D1484751

MODERN
MULTIDIMENSIONAL
CALCULUS

MARSHALL EVANS MUNROE

DOVER PUBLICATIONS, INC.
MINEOLA, NEW YORK

Bibliographical Note

This Dover edition, first published in 2019, is an unabridged republication of the first edition of the work, originally published by Addison-Wesley Publishing Company, Reading, Massachusetts, in 1963.

Library of Congress Cataloging-in-Publication Data

Names: Munroe, M. Evans (Marshall Evans), 1918– author.
Title: Modern multidimensional calculus / Marshall Evans Munroe.
Description: Dover edition. | Mineola, New York : Dover Publications, Inc., 2019. | Originally published: Reading, Massachusetts : Addison-Wesley Publishing Company, 1963. | Includes index.
Identifiers: LCCN 2018039376| ISBN 9780486834023 | ISBN 0486834026
Subjects: LCSH: Mathematical analysis—Textbooks. | Calculus—Textbooks. | Algebras, Linear—Textbooks.
Classification: LCC QA300 .M77 2019 | DDC 515/.14—dc23
LC record available at https://lccn.loc.gov/2018039376

Manufactured in the United States by LSC Communications
83402601   2019
www.doverpublications.com

PREFACE

This is a second-year calculus text devoted primarily to topics in multi-dimensional analysis. These topics have been developed in accordance with three major guidelines:

1. A first course in multidimensional calculus becomes far too heavy for most students if it is burdened down with rigorous proofs of all pertinent results. Therefore, the emphasis here is on concepts and methods. Where proofs are required but omitted, this is clearly pointed out.

2. When properly used the differential becomes an extremely elegant tool in the development of calculus, particularly in a multidimensional setting. Thus, the notion of differential is developed in its most potent, modern form and then used as extensively as possible.

3. A major contribution of differential calculus is to reduce nonlinear transformations to linear ones, thus making linear algebra an important part of the subject. Matrix methods are the most efficient in the study of linear transformations and hence are used extensively.

Certain innovations are necessary to accomplish these ends. This is particularly true in laying a proper foundation for an intellectually respectable and yet maximally useful theory of the differential.

The first step in this program is to insist on a careful distinction between a point in the plane and the ordered pair of real numbers consisting of its rectangular coordinates. Indeed, coordinate variables, such as x and y, must be recognized as mappings. For example, x is a symbol for the mapping that carries each point p into its abscissa $x(p)$. Mappings of this type are the major concern here; so they must be clearly recognized.

Another type of mapping that plays a completely different role is a mapping that carries n-tuples of numbers into numbers. Suppose that f maps number pairs into numbers and x, y, and z, map points of the plane into numbers. Then, the equation $z - f(x, y)$ states that the mapping z is the same as the composite mapping consisting of the ordered pair of mappings x, y followed by the mapping f. This is a typical relation in multidimensional calculus and clearly x, y, and z belong in one category here with f in another. To distinguish between these two types of mappings, the following terminology is introduced. Point-to-number mappings such as x, y, and z are called variables. This is inspired by the phrase "coordinate variable," although the word variable assumes a more general meaning here. Connecting mappings, such as f above, are called functions.

It is possible to develop a reasonably complete calculus of functions in which variables do not appear at all. However, the variables connect calculus with geometry, physics, etc.; so a calculus of functions only is strictly an "ivory tower" discipline. More germane to the applications is a calculus of variables in which functions play an essential implicit role but seldom need to be mentioned explicitly. This latter point of view is the one adopted here.

Differentials, then, need to be defined for variables. Chevalley has given a definition that accomplishes exactly this, and a suitable specialization of his development of the subject is employed here. It is principally in this respect that this calculus book differs from others of recent vintage. It has been recognized for some time now that the "$dx = \Delta x \cdots$" approach leads to nothing but double talk, but the usual procedure in making the differential respectable is to employ the Frechet definition. Though it involves a more sophisticated procedure, the Chevalley approach is used here because it defines directly the differential of a variable, whereas the Frechet definition yields the differential of a function.

The first three chapters contain more or less introductory material. Chapter 1 establishes the hierarchy of mappings to be studied; that is, variables connected by functions. Chapter 2 develops the one-dimensional case of the differential. The Chevalley definition of a differential is an algebraic maneuver in which the dimension of the domain of the variables makes a very minor difference. The purpose of Chapter 2 is to introduce this algebraic procedure in a setting in which each step can be easily interpreted in terms of a concrete geometric picture. The algebra is then repeated in Chapter 5 for the n-dimensional case where pictures are not so practical. Chapter 3 introduces vectors in the plane and gives some illustrations of procedures that make full use of an effectively defined differential. Topics covered in these three chapters are normally included in an introductory calculus course. The revised treatment of selected topics given here should serve to give a picture of one-dimensional calculus consistent with the development to follow for the multidimensional case.

The serious study of multidimensional calculus begins in Chapter 4 with the development of the matrix algebra that will be needed. Chapter 5 introduces the differential. Chapters 6 and 7 present topics in differential calculus, notably max-min problems, transformations and chain rules, and vector derivative operators. Chapter 8 should be review. It treats iterated integrals essentially as they are treated in any introductory calculus book. Chapters 9 and 10 are concerned with multiple integrals. Here again the differential is used as much as possible. The exterior product of differentials is defined concretely and used in the definition of the integral. Exterior algebra is thus not merely a way of systematizing known results. Instead, it becomes an effective tool in developing such things as the substitution theorem and a generalized Stokes' theorem that specializes to the classical theorems of Green, Gauss, and Stokes.

M. E. M.

Durham, N. H.
October, 1962

CONTENTS

CHAPTER 1

FUNCTIONS AND VARIABLES

1-1 Elementary functions. A *function* is defined as a set of ordered pairs of numbers no two of which have the same first entry. The set of all first entries in the ordered pairs is called the *domain* of the function; the set of all second entries is called its *range*. If a function is displayed in a two-column table, the ordered pairs are read across. The domain appears as the first column of the table and the range as the second column.

If f is a function and (a, b) is one of its ordered pairs, the common practice is to indicate this fact by writing

$$b = f(a).$$

The range entry b that corresponds to the domain entry a is called the *value* of f at a. The symbol $f(a)$ is frequently read, "f of a," or in better modern usage, "f at a." A convenient way of describing a specific function f is to give a formula which yields $f(a)$ in terms of a for each a in the domain. For example,

$$f(a) = a^2 - 3a + 2$$

describes a simple polynomial function f.

Addition, subtraction, multiplication, and division of functions are described by the general rule, "Pair off equal domain entries and perform the indicated operation on the corresponding range entries." More specifically, these operations are defined by the following formulas:

$$(f \pm g)(a) = f(a) \pm g(a),$$

$$(fg)(a) = f(a)g(a),$$

$$\frac{f}{g}(a) = \frac{f(a)}{g(a)} \qquad [g(a) \neq 0].$$

It is also convenient to define a real number times a function by the formula

$$(cf)(a) = cf(a).$$

Note that if f and g are functions and c is a number, then $f + g$, $f - g$, fg, f/g, and cf are functions.

With regard to this last point, multiplication by a number coupled with division can lead to some confusion. Suppose that

$$g = cf, \tag{1}$$

then one is tempted to write

$$\frac{g}{f} = c.$$

This, however, equates a function and a number and cannot be correct. A correct deduction from (1) is that

$$\frac{g}{f} = c\frac{f}{f}.$$

The function f/f is a unit constant function. Each of its range entries is equal to 1. Let

$$f^0$$

(f to the zero power) be defined as the set of all ordered pairs $(a, 1)$ where (a, b) is a member of f. Strictly speaking f^0 and f/f may be different in that the domain of f^0 may contain numbers a for which $f(a) = 0$ while that of f/f does not. It is a useful convention to agree that given (1),

$$\frac{g}{f} = cf^0.$$

The *inverse* of a function is obtained by reversing each of its ordered pairs. Since this may produce duplicate first entries, the inverse of a function may not be a function. The usual practice is to delete duplicate domain entries in some systematic way to obtain a *principal inverse* which is a function. For commonly used functions there are established conventions for constructing principal inverses. For example, the principal inverse of the square function is the nonnegative square root function. There are well-known conventions for defining principal inverses of the trigonometric and hyperbolic functions.

Given functions f and g, the *composite* function

$$f \circ g$$

(read f circle g) is defined as the set of all ordered pairs (a, c) such that for some b, (a, b) is in g and (b, c) is in f. The composite is an iterated mapping process shown schematically in Fig. 1–1.

In terms of a formula for function values, $f \circ g$ may be defined as follows:

$$(f \circ g)(a) = f[g(a)].$$

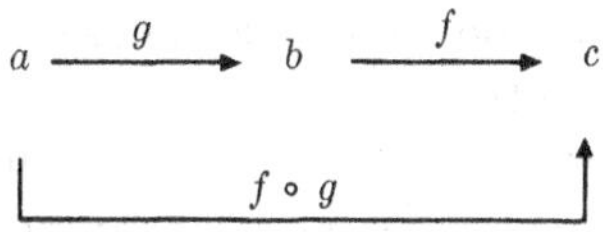

$$a \xrightarrow{\quad g \quad} b \xrightarrow{\quad f \quad} c$$

$$\underline{\qquad\qquad f \circ g \qquad\qquad}$$

FIGURE 1–1

Composition is associative; that is,

$$(f \circ g) \circ \phi = f \circ (g \circ \phi),$$

because each of these has for its value at a, $f\{g[\phi(a)]\}$. It is not commutative; as a rule

$$f \circ g \neq g \circ f.$$

Composition is distributive over the algebraic operations *from the right:*

$$(f \pm g) \circ \phi = (f \circ \phi) \pm (g \circ \phi),$$

$$(fg) \circ \phi = (f \circ \phi)(g \circ \phi),$$

$$\frac{f}{g} \circ \phi = \frac{f \circ \phi}{g \circ \phi}.$$

It is not distributive from the left; in general

$$\phi \circ (f \pm g) \neq (\phi \circ f) \pm (\phi \circ g),$$

$$\phi \circ (fg) \neq (\phi \circ f)(\phi \circ g),$$

$$\phi \circ \frac{f}{g} \neq \frac{\phi \circ f}{\phi \circ g}.$$

Because of the associative law, it is unambiguous to write

$$f \circ g \circ \phi.$$

To find a value of this function one applies ϕ, then g, then f. Note, however, to differentiate it, one proceeds from left to right. The chain rule for differentiating $f \circ g \circ \phi$ may be written

$$(f \circ g \circ \phi)' = (f' \circ g \circ \phi)(g' \circ \phi)\phi'.$$

The *identity function*, I, is defined by saying that

$$I(a) = a$$

for every real number a; that is, I pairs each real number with itself.

It is natural to introduce positive integers as exponents on function symbols in terms of multiplication; that is, $f^2 = ff$; $f^3 = fff$; and in general $f^n = f^{n-1}f$. This notation and the notation for the identity function furnish self-explanatory symbols for all polynomial functions. For example, the function f defined by

$$f(a) = a^2 - 3a + 2$$

may be written

$$I^2 - 3I + 2I^0.$$

Fractional powers are defined by saying that $I^{1/n}$ is the principal inverse of I^n.

Calculus generally deals with the *elementary functions*. These are the functions generated by the identity function, the sine function, and the natural logarithm function through the operations of addition, subtraction, multiplication, division, multiplication by numbers, inversion, and composition.

Roughly speaking, elementary functions are those functions for which there are standard symbols. As noted above, the symbol I furnishes polynomial symbols. The trigonometric and hyperbolic functions (sin, cos, sinh, cosh, etc.) and their inverses (arcsin, arcsinh, etc.) have well-known symbols. The usual symbol for the natural logarithm is ln. The exponential and absolute-value functions are familiar ones, but symbols for them are not commonly used. Let exponential (exp) and absolute (abs) be defined by

$$\exp(a) = e^a$$

and

$$\operatorname{abs}(a) = |a|,$$

respectively.

Derivative formulas are specifically designed to fit the elementary functions. The standard formulas for derivatives of combinations are:

$$(f \pm g)' = f' \pm g',$$

$$(fg)' = f'g + g'f,$$

$$\left(\frac{f}{g}\right)' = \frac{gf' - fg'}{g^2},$$

$$(cf)' = cf',$$

$$g' = \frac{I^0}{f' \circ g} \qquad (f \text{ and } g \text{ inverses}),$$

$$(f \circ g)' = (f' \circ g)g'.$$

These, together with the three specific formulas,

$$I' = I^0,$$

$$\sin' = \cos,$$

$$\ln' = \frac{I^0}{I},$$

will yield by direct substitution all the standard derivative formulas.

Of more basic importance than the organization of formulas is the fact that the structure of the class of elementary functions leads to a concise description of the continuity and differentiability properties of these functions.

A function f is *continuous* at a provided that

$$\lim_{h \to a} f(h) = f(a).$$

It is *differentiable* at a provided that

$$\lim_{h \to 0} \frac{f(a + h) - f(a)}{h}$$

exists. Many results in calculus depend heavily on the continuity and differentiability properties of the functions involved. To apply these results with confidence, one must be able to verify that the required conditions hold, but for most practical purposes, this does not require an exhaustive study of the limit concept. Since specific applications generally involve elementary functions, it usually suffices to know about them. The necessary information is easily outlined in terms of the three basic functions and seven operations that generate the elementary functions.

The functions I and sin are continuous everywhere, and ln is continuous *except* at 0. Addition, subtraction, multiplication, composition, and multiplication by numbers all preserve continuity, and for functions defined on intervals, inversion preserves continuity. Division preserves continuity *except* where the denominator is zero.

A similar set of statements describes the differentiability properties of the elementary functions. For I and sin, the derivative exists everywhere; for ln, it exists everywhere *except* at 0. Addition, subtraction, multiplication, composition, and multiplication by numbers all preserve differentiability. If f and g are inverses and f is differentiable, then g is also differentiable, *except* where $f' \circ g$ is zero. Division preserves differentiability *except* where the denominator is zero.

In a nutshell, elementary functions are continuous except where the formula for function values indicates division by 0 or ln 0, and they are differentiable, except where the derivative formula indicates division by 0 or ln 0.

This brief summary may be a little deceptive. For example, tan is discontinuous at $\pi/2$, but the division by zero appears only if one writes it as sin/cos. However, the complete statements preceding the summary give precise criteria for determining continuity and differentiability properties of specific elementary functions.

EXAMPLES

1. Let $f = I^2$ and $g = I^3 + I^0$. Find $f(0)$, $f(-1)$, $g(0)$, $g(-1)$. As noted above, this is done by substitution; thus

$$f(0) = 0^2 = 0, \qquad f(-1) = (-1)^2 = 1,$$

$$g(0) = 0^3 + 1 = 1, \qquad g(-1) = (-1)^3 + 1 = 0.$$

2. Let $f = I^2 - I^{1/2}$. In terms of an arbitrary positive number a, find $f(a^2)$, $f(a^{1/2})$, $[f(a)]^{1/2}$, $[f(a)]^2$. Again, substitution is called for; hence

$$f(a^2) = (a^2)^2 - (a^2)^{1/2} = a^4 - a,$$

$$f(a^{1/2}) = (a^{1/2})^2 - (a^{1/2})^{1/2} = a - a^{1/4},$$

$$[f(a)]^{1/2} = [a^2 - a^{1/2}]^{1/2} \qquad (a > 1),$$

$$[f(a)]^2 = [a^2 - a^{1/2}]^2 = a^4 - 2a^{5/2} + a.$$

3. In words describe the rule for computing values of the function f in Example 2. Take the square of a number, then take its square root, and finally subtract the second of these two results from the first.

4. Write

$$[I + (I - 2I^0)^3]^{1/2},$$

using $\circ$ to denote composition. *Answer.* $I^{1/2} \circ \{I + [I^3 \circ (I - 2I^0)]\}$.

5. Write

$$I^5 \circ \frac{I^{1/2} \circ (I + I^2)}{I^2 \circ (I - I^{1/2})}$$

in a form without the $\circ$ symbol. *Answer.* $[(I + I^2)^{1/2}/(I - I^{1/2})^2]^5$.

6. Locate the discontinuities of

$$\ln \circ \text{abs} \circ \left(\frac{I - I^0}{I + I^0}\right).$$

This is an elementary function and has two discontinuities. It is discontinuous at -1 because division by zero is indicated there, and it is discontinuous at 1 because ln 0 is indicated there.

6. Write each of the following without the symbol $\circ$.

(a) $I^{1/2} \circ (I^2 - I^0)$

(b) $I^3 \circ (I + I^{1/2})$

(c) $I^2 \circ (I^0 + I^{1/2}) \circ (I^2 + I^0)$

(d) $I^4 \circ (I^0 - I) \circ (3I^0 + I^{1/2})$

(e) $3I^5 \circ (I^2 - 2I)$

(f) $I^{1/2} \circ (I^3 + 3I^2)$

(g) $I^{1/2} \circ (I^0 - I) \circ I^{1/2} \circ (I^0 - I)$

(h) $I^3 \circ (I^0 - 2I^{1/2}) \circ (3I^0 + I)$

7. Rewrite each of the following, using $\circ$ to denote composition.

(a) $(I + I^2)^5$

(b) $(I - I^{1/2})^3$

(c) $(I - I^3)^{1/2}$

(d) $[(I - I^0)^{1/2} - (I + I^0)^{1/2}]^3$

(e) $[I^0 + (I^0 + I)^2]^2$

(f) $[I + (I - I^0)^{1/2}]^{1/2}$

(g) $\{I^0 + [I^0/(I^0 - I^2)]\}^3$

(h) $\{I + [I + (I - I^0)^{1/2}]^{1/2}\}^{1/2}$

8. Functions are often informally described as machines; that is, one takes a set of numbers, feeds it into the machine, and gets another set of numbers out at the other end. In their simplest uses, modern computers are just such machines. One gives a computer instructions for processing data, then feeds in numbers and gets others out. For the computing-machine model, use the words "input," "instructions," and "output" to describe each of the following:

(a) function, as defined in this section,

(b) rule for computing function values,

(c) domain,

(d) range,

(e) composition of functions.

9. Let f be a function whose domain is the entire real number system. Discuss the relations among

$$I^0, \qquad I^0 \circ f, \qquad f^0.$$

10. A standard book of numerical tables will contain tables of logarithms, logarithms of trigonometric functions, and natural trigonometric functions. These tables display functions f, g, and ϕ such that

$$\phi = f \circ g.$$

Which function comes from which table? Obtain a book of tables and check your answer numerically.

11. In general $f \circ g$ and $g \circ f$ are quite different functions; so the phrase "composite of f and g" is ambiguous. Commonly used terminology is as follows: $f \circ g$ is referred to as "the composite function consisting of g followed by f." Discuss this terminology in light of (a) the definition, and (b) the notation.

12. Locate the discontinuities for each of the following functions.

(a)　ln ∘ abs ∘ sin

(b)　$\exp \circ \dfrac{-I^0}{I}$

(c)　cot ∘ $2I$

(d)　$\dfrac{I - I^0}{I^2 - 4I^0}$

(e)　$I - \dfrac{I^0}{I}$

(f)　$2I + \tan$

(g)　tanh

(h)　$\dfrac{I^0}{I^0 - \exp}$

(i)　ln ∘ ln ∘ abs

(j)　$\dfrac{I^0}{I^0 + \sin}$

(k)　$\dfrac{I}{\text{abs}}$

(l)　ln ∘ abs ∘ $(I^0 - \ln)$

(m)　arctan ∘ $\dfrac{I^0}{I}$

(n)　ln ∘ abs ∘ tan

(o)　ln ∘ abs ∘ $\left[I^0 - \dfrac{I^0}{I^2} \right]$

(p)　$I(I^0 - I)^{1/3}$

(q)　$I(I^0 - I)^{-1/3}$

(r)　$I^{1/2} \circ (I^4 - I^2)$

(s)　ln ∘ $(I^4 - I^2)$

(t)　exp ∘ ln ∘ abs

13. The function abs is not differentiable at 0. (a) Form the difference quotient and show that it has different one-sided limits. (b) Deduce this from the master rule on differentiability of elementary functions. [*Hint:* abs $= I^{1/2} \circ I^2$.]

14. How many derivatives does $I^{m/n}$ have at 0?

1–2 Coordinate variables. A function is often referred to as a mapping of its domain onto its range. Here the word "mapping" will be used to denote a generalization of the notion of function. Define a *mapping* as any set of ordered pairs no two of which have the same first entry. A function is then a mapping with domain and range each a set of numbers.

The foundations of analytic geometry involve certain important mappings that are not functions by this definition. Specifically, the coordinate variables x and y are mappings. If p is a point of the plane, then $x(p)$ is its abscissa, and the symbol x, itself, stands for the set of all ordered pairs (point, abscissa). Similarly, y is a mapping that pairs each point with its ordinate.

A *variable* is thus defined as a mapping whose domain is a set of geometric or physical entities and whose range is a set of numbers. The abscissa and ordinate variables are the prime examples, although many others will appear as physical and geometric applications of calculus are discussed.

If x is the abscissa variable and f is a function, the composite mapping $f \circ x$ is defined, and a basic notion in analytic geometry is that of the locus of

$$y = f \circ x. \tag{2}$$

This is the set of all points p such that

$$y(p) = (f \circ x)(p).$$

Here a word about notation is in order. Equation (2) is usually written $y = f(x)$ with no precise explanation of what x and y stand for. Presumably, $f(x)$ means the value of f at the number x. However, parentheses are also used to denote composition, so $f(x)$ could mean the composite denoted here by $f \circ x$. In adapting the modern theory of the differential to elementary calculus it is essential that the coordinate variables be recognized as mappings. Thus, in the present context whether one writes $f \circ x$ or $f(x)$, it must be understood that the relation between f and x is that of composition. To emphasize this fact, the unambiguous circle notation for composition will be used throughout this introductory chapter. Later, parentheses for composition will be introduced, and the notation will assume a more familiar form, although the meanings attached to standard symbols here must be retained.

The question of notation in (2) is very superficial. A much more basic question is the following. Is equation (2) to be taken by itself and regarded as an assertion that y and $f \circ x$ are two symbols for the same mapping? Unfortunately, the answer is "no" in analytic geometry and "yes" in calculus.

In analytic geometry each of the variables x and y has the entire plane for its domain, and the basic idea is that of the "locus of $y = f \circ x$." The entire phrase in quotes is defined immediately following equation (2), and as long as the phrase in quotes is considered as a unit, there is no trouble.

Suppose the locus of $y = f \circ x$ is found to be a curve C. In calculus one wants to redefine the symbols x and y so that they still map points into their abscissas and ordinates, respectively, but in such a way that each mapping has only the curve C for its domain. Once this operation is performed, it becomes quite literally true that $y = f \circ x$. Now, in calculus one wants equations to be simple assertions of fact like this because one of the principal manipulative tools in calculus is substitution, and substitution of one symbol for another leaves meanings unchanged only if the two symbols stand for the same thing.

Calculus equations generally involve variables, and the most polished way to introduce them is as it was done in the preceding paragraph. That is, define the variables carefully (including specification of domain) in advance, then make a valid statement about them in the form of an equation. In practice this becomes too involved. Each change of domain introduces a new variable and should call for a new symbol. Clearly, this is not practical. What one can do is to write statements of the form

$$y = f \circ x \quad \text{on } C. \tag{3}$$

Agree that x and y always mean abscissa and ordinate, respectively; then (3) means that, if x and y have domain C, $y = f \circ x$ is a statement of fact and may be used for purposes of substitution.

Abscissas and ordinates are not the only useful coordinate variables on the plane. Polar coordinates must certainly be considered, and as the study of multidimensional analysis progresses, other pairs will be introduced. The polar-coordinate variables r and θ are mappings just as x and y are. For each point p in the plane, $r(p)$ is the distance from the origin to p, and $\theta(p)$ is the angle from the positive x-direction to that of the radial line through p. It is readily verified that the equations

$$x = r \cos \theta, \qquad y = r \sin \theta$$

hold over the entire plane. Therefore, substitutions from them are valid on any figure in the plane.

It is well to consider two different coordinate systems in the plane by pointing out that, in the description of a locus, the variables involved play just as important a role as the functions. For example,

$$y = x^2$$

is the equation of a parabola, while

$$r = \theta^2$$

is the equation of a spiral. These equations can be written

$$y = I^2 \circ x$$

and

$$r = I^2 \circ \theta,$$

respectively. The function involved is the same, although the loci are quite different.

EXAMPLES

1. Rewrite

$$\sqrt{1 + x^2}$$

in a form that displays the structure of the mapping. *Answer.*

$$I^{1/2} \circ (I^0 + I^2) \circ x.$$

2. Rewrite

$$I^3 \circ (I^0 + I^{1/2}) \circ (I^0 + I) \circ x$$

in standard notation for variables. *Answer.* $(1 + \sqrt{1 + x})^3$.

EXERCISES

1. Rewrite each of the following in a form that displays the structure of the mapping.

(a) $(1 + \sqrt{x})^2$

(b) $\sqrt{1 - x^3}$

(c) $3 + (2 - x)^4$

(d) $(3 + \sqrt{2 - x})^4$

(e) $2 + (1 - \sqrt{x + 4})^3$

(f) $\sqrt{1 - (4 + x^2)^3}$

(g) $(2 - \sqrt{1 - x^3})^2$

(h) $(1 - x\sqrt{1 - x^2})^2$

(i) $\sqrt{x - \sqrt{1 - x}}$

(j) $x\sqrt{1 - 2\sqrt{x}}$

2. Rewrite each of the following in standard notation. (See Example 2 for an illustration of "standard" notation.)

(a) $I^{1/2} \circ (I^0 + I) \circ x$

(b) $I^3 \circ (I^{1/2} - I^2) \circ x$

(c) $I^2 \circ (2I - I^2) \circ (I^0 - I) \circ x$

(d) $I^{1/2} \circ (I^0 + I^{1/2}) \circ x$

(e) $(I^0 - I^2) \circ (I - I^2) \circ x$

(f) $(I - I^3) \circ (I - I^{1/2}) \circ x$

(g) $(I^{3/2}) \circ (I^0 - I) \circ x$

(h) $\{I[I^{1/2} \circ (I^0 - I)]\} \circ x$

(i) $\{[I^3 \circ (I^3 - I^2)](3I^2 - 2I)\} \circ x$

(j) $I^{1/2} \circ (I^0 + I) \circ I^{1/2} \circ (I^0 + I) \circ x$

3. Let x and y be the usual coordinate variables; let f and g be functions; let p and q be points of the plane; let a and b be numbers. Assume that each of the functions f and g has the entire real number system for both domain and range. Each of the following symbols falls in exactly one of four categories: a number, a function, a variable, meaningless. Classify each of them.

(a) $f \circ x$ (b) $g[y(p)]$ (c) $x \circ g$ (d) $p \circ x$

(e) $f(a)$ (f) $y \circ x$ (g) $f[x(a)]$ (h) $g[f(b)]$

(i) $a \circ b$ (j) $f \circ g \circ x$ (k) $y(b)$ (l) $x[y(p)]$

(m) $f \circ g$ (n) $p \circ g$ (o) $x(q)$ (p) $q(p)$

(q) $g \circ f \circ x$ (r) $f \circ y$ (s) $b(p)$ (t) $f\{g[y(q)]\}$

(u) $g\{x[f(a)]\}$ (v) $x\{g[y(p)]\}$ (w) $g(b)$ (x) $g \circ y$

(y) $g(q)$ (z) $g \circ f$

4. Write each of the following expressions in symbols.
  (a) the sum of the abscissa and ordinate variables;
  (b) the product of the abscissa and ordinate variables applied to the point p;
  (c) the product of the abscissa of p and the ordinate of p;
  (d) the composite consisting of the abscissa variable followed by the cube of the identity function; [*Note:* See Exercise 11, Section 1–1.]
  (e) the value at the number a of the sum of the cube of the identity function and the square of the identity function;
  (f) the composite consisting of the sum of the abscissa and ordinate variables followed by the square of the identity function;
  (g) the sum of the two composites consisting of the abscissa and ordinate variables, each followed by the square of the identity function;
  (h) the square root of the sum of the unit constant function and the identity function;
  (i) the sum of the unit constant function and the square root of the identity function.

5. Each of the following lists a pair of variables (with explanations of the symbols following). In each case name a function that maps the first variable into the second. Note that the three answers are the same. One function may appear in many different connections.
  (a) s, A; s measures the length of a side and A the area, both applied to squares.
  (b) r, A/π; r measures the radius and A the area, both applied to circles.
  (c) t, $2s/g$; t measures elapsed time and s measures distance traveled, both applied to freely falling bodies which start from rest; g is the constant of the acceleration due to gravity.

1–3 Derivatives and integrals. If f is a function defined on an interval containing a, the derivative of f at a is defined by

$$f'(a) = \lim_{h \to 0} \frac{f(a + h) - f(a)}{h}$$

whenever this limit exists. The derivative of f is another function f' consisting of all ordered pairs $[a, f'(a)]$ for which $f'(a)$ is defined. As indicated in Section 1–1, a systematic set of formulas can be compiled to yield the derivative of any elementary function.

All these considerations belong to what might be called the calculus of functions. That is, the notion of derivative and the entire battery of formulas can be studied without any mention of the notion of variable as defined in Section 1–2.

The notion of integral may also be defined strictly within the framework of the calculus of functions. Suppose f is continuous on $[a, b]$. Take numbers c_i such that

$$a = c_0 < c_1 \cdots < c_n = b.$$

It can be shown that there is a unique limit,

$$\lim \sum_{i=1}^{n} f(c_i)(c_i - c_{i-1}),\qquad\qquad(4)$$

where the limit is taken over any sequence of sets $\{c_i\}$ for which the maximum difference, $c_i - c_{i-1}$, tends to zero. This limit depends on a, b, and f. It is denoted by $\int_a^b f$ and called the integral from a to b of f.

The fundamental theorem of calculus may also be embedded in the calculus of functions. Essentially, the theorem says that if f' is continuous, then

$$\int_a^b f' = f(b) - f(a).$$

Because of this theorem, derivative formulas (which are fairly easy to come by) yield a useful list of integral formulas.

If the functions involved are used in conjunction with variables as indicated in Section 1–2, a calculus of variables begins to emerge. First let us consider derivatives. Let u and v be variables, and let f be a function. Suppose that on some domain

$$v = f \circ u.$$

Given this situation, the *derivative of v with respect to u* is denoted by $D_u v$ and defined by

$$D_u v = f' \circ u.$$

The symbol D_u stands for what is called a *derivative operator*. This is a mapping; it maps variables into variables. Specifically, for example,

$$D_u u^n = n u^{n-1};$$

hence D_u maps u^n into nu^{n-1}.

From this point of view, successive differentiations mean successive applications of the operator D_u. So, the second derivative of v with respect to u is

$$D_u D_u v = f'' \circ u,$$

and this is commonly written

$$D_u^2 v.$$

The nth derivative of v with respect to u is then

$$D_u^n v.$$

With regard to integrals, the situation is similar but the explanation is a little more involved. The general notion of integration of a variable will be presented in Section 2–5. Consider for the present an important special case. Let x be the abscissa variable with the horizontal axis only as a domain. Let f be a function, and think of $f \circ x$ as measuring vertical distances from the domain of x to the locus of $y = f \circ x$. That is, for each point p on the x-axis, $x(p)$ is the abscissa of p, and $f[x(p)]$ is the vertical distance from p to the curve.

For each p in the domain of x there is defined a variable denoted by dx_p and called the *differential of x at p*. A comprehensive definition of differentials will appear later (Section 2–2), but for the specific model being considered here, the following is an accurate description of dx_p. Its domain is the horizontal axis, and for each q in this domain, $dx_p(q)$ is the abscissa of q with respect to p; in other words, $dx_p(q)$ is the directed distance from p to q.

In general, differentials measure distances like this, and different differentials measure them to different scales. Roughly speaking, the theory developed in Section 2–2 shows how a variable u determines the scale to which du_p measures distances from p. Suffice it to say for the present that for the abscissa variable on a horizontal line, the differential measures distances to the same scale as the original variable.

Now, take the interval on the x-axis from $x = a$ to $x = b$, and partition it by points $p_0,\ p_1,\ p_2, \cdots ,\ p_n$ with $x(p_0) = a$ and $x(p_n) = b$. Form the sum

$$\sum_{i=1}^{n} f[x(p_i)]\, dx_{p_{i-1}}(p_i). \tag{5}$$

The limit of this sum, as the number of partition points tends to infinity and the maximum distance between adjacent ones tends to zero, is denoted by

$$\int_a^b f \circ x\, dx$$

and is called the *integral from a to b of $f \circ x$ with respect to x*.

An integral with respect to a different variable would employ a different differential; thus, it would measure the distances from p_{i-1} to p_i to a different scale.

It is easily seen that the sums (4) and (5), above are the same. That is, set

$$c_i = x(p_i);$$

then

$$f[x(p_i)] = f(c_i), \qquad \text{and} \qquad dx_{p_{i-1}}(p_i) = c_i - c_{i-1}.$$

Thus, it appears that

$$\int_a^b f \circ x\, dx = \int_a^b f.$$

This very specialized picture of a specific differential suffices to define one basic integral form in the calculus of variables. A full development of the calculus of variables depends on a complete theory of the differential. This is given for the one-dimensional case in Sections 2–2 and 2–3, and the multidimensional case is discussed in Chapter 5.

Of what does a theory of the differential consist? Very briefly, it is as follows. Note that in the above discussion it is x that has a differential, not f. Differentials are associated with variables, not with functions. It was noted that if a variable other than x were involved, its differential would measure distances to a different scale. The significant thing is that a differential dv should depend on the variable v alone, not on other variables or other differentials. Finally, the theory should lead to the formula

$$dv = D_u v \, du. \tag{6}$$

That is, du, dv, and $D_u v$ are all defined independently but in such a way that this relation holds.

Clearly, the relation (6) is the basis for integration by substitution. In this connection (6) is sometimes presented only as a convenient way of remembering the right answer. With an adequate theory of the differential, however, it is possible to define the integral of a variable in such a way that integration by substitution is just exactly that. This will be done in Section 2–5.

An immediate corollary of (6) is that

$$\frac{dv}{du} = D_u v.$$

The derivative of a variable is a quotient of differentials. This, too, is sometimes regarded as fiction, but it can be made quite literally true with the proper foundation. A precise interpretation is given in Section 2–2.

EXERCISES

1. If f is a function and y is a variable, one speaks of the "derivative of f" but of the "derivative of y with respect to x," where x and y are related variables. Let y map points into numbers. Why cannot y' be defined by direct analogy with f'? Set up a "difference quotient" leading to y'; what part of this difference quotient is meaningless?

2. With no reference to the connecting function, formulate a description of $D_u v$ as follows: Form a quotient of differences based on values of u and v at points p and q and give the value of $D_u v$ at p as a limit of this quotient as $q \to p$ (these latter symbols mean that p is fixed and the distance between p and q tends to zero).

3. The variable $D_u v$ is usually described as measuring the "rate of change of v with respect to u." Discuss this in the light of Exercise 2.

1–4 Polynomial approximations. The approximation

$$f(a+h) \sim f(a) + f'(a)h + f''(a)h^2 + \cdots + \frac{1}{n!}f^{(n)}(a)h^n$$

appears in many connections in calculus. If the right-hand side is a partial sum of a convergent Maclaurin series, the approximation improves as $n \to \infty$. There are various theorems giving the order of magnitude of the error as $h \to 0$. There is Taylor's formula with remainder which involves no limiting processes but gives an error estimate in terms of $f^{(n+1)}(a + \theta h)$ for an undetermined θ between 0 and 1.

A key step in the development of the differential (Section 2–2) depends on this same polynomial approximation formula with an error term slightly different from that in Taylor's remainder formula. Specifically, if f has an $(n + 2)$ derivative at a, there is a differentiable function g such that

$$f(a+h) = f(a) + f'(a)h + \cdots + \frac{1}{n!}f^{(n)}(a)h^n + g(a+h)h^{n+1}.$$

To prove this, solve the equation for $g(a + h)$:

$$g(a+h) = \frac{f(a+h) - f(a) - f'(a)h - \cdots - (1/n!)f^{(n)}(a)h^n}{h^{n+1}}.$$

Let this define $g(a + h)$ for $h \neq 0$, and set

$$g(a) = \frac{1}{(n+1)!}f^{(n+1)}(a).$$

Clearly, $g'(a + h)$ exists provided that $f'(a + h)$ does, except perhaps for $h = 0$. Now,

$$g'(a) = \lim_{h \to 0} \frac{g(a+h) - g(a)}{h}$$

$$= \lim_{h \to 0} \frac{f(a+h) - f(a) - f'(a)h - \cdots - \frac{f^{(n)}(a)h^n}{n!} - \frac{f^{(n+1)}(a)h^{n+1}}{(n+1)!}}{h^{n+2}}.$$

This is an indeterminate form, and $n + 1$ applications of L'Hôpital's rule reduce it to

$$\lim_{h \to 0} \frac{f^{(n+1)}(a+h) - f^{(n+1)}(a)}{(n+2)!h}.$$

By definition, this limit is $f^{(n+2)}(a)/(n + 2)!$, and this was assumed to exist.

The following is the guise in which this theorem will be used. Suppose u and v are variables such that $v = f \circ u$ on some curve. Let p be a fixed point on the curve, and let q be any other point on the curve. Use the above theorem with $n = 1$, and make the following substitutions:

$$a = u(p),$$
$$h = u(q) - u(p),$$
$$a + h = u(q),$$
$$f(a) = f[u(p)] = v(p),$$
$$f(a + h) = f[u(q)] = v(q).$$

The theorem says that if $f'''[u(p)]$ exists, then there is a differentiable function q such that

$$v(q) = v(p) + f'[u(p)][u(q) - u(p)] + g[u(q)][u(q) - u(p)]^2.$$

Since this holds for all q on some portion of the curve, it follows that on an appropriate domain

$$v = v(p) + f'[u(p)][u - u(p)] + (g \circ u)[u - u(p)]^2. \tag{7}$$

CHAPTER 2

DIFFERENTIALS

2–1 Differentials: geometric picture. Let x and y be the rectangular coordinate variables restricted to a curve C, and suppose that $y = f \circ x$. As noted in Section 1–3, dx_p and dy_p should be so defined that

$$dy_p = f'[x(p)] \, dx_p. \tag{1}$$

The following geometric description of dx_p and dy_p is quite consistent with the general algebraic definition of Section 2–2, and it yields immediately the relation (1). Let T_p be the tangent line to C at p. Each of the variables dx_p and dy_p will have T_p for its domain; and for each point q on T_p, $dx_p(q)$ and $dy_p(q)$ will be the directed distances shown in Fig. 2–1. In words, $dx_p(q)$ is the abscissa of q with respect to p as the origin, and $dy_p(q)$ is the ordinate of q with respect to p as the origin.

Since these relative abscissas and ordinates are defined *on T_p only*, it is clear from Fig. 2–1 that for every admissible q,

$$\frac{dy_p(q)}{dx_p(q)} = \tan \alpha = f'[x(p)];$$

hence the identity (1) is established.

Figure 2–1 shows another differential, ds. This comes from the variable s that measures arc length on C from some fixed reference point. From the integral formula for arc length it is clear that at p,

$$D_x s = \sqrt{1 + (D_x y)^2} = \sqrt{1 + \tan^2 \alpha} = \sec \alpha;$$

so this picture yields the required relation between ds_p and dx_p.

As indicated by the notation, each differential is applied to an ordered pair of points. However, this domain (set of point pairs) has a very special structure. The first entry from such a pair (the one that goes into the subscript) is to be a point of C. The second entry (the one in parentheses) is to be on the tangent line determined by the first. The set of pairs (p, q) restricted and related in this way forms what is called the *tangent bundle* for C (see Fig. 2–2).

In the present discussion, then, x, y, and s are variables on C, and dx, dy, and ds are variables on the tangent bundle for C. More generally (Section 2–2), each variable u of a large class of variables on C will generate

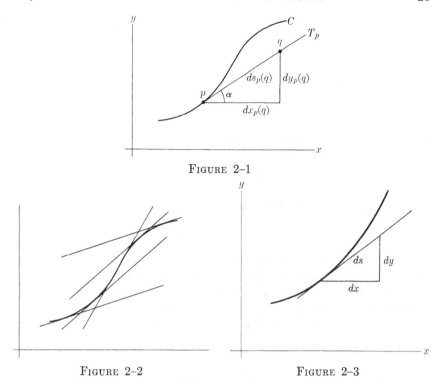

FIGURE 2–1

FIGURE 2–2   FIGURE 2–3

a variable du on the tangent bundle. Note that the establishment of (1) as an identity on T_p for *every* p on C establishes the broader identity

$$dy = D_x y \, dx$$

on the tangent bundle. This is, for dx and dy, the fundamental theorem identity that is the chief working tool in the theory of differentials. It will be shown in Section 2–2 that for any u and v to which the theory applies,

$$dv = D_u v \, du.$$

This is the formula used in formal substitution in an integral. It is probably more important, however, for yielding the representation of a derivative as a quotient of differentials:

$$\frac{dv}{du} = D_u v. \tag{2}$$

For example, to find $D_s x$ on a curve it is usually futile to look for a relation $x = f \circ s$ that will yield this derivative directly. Instead, use (2) and find $D_s x$ by finding dx/ds. Figure 2–3 is the standard starting point.

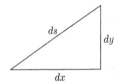
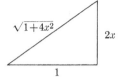

FIGURE 2–4

It shows the geometric significance of the differentials, and (see Example 1 below) for a specific curve, it is easy to find a triangle similar to the differential triangle of Fig. 2–3. Once similar triangles are noted, ratios of sides are obtained immediately.

From Fig. 2–3 comes the important relation

$$ds^2 = dx^2 + dy^2.$$

Working from this to find derivatives of (or with respect to) s, one must take a square root; and the question of sign should be noted. Positive dx and dy have the obvious geometric interpretation, but positive ds may go either way. In any given problem, the positive arc-length direction is established by the choice of sign when a radical is introduced.

EXAMPLES

1. On the locus of $y = x^2$ find $D_s x$ and $D_s y$. Proceed as follows:

$$dy = 2x\, dx;$$

so, the triangles shown in Fig. 2–4 are similar, and

$$ds = \sqrt{1 + 4x^2}\, dx, \tag{3}$$

$$D_s x = \frac{dx}{ds} = \frac{1}{\sqrt{1 + 4x^2}},$$

$$D_s y = \frac{dy}{ds} = \frac{dx}{ds}\frac{dy}{dx} = \frac{2x}{\sqrt{1 + 4x^2}}.$$

These results depend on the direction convention established by (3), namely that the positive direction is that of increasing x.

2. Find expressions for ds on the tangent bundle for the unit circle, and discuss positive direction conventions. On the unit circle,

$$x^2 + y^2 = 1 \tag{4}$$

or

$$y = \pm\sqrt{1 - x^2};$$

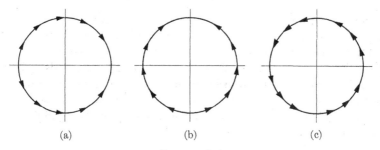

(a)   (b)   (c)

FIGURE 2–5

so, over the tangent bundle,

$$dy = \frac{-x\, dx}{\pm\sqrt{1 - x^2}} = \frac{-x\, dx}{y}.$$ (5)

Thus, recalling (4), one finds that

$$ds^2 = dx^2 + dy^2 = \left(1 + \frac{x^2}{y^2}\right) dx^2 = \frac{dx^2}{y^2}.$$

In a similar way it follows that

$$ds^2 = \frac{dy^2}{x^2}.$$

In taking square roots there is a more or less arbitrary choice of sign, and each of the following is correct for the direction convention shown in the indicated figure:

$$ds = \frac{dx}{|y|} \quad \text{(Fig. 2–5a)},$$

$$ds = \frac{dy}{|x|} \quad \text{(Fig. 2–5b)},$$

$$ds = \frac{-dx}{y} = \frac{dy}{x} \quad \text{(Fig. 2–5c)}.$$

3. A common procedure for deriving equation (5) from (4) above is as follows. Start with

$$x^2 + y^2 = 1,$$

and take differentials:

$$2x\, dx + 2y\, dy = 0.$$ (6)

From this (5) follows at once. While the only differentials that appear explicitly are dx and dy, a complete analysis of this process involves concepts not introduced in this section. Specifically, from $x^2 + y^2 = 1$, one concludes that $d(x^2 + y^2) = 0$, but what does $d(x^2 + y^2)$ mean? Presumably it expands as in (6), but it should be defined first.

The point to this discussion is that a "theory of differentials" with only three differentials in it is quite inadequate. Furthermore, it is obviously futile to introduce others one at a time. For this reason, one uses the algebraic approach in Section 2–2. It furnishes a blanket definition that includes all needed differentials.

EXERCISES

1. In each of the following find two expressions for ds. From each of these find $D_s x$ and $D_s y$. Sketch the locus, and discuss positive direction conventions for each result.

(a) $x^2 + y^2 = 4$	(b) $xy = 1$
(c) $xy = -1$	(d) $y = x^3$
(e) $y = 2x - x^2$	(f) $y = 3x - x^3$
(g) $x^2 - y^2 = 1$	(h) $y = \sin x$
(i) $y^2 = x$	(j) $y^2 = -x$
(k) $x = \sin y$	(l) $y = \arcsin x$

2. Let C be a parabola, the locus of $y = x^2$. Define x, y, and s on C in the usual way with the direction of increasing s being that of increasing x. Each of the following is in the form $du_p(q)$, where u is x, y, or s and where p and q are denoted by their rectangular coordinates. In each case, verify that (p, q) is on the tangent bundle (i.e., that p is on C and q is on T_p) and evaluate the expression given.

(a) $dx_{(1,1)}(2, 3)$	(b) $dy_{(1,1)}(3, 5)$
(c) $ds_{(-2,4)}(-1, 0)$	(d) $dx_{(0,0)}(3, 0)$
(e) $dx_{(-2,4)}(-3, 8)$	(f) $dy_{(0,0)}(2, 0)$
(g) $ds_{(0,0)}(-1, 0)$	(h) $dx_{(1,1)}(-1, -3)$
(i) $dy_{(1,1)}(0, -1)$	(j) $ds_{(-2,4)}(0, -4)$
(k) $dx_{(1,1)}(-1, -3)$	(l) $dy_{(-2,4)}(-1, 0)$
(m) $ds_{(1,1)}(3, 5)$	(n) $ds_{(1,1)}(-1, -3)$
(o) $dx_{(0,0)}(-4, 0)$	(p) $dy_{(-2,4)}(-3, 8)$

3. Let x, y, and s be as in Exercise 2. In each of the following find the rectangular coordinates of q.

(a) $dx_{(1,1)}(q) = 2$	(b) $dy_{(-1,1)}(q) = 3$
(c) $ds_{(2,4)}(q) = 17$	(d) $dy_{(-2,4)}(q) = -2$
(e) $dx_{(0,0)}(q) = -5$	(f) $ds_{(1,1)}(q) = -5$
(g) $dy_{(1,1)}(q) = 3$	(h) $dx_{(-2,4)}(q) = 2$
(i) $ds_{(0,0)}(q) = 4$	(j) $ds_{(-1,1)}(q) = -5$

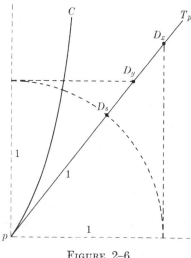

FIGURE 2–6

2–2 Differentials: algebraic theory. To define the differential is to describe a relation giving du from u. Now, du is supposed to be applied to point pairs (p, q), and the secret is to identify q, the point on the tangent line, with a mapping and use this latter mapping to generate (in a way to be described shortly) the u-to-du relation.

The mappings with which points on the tangent lines are to be identified are derivative operators, such as D_x, D_y, etc. As for the question, Why derivative operators?, the best answer is, It works. Remember, it took 250 years to discover this! The actual scheme of identification is easy to describe geometrically for the operators D_x, D_y, and D_s. This is shown in Fig. 2–6. Put D_s at the point one unit out the tangent line from the point of tangency p and draw a circle through D_s with center at p. From the point where this circle intersects the horizontal line through p, project vertically onto the tangent line to get D_x. Similarly, from the intersection of the circle and the vertical line through p, project horizontally onto the tangent line to locate D_y.

Three derivative operators are not enough. One needs a whole "line full" of them. So, at this stage it is best to abandon the geometric approach. Here is the plan. Derivative operators are defined in general by a pair of postulates. The postulates are not at all surprising. They endow derivative operators with familiar properties of differentiation. From these postulates it is fairly easy to prove that the set of derivative operators at a point has the algebraic structure of a line, so the above mentioned identification with the tangent line is justified. Then, to define a differential one must specify what it does to a derivative operator. The solution is to say, in effect, that du does to the operator D what D does to u.

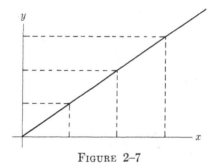

FIGURE 2-7

This last maneuver is a common one in modern algebra. Perhaps it is well to give an illustration of it in a more concrete setting. Consider the family of all lines through the origin. A fixed line from this family maps points on the x-axis into points on the y-axis (Fig. 2-7). On the other hand, a fixed point on the x-axis maps lines into points on the y-axis (Fig. 2-8). In terms of symbols for mappings this idea is expressed as follows. Take a set A and a set F of mappings, each carrying A into B. By means of the relation

$$\alpha(f) = f(a),$$

each a in A generates a mapping α which carries F into B. That is, when f operates on a, then a generates an α that operates on f. The modern definition of a differential uses this basic idea with a replaced by u (a variable on the curve), f by D (a derivative operator), and α by the differential du. So, the relation defining du takes the form

$$du(D) = Du.$$

This is the general outline of procedure. Now to fill in details, let D map certain variables on a curve into other variables on this curve. (The question of what kind of curve and just what variables is postponed to Section 2-3.) Let Du denote the map of u by D. If p is a point on the curve, denote the value of Du at p by $Du(p)$. Such a mapping D is called a *derivative operator at p* provided that

(i) $\ D(au + bv)(p) = a\,Du(p) + b\,Dv(p),$

(ii) $\ D(uv)(p) = u(p)\,Dv(p) + v(p)\,Du(p).$

Here u and v are arbitrary variables in the domain of D, and a and b are numbers.

The first step is to show from these postulates that D applied to any constant yields zero. In line with the usual practice here concerning

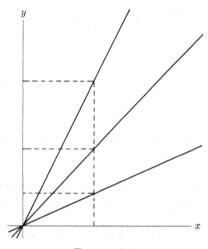

FIGURE 2-8

variables, this result will be written

$$Da(p) = 0. \tag{7}$$

More precisely, what is meant is that

$$D(au^0)(p) = 0.$$

The precise form helps in following the derivation. After that, the colloquial form (7) will be used. Note that for any v,

$$au^0v = av. \tag{8}$$

First, one applies postulate (ii) to the left side of (8):

$$D(au^0v)(p) = au^0(p)\,Dv(p) + v(p)\,D(au^0)(p)$$
$$= a\,Dv(p) + v(p)\,D(au^0)(p).$$

Now, one applies postulate (i) to the right side of (8):

$$D(av)(p) = a\,Dv(p).$$

If v is chosen so that $v(p) \neq 0$, a comparison of the two results shows that $D(au^0)(p) = 0$.

It also follows from these postulates that squares differentiate as one would expect. Set $u = v$, and postulate (ii) reads

$$D(u^2)(p) = 2u(p)\,Du(p). \tag{9}$$

In Section 2–3 it will be pointed out that this theory is to apply to some parent variable u and to all other variables v such that

$$v = f \circ u \tag{10}$$

for some thrice differentiable function f. The definition

$$D_u v = f' \circ u \tag{11}$$

given in Section 1–3 serves to define an operator D_u on this set of variables, and it is easily checked that D_u satisfies postulates (i) and (ii).

There are other derivative operators, but they are all very closely related to D_u. One can show the relation by introducing the polynomial approximation formula [(7), Section 1–4] for the function f in (10) and writing

$$v = v(p) + f'[u(p)][u - u(p)] + (g \circ u)[u - u(p)]^2. \tag{12}$$

Now, operate on (12) with D at p, and use (7) and (9) and the postulates. The result is

$$Dv(p) = f'[u(p)] \, Du(p). \tag{13}$$

Each of the other terms on the right vanishes either because of (7) or because it contains the factor $[u(p) - u(p)]$. In the light of (11), (13) may be rewritten

$$Dv(p) = Du(p) \, D_u v(p); \tag{14}$$

equation (14) displays the linear structure of the set of derivative operators. That is, since $Du(p)$ does not depend on v, (14) says that on every v, D is the same multiple of D_u. At p, the derivative operators are D_u and constant multiples thereof, which justifies the identification of this set of operators with the tangent line, T_p.

Consider that the identification is made, and define the *tangent bundle* for the given curve as the set of all ordered pairs (p, D), where p is a point of the curve, and D is a derivative operator at p. Clearly, this is an abstraction of the notion presented in Section 2–1.

Finally, the *differential* is defined in general as follows: For v a variable on the curve, dv is a variable on the tangent bundle. Its value at (p, D) is denoted by $dv_p(D)$, and dv is defined by

$$dv_p(D) = Dv(p). \tag{15}$$

Note that this definition gives dv in terms of v alone. For example, the variable u and the function f of equation (10) play no direct role in the definition of dv. Indeed, if (10) were replaced by $v = g \circ w$, dv would not be changed at all. The purpose of the entire discussion from the postulates through equation (14) is to justify the choice of domain for dv; i.e., ordered pairs (p, D) rather than (p, q) as in Section 2–1. Once this domain is chosen, the definition falls in place immediately.

If, now, $Dv(p)$ and $Du(p)$ in (14) are replaced by appropriate values of differentials, the result reads

$$dv_p(D) = D_u v(p) \, du_p(D). \tag{16}$$

Equation (16) holds for any p on the curve and any D on T_p; so,

$$dv = D_u v \, du \tag{17}$$

over the tangent bundle. The *fundamental theorem on differentials* states that given suitable conditions on u and v (to be described in Section 2–3) the relation (17) holds over the tangent bundle.

The right side of (17) is the product of a variable on the curve and one on the tangent bundle. This particular multiplication of dissimilar mappings is very simply defined; (17) means that (16) holds for all p and D. However, just as with the product of a number and a function, division can lead to trouble; and as pointed out in Section 2–1, one should divide (17) by du. So, let it be noted that the scrupulously correct way is to write

$$\frac{dv}{du} = D_u v \, du^0. \tag{18}$$

Normally (18) will be written

$$\frac{dv}{du} = D_u v; \tag{19}$$

that is, as usual, with variables, colloquial notation is recommended. The form (19) is probably best for solving problems, but for a complete understanding, one must recall that (18) gives the real meaning.

In Section 2–1 only three differentials appeared: dx, dy, and ds. Each of these was given a separate, geometric definition; but certain relations among these differentials were proved. Here a general definition of differential has been given (equation 15), applicable to any variable on the curve, and equation (17) gives a general form relating any two of these.

To be usable, this algebraic theory must have a geometric interpretation; and this is furnished, in outline form, by Fig. 2–6. Note that with D_x, D_y, and D_s representing the points shown in Fig. 2–6 and with dx, dy, and ds described as in Fig. 2–1,

$$dx(D_x) = dy(D_y) = ds(D_s) = 1.$$

This is in accord with the theory of the present section.

Actually, the algebraic structure is given a unique, geometric representation if one derivative operator is identified with a specific point. Let it be agreed that D_s will always be the point one unit out. This determines the geometric significance of ds and, through (17), of all other differentials. It is left to the student to check that Figs. 2–1 and 2–6 are consistent with this convention.

EXAMPLES

1. Prove that

$$d(uv) = u \, dv + v \, du.$$

For each (p, D) on the tangent bundle

$$
\begin{aligned}
d(uv)_p(D) &= D(uv)(p) &&\text{by (15)} \\
&= u(p) \, Dv(p) + v(p) \, Du(p) &&\text{by (ii)} \\
&= u(p) \, dv_p(D) + v(p) \, du_p(D) &&\text{by (15).}
\end{aligned}
$$

2. Show that the slope of the curve whose equation is

$$x^3 - xy + y^2 = 0$$

is given by

$$\frac{3x^2 - y}{x - 2y}.$$

On this locus the variable $x^3 - xy + y^2$ is constant; thus, by (7), its differential is identically zero. This differential may be expanded as follows:

$$d(x^3 - xy + y^2) = 3x^2 \, dx - y \, dx - x \, dy + 2y \, dy = 0. \tag{20}$$

Solve this equation for dy/dx to get the desired result. This process is sometimes called *implicit differentiation*. Actually, it is just a bit of routine manipulation with differentials.

Note the ideas and results involved in obtaining the relation (20). In addition to dx and dy, there appear (at least tacitly) differentials of $x^3 - xy + y^2$, x^3, xy, and y^2. To justify (20), one must use Exercise 2(b) below, equation (17), equation (7), and Example 1 above. Just where is each of these used?

3. Suppose that the base curve is the x-axis. Then x is the arc-length variable, and dx measures distances on the tangent lines. In this case, however, the tangent lines are superposed on the curve itself. They are distinguished one from the other by the positions of their origins. Now, let $y = x^2$ on this curve. This equation defines y; no geometric significance is prescribed for it, and there are no parabolas involved. By (17),

$$dy = 2x \, dx.$$

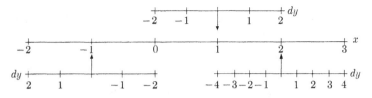

FIGURE 2-9

The significance of dx is known; so, the significance of dy is hereby determined. It measures distances to different scales on different tangent lines. A few samples are shown in Fig. 2-9.

This example is designed to illustrate two points. First, the relations among variables on a curve in this theory need have nothing to do with the equation of the curve in the sense of analytic geometry. Second, it is often convenient to introduce a coordinate axis as the base curve in this theory. In this case the tangent bundle has the usual algebraic structure, but it is hard to draw because each tangent line is merely a translation of the base curve.

EXERCISES

1. Let x and y be rectangular coordinate variables on the parabola such that $y = x^2$. Let the direction of increasing s be that of increasing x. Find the value of each of the following.

(a) $dx_{(1,1)}(D_y)$

(b) $dy_{(1,1)}(D_x)$

(c) $dx_{(-1,1)}(D_s)$

(d) $dy_{(-1,1)}(D_x)$

(e) $ds_{(2,4)}(D_x)$

(f) $dy_{(-2,4)}(D_y)$

(g) $ds_{(0,0)}(D_x)$

(h) $dx_{(2,4)}(D_s)$

(i) $dx_{(-1,1)}(D_y)$

(j) $dy_{(2,4)}(D_s)$

(k) $ds_{(-2,4)}(D_y)$

(l) $ds_{(-2,4)}(D_x)$

(m) $dx_{(1,1)}(D_z)$

(n) $dx_{(-1,1)}(D_x)$

(o) $dy_{(-1,1)}(D_s)$

(p) $ds_{(-1,1)}(D_y)$

[*Note:* Do each of these problems in two ways. (a) Apply (15) directly; each of these values of a differential is defined as the value of an appropriate derivative. To find the needed derivatives quickly, refer to Example 1, Section 2-1. (b) Solve them geometrically. Draw the required tangent lines and locate D_x, D_y, and D_s as in Fig. 2-6. Note the geometric definitions of dx, dy, and ds in Section 2-1; and obtain answers by considering similar triangles.]

2. Derive each of the following formulas from the postulates and definitions of this section.

(a) $da = 0$

(b) $d(au + bv) = a\,du + b\,dv$

(c) $d(u/v) = (v\,du - u\,dv)/v^2$

(d) $df(u + v) = f'(u + v)(du + dv)$

(e) $df(uv) = f'(uv)(u\,dv + v\,du)$

3. In each of the following find $D_x y$ in terms of x and y. Point out where the various parts of Exercise 2 are used.

(a) $x^2 + y^2 = 1$ (b) $x^2 y^3 = 1$

(c) $\sin xy = \cos(x + y)$ (d) $xy - \sqrt{y} = \sqrt{x}$

(e) $y = \ln(x + y)$ (f) $x^2 y - 2\sqrt{xy} = 2$

(g) $x^2 y - (y/x^2) = x$ (h) $x \sin y = \sin xy$

(i) $\sqrt{xy} + \sqrt{x/y} = \sqrt{y}$ (j) $xe^{xy} = y$

4. In terms of differentials write (a) the chain rule, and (b) the formula for derivative of an inverse function.

5. Use differentials to derive each of the following results.

(a) $D_{uv}w = \dfrac{D_v w}{u + v \, D_v u} = \dfrac{D_u w}{v + u \, D_u v}$

(b) $D_{u/v}w = \dfrac{v^2 \, D_v w}{v \, D_v u - u} = \dfrac{v^2 \, D_u w}{v - u \, D_u v}$

(c) If $D_x y = u$ and $D_x u = v$, then $D_y u = v/u$.

6. Suppose $u = av$, where a is a constant. (a) How are du_p and dv_p related as measuring devices on T_p? (b) How are D_u and D_v related as points of T_p? (c) Find $du(D_v)$ and $dv(D_u)$.

7. Draw tangent lines to the parabola on which $y = x^2$ at $(-1, 1)$, $(1, 1)$, and $(2, 4)$. On each of these mark off the du-scale of measurement, where u is defined on the parabola by

(a) $u = x^3$ (b) $u = \ln x$

(c) $u = x \ln y$ (d) $u = x/y$

2–3 Manifolds. Clearly, the underlying structure for the theory of Section 2–2 is a set of points and a set of variables defined thereon. In setting up the technical requirements for such a structure there are three main points to be considered.

I. The theory is based on the variables and their values. For these to represent fairly the geometry of the underlying point set, the variables should be one-to-one and in some sense continuous.

II. Equation (11), relates the notion of derivative operator to that of derivative in the function-theoretic sense. For a function to have a derivative its domain must be an interval; therefore, the range of each variable should be an interval.

III. To set up (12), one should require that each pair of variables considered be related by a thrice differentiable function.

Let M be a subset of some space in which distance is defined. Examples will be confined to subsets of Euclidean spaces, although it is possible to generalize the idea considerably. Given a number $a > 0$ and a point p

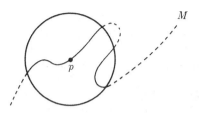

FIGURE 2–10

of M, the set of all points of M whose distances from p are less than a is called a *neighborhood of p in M*. In Fig. 2–10 the pieces of the curve lying inside the circle constitute a neighborhood of p. A variable u defined on M is *continuous at p* provided that, given $\epsilon > 0$, there is a neighborhood N of p in M such that $|u(p) - u(q)| < \epsilon$ whenever q is in N. Continuity of a mapping whose range is M is defined similarly.

Let M be a point set with neighborhoods defined in it as above, and let p be a point of M. A variable u is called a *coordinate in M at p* provided that

(a) The domain of u is a neighborhood of p in M.

(b) The range of u is an interval in the real number system.

(c) u is continuous.

Note that one speaks of a coordinate *at p*, but the coordinate must be defined over a neighborhood of p in M.

A *one-dimensional manifold* is a structure consisting of a point set M and a set V of variables defined on neighborhoods in M for which the following conditions are satisfied.

(d) For each point p of M, some variable in V is a coordinate in M at p. This coordinate is one-to-one, and its inverse is continuous.

(e) Any two points of M can be connected by a finite chain of overlapping neighborhoods each of which is the domain of one of these coordinate variables.

(f) If u and v are two variables from V with overlapping domains, then there is a thrice differentiable function f such that $v = f \circ u$ identically on the common part of their domains.

It may (and frequently does) happen that there is a single variable u defined on the whole of M so that u will qualify as a coordinate at every point. In this case, an informal description of the situation would be that M is a continuous image of a linear interval. More generally, a one-dimensional manifold is based on a set that can be broken into overlapping pieces each of which is mapped continuously onto a linear interval in such a way that these mappings are related by thrice differentiable functions on the overlapping parts.

Let (M, V) be a one-dimensional manifold. In addition to the variables in V, others are also considered, and these fall into three classes. Let p be a point of M, and let u be the coordinate at p taken from V. Consider variables v, each defined on some neighborhood of p in M, and such that $v = f \circ u$ on the common part of the domains of u and v. Classify these variables by the properties of the connecting function f as follows.

    A: f is differentiable,

    B: f is thrice differentiable,

    C: f is thrice differentiable, and $f'[u(p)] \neq 0$.

If v is in class C, then it can be shown that in some neighborhood of p, $u = \phi \circ v$ where ϕ is thrice differentiable and $\phi'[v(p)] \neq 0$. Furthermore, $F \circ \phi$ is differentiable or thrice differentiable according as F is; so the classes A, B, and C remain the same when the coordinate u is replaced by any variable of class C in their definition. Such a change of coordinate may change the neighborhoods of p on which pertinent relations hold, but this is immaterial so long as there still are such neighborhoods.

For obvious reasons, then, variables of class C will be called *admissible coordinates at* p; and for reasons soon to appear, variables of class B will be called *differentiable at* p.

In the light of these definitions, reexamine the development in Section 2-2. The extrapolation formula (12) is valid if u is an admissible coordinate at p and v is differentiable at p. The variable $g \circ u$ appearing in (12) will be of class A. Thus, if D is defined on class A, equation (13) may be derived from formula (12). This shows that if derivative operators are defined on class A, then they obey the chain rule (13) at least on class B. On the differentiable variables the derivative operators exhibit all the properties that one associates with a derivative. This is the reason for the definition of differentiable variable as given above.

Now a differential may be defined for any variable of class A by equation (15); but since equation (17) depends on equation (13) the fundamental theorem on differentials will be restricted to the differentiable variables. It may be stated in detail as follows: Let (M, V) be a one-dimensional manifold, and let M' be a subset of M. Let u be an admissible coordinate at each point of M', and let v be differentiable at each point of M'. Then,

$$dv = D_u v \, du$$

over that portion of the tangent bundle corresponding to points in M'.

This completes the basic theory. A number of comments are now in order. Observe first that a manifold is not merely a set of points. It is a point set plus a coordinate structure. One point set may admit two different coordinate structures and thus two different systems of derivative operators and differentials. (See Example 2 below.)

What does the term "one-dimensional" signify? Geometrically, it means that each key neighborhood is a one-to-one continuous image of a line interval. Analytically, it means that within its domain a single coordinate characterizes points uniquely so that any two admissible coordinates are connected by a function of one argument. One may, and frequently does, introduce relations of the form $w = \phi \circ (u, v)$ on a one-dimensional manifold. Ultimately, w depends on only one coordinate because u and v must be related here. However, in many instances it is unnecessary to express the (u, v) relation [see Exercises 1(d) and 1(e), Section 2–2].

Manifolds are defined by postulates—(a) through (f) above. As pointed out in this section, these postulates have been so contrived that, for appropriate variables, the algebraic derivative operators and equivalent quotients of differentials represent derivatives in the function-theoretic sense. That is, questions concerning the existence of derivatives are answered *a priori* in the postulates so that one may proceed with confidence to derive results by purely algebraic manipulation with differentials.

All this presupposes, of course, that one is operating on a manifold; so the question arises, Which structures are manifolds and which are not?

The remarks immediately following the postulates suffice to give the usual intuitive picture of geometric models. In this connection, it should be mentioned that a curve that crosses itself cannot form a one-dimensional manifold, and neither can a figure that contains the entire interior of a circle. This statement stems from postulates (b) and (d), of course. It seems clear (though it is not trivial to prove) that neither a cross nor the interior of a circle is a one-to-one, continuous image of a line interval.

A more pertinent form of the question, What sets can form one-dimensional manifolds? is the following. Assuming a coordinate structure on the plane in the sense of analytic geometry, what equations in these coordinates have loci that can form one-dimensional manifolds? Briefly, if an equation in x and y is built up by elementary operations, that is addition, multiplication, ln, sin, inversion, and composition, then selected portions of its locus can form one-dimensional manifolds. The portions must be selected to avoid discontinuities and crosses. If, in addition, the portions are selected to avoid corners, then both x and y can be admissible coordinates.

In many physical problems the manifold itself does not actually appear. What does appear is a set of physical variables, and it is tacitly assumed in setting up the problem that these are all related to some coordinate (frequently the time variable t) in such a way that the present theory is applicable.

EXAMPLES

1. Show that a circle can form a one-dimensional manifold. The problem is to define a set of coordinates in such a way that the postulates are satisfied. Observe, first, that the rectangular coordinates restricted to the circle will not qualify, because they are not one-to-one. A variable measuring

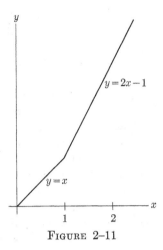

$$y = 2x - 1$$

$$y = x$$

FIGURE 2–11

central angles from a standard position seems the best bet, but one of these will not suffice. Suppose α is such an angle variable mapping the usual point into 0. By postulate (a) its domain must extend on both sides of the zero point. Now, swing around toward the 2π-position. If the ends of the domain of α overlap, then α will be double valued, violating postulate (d). If not, at least one point of the circle will be left out, violating postulate (d). One solution is to introduce two angle variables α and β, both measuring positive angles in the usual way. However, restrict α to a range $(0, 3\pi/2)$, and restrict β to a range $(\pi, 5\pi/2)$. The domains overlap in the third and first quadrants. In the third quadrant $\alpha = \beta$, and in the first quadrant $\alpha = \beta - 2\pi$. Thus, postulate (f) is satisfied, and the others are easily checked.

　　2. Let M be the locus of

$$y = \begin{cases} x & \text{for} & 0 \le x \le 1, \\ 2x - 1 & \text{for} & 1 \le x \le 2. \end{cases}$$

A sketch is shown in Fig. 2–11. With either x or y as a single coordinate, M is made into a one-dimensional manifold; each of these variables maps M by projection onto a line interval. What is the tangent line at the corner point $(1, 1)$? The trouble is that x and y are not both admissible coordinates at $(1, 1)$; one may use either, but not both. Choosing either of these as the coordinate, one gets the usual algebraic structure, but the familiar geometric picture of the tangent bundle often breaks down when the rectangular coordinates are not both admissible. It is entirely too restrictive to rule out manifolds with corners, but generally speaking, consideration in this book will be limited to those for which both x and y are admissible coordinates at all but a finite number of points.

EXERCISES

1. Show that a square can form a one-dimensional manifold.

2. Show that if f is continuous and has an interval for a domain, then the locus of $y = f(x)$ forms a one-dimensional manifold with x as a single coordinate.

3. It is possible (although complicated) to define on the interval $[0, 1]$ a function that is continuous, but nowhere differentiable. Let this be the function in Exercise 2, and discuss the tangent bundle for the resulting manifold.

4. Why (intuitively) cannot the graph of $y = \sin(1/x)$, together with the segment of the y-axis from -1 to 1, form a one-dimensional manifold?

5. Figure 2–12 shows a number of sketches. Which of these sets can form one-dimensional manifolds? For those that cannot, which postulates give trouble?

6. Prove that if u is an admissible coordinate, then du is not identically zero. At what points is $du = 0$?

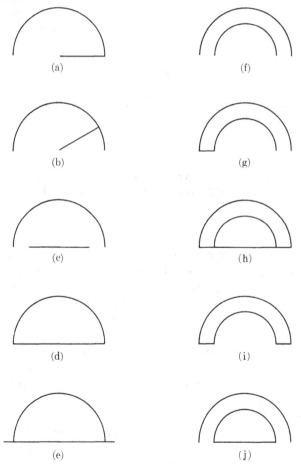

FIGURE 2–12

7. An obvious and enlightening corollary to the fundamental theorem on differentials is that, given the hypotheses of that theorem, $dv/du = D_u v$. Discuss the question of possible division by zero here.

8. Let M be the locus of $y = x^3$. (a) With x chosen as the coordinate, and (b) with y chosen as the coordinate, for what values of α is x^α differentiable? an admissible coordinate?

9. Define continuity for a mapping whose domain is part of the real number system and whose range is a set of points in a space in which distance is defined.

10. Prove that if (M, V) is a one-dimensional manifold, then each variable in V is an admissible coordinate at each point of its domain.

2–4 Higher-order derivatives. Let (M, V) be a one-dimensional manifold. Let u be an admissible coordinate over M, and let v be a differentiable variable over M with $v = f \circ u$. Then, $D_u v$ is a variable on M, and $D_u v = f' \circ u$. Thus, if f' is differentiable, $D_u v$ is susceptible to mapping by the operator D_u, yielding a variable

$$D_u(D_u v) = D_u^2 v.$$

Similarly, $D_u^2 v$ is a variable on M, and this process can be repeated as long as differentiability conditions hold to define

$$D_u^n v,$$

the nth *derivative of v with respect to u.* It follows at once by repeated applications of formula (12), that if f has $n + 2$ derivatives, then

$$D_u^n v = f^{(n)} \circ u$$

over M.

The situation with differentials is quite different, and the reason for this appears right at the start. The differential dv is not a variable on M; it is a variable on the tangent bundle for M. Now, differentials are defined for variables on M; so according to this definition the symbol $d(dv) = d^2 v$ is meaningless, and throughout this book it will remain so. That is, *higher-order differentials will not be defined.*

This categorical statement should serve to close the discussion, but in all fairness it should be added that, in the literature, one will find definitions of higher-order differentials and, more commonly, the derivative notation $d^n v/du^n$ (meaning $D_u^n v$) implying indirectly that $d^n v$ means something. The real difficulty (that dv is a variable on the wrong domain) is pointed out above, but Example 1 and Exercise 6 below show some of the additional problems that arise if one ignores the basic difficulty and tries to define higher-order differentials merely by algebraic forms, using relations modeled after the fundamental theorem of Section 2–3 as definitions.

Examples

1. From the basic relation

$$\frac{dv}{du} = D_u v$$

(corollary to the fundamental theorem, Section 2–3) there comes the temptation to write

$$\frac{d}{du} = D_u,$$

making d/du into a derivative operator. If this is done, then it seems natural to write

$$D_u^2 v = \left(\frac{d}{du}\right)^2 v = \frac{d^2 v}{du^2}.$$

In this case, the parallel to the relation in the fundamental theorem would be

$$d^2 v = D_u^2 v \, du^2, \tag{21}$$

and the idea occurs, "Let this be the definition of d^2v."

The difficulty here is that, unlike the relation

$$du = D_u v \, du,$$

the relation (21) is not invariant under change of coordinate. Consider the following simple example. Let u be an admissible coordinate on some manifold, and let v and w be defined on this same manifold by setting

$$v = u^6, \qquad w = u^2;$$

then

$$v = w^3.$$

By the fundamental theorem,

$$dv = 6u^5 \, du, \qquad dw = 2u \, du, \qquad dv = 3w^2 \, dw,$$

and substitution shows that the two expressions for dv are identical, as, indeed, they always will be; the fundamental theorem applies to any two suitably related variables. Now, try this for second-order differentials:

$$d^2 v = 30u^4 \, du^2, \qquad d^2 v = 6w \, dw^2;$$

substitute $dw = 2u \, du$ into the second of these to obtain

$$d^2 v = 6u^2 (2u \, du)^2 = 24u^4 \, du^2.$$

Thus, the values of d^2v depend on whether its definition is given in terms of u and du or in terms of w and dw. If second-order differentials are defined by (21), no coordinate changes are allowable.

2. Given $x^2 + y^2 = 1$, use differentials to find $D_x^2 y$. Set the differential of $x^2 + y^2$ equal to zero:

$$2x \, dx + 2y \, dy = 0.$$

From the equation above, it follows that

$$D_x y = \frac{dy}{dx} = \frac{-x}{y}.$$

From this the differential of $D_x y$ can be computed:

$$d(D_x y) = \frac{-(y \, dx - x \, dy)}{y^2}.$$

Thus,

$$D_x^2 y = \frac{d(D_x y)}{dx} = -\frac{y - x(dy/dx)}{y^2} = -\frac{y - x(-x/y)}{y^2}$$

$$= -\frac{x^2 + y^2}{y^3} = -\frac{1}{y^3}.$$

3. Show that

$$D_w^2 v = D_u^2 v \, (D_w u)^2 + D_u v \, D_w^2 u.$$

First, note that

$$D_w v = \frac{dv}{dw} = \frac{dv}{du} \frac{du}{dw} = D_u v \, D_w u;$$

so

$$d(D_w v) = D_u v \, d(D_w u) + d(D_u v) \, D_w u = D_u v \, d(D_w u) + \frac{d(D_u v)}{du} \frac{du^2}{dw}.$$

Thus,

$$D_w^2 v = \frac{d(D_w v)}{dw} = D_u v \, \frac{d(D_w u)}{dw} + \frac{d(D_u v)}{du} \left(\frac{du}{dw}\right)^2,$$

and this is equal to the required expression.

EXERCISES

1. Identify each of the following as a number or a variable and specify the domain of each variable. As usual, u and v are variables on M; f and g are functions; p is a point of M.

    (a) $D_u v$ (b) $f \circ u$ (c) $f[v(p)]$ (d) du_p

    (e) $du_p(D_v)$ (f) $d(f \circ u)_p$ (g) $d(D_u v)_p$ (h) $D_u(f \circ u)$

(i) du (j) $f' \circ u \, du$ (k) $d(f \circ u)$ (l) dv/du

(m) $D_u^2 v(p)$ (n) $f \circ g \circ u$ (o) $D_u(f \circ u)(p)$ (p) $D_u v(p) \, du_p$

(q) $D_u v \, du$ (r) dv_p/du_p

2. In each of the following find $D_x^2 y$.

(a) $x^2 - y^2 = 1$ (b) $x^2 + xy + y^2 = 1$

(c) $e^{xy} = x$ (d) $y \sin x = x$

(e) $y = \ln (xy)$ (f) $x^3 + y^2 = y$

(g) $\sin (x + y) = 2x$ (h) $e^{y/x} = y$

(i) $e^{x/y} = y$ (j) $x^2 + \sqrt{xy} - y = 0$

3. Find $D_y^2 x$ in terms of $D_x y$ and $D_x^2 y$.

4. Show that the differential equation

$$b_n x^n \, D_x^n y + b_{n-1} x^{n-1} \, D_x^{n-1} y + \cdots + b_1 x \, D_x y + b_0 y = 0$$

is transformed by the substitution $z = \ln x$ into

$$a_n \, D_z^n y + a_{n-1} \, D_z^{n-1} y + \cdots + a_1 \, D_z y + a_0 y = 0.$$

The a's and b's are constants. [*Hint:* Show by induction that

$$x^k \, D_x^k = D_z(D_z - 1)(D_z - 2) \cdots (D_z - k + 1).]$$

5. Show that the substitution $z = \cos x$ transforms

$$D_x^2 y + \cot x \, D_x y + n(n + 1)y = 0$$

(where n is a constant) into

$$(1 - z^2) \, D_z^2 y - 2z \, D_z y + n(n + 1)y = 0.$$

The latter expression is *Legendre's differential equation.*

6. Suppose that d is regarded as an operator operating indiscriminately on derivatives and differentials and satisfying a product law. That is, purely formally, let $d(uv) = u \, dv + v \, du$ regardless of what these symbols mean.

(a) **Start** with the fundamental theorem, $dv = D_u v \, du$; operate on this equation with d; reapply the fundamental theorem; and obtain the formal result

$$d^2 v = D_u^2 v \, du^2 + D_u v \, d^2 u.$$

Consider this as a definition of $d^2 v$.

(b) Show that the form in part (a) is invariant under coordinate changes; that is,

$$D_w^2 v \, dw^2 + D_w v \, d^2 w = D_u^2 v \, du^2 + D_u v \, d^2 u,$$

provided that du, dw, $d^2 u$, and $d^2 w$ are related as indicated.

(c) With this definition of second differential, try to express a second derivative as a quotient of differentials.

(d) Following this pattern, what would a third order differential look like?

2-5 Line integrals. Let (M, V) be a one-dimensional manifold with M having finite arc length. The purpose of this section is to introduce, for appropriate variables u and v on M, an integral to be written

$$\int_{\mathbf{M}} u \, dv. \tag{22}$$

Form (22) represents what is commonly called a *line integral*. The classical theory of line integrals will appear much later, but the basic idea will be introduced here because the standard treatment of many topics in calculus involves the use of line integrals, though these are not generally recognized as such in the literature.

The meaning of the integral (22) may be presented informally as follows. Partition M by points $p_0, p_1, p_2, \ldots, p_n$. For each i, let q_i be the point on the tangent line T_{p_i} whose distance from p_i is the length of the arc $p_i p_{i+1}$ on M. That is, unroll each increment of arc onto a tangent line (see Fig. 2-13). Now, form sums

$$\sum_{i=1}^{n-1} u(p_i) \, dv_{p_i}(q_i), \tag{23}$$

and define (22) as the limit of these sums as $n \rightarrow \infty$ and the distance between successive partition points tends to zero. This limit will exist provided that u is continuous and v is differentiable over the whole of M. The proof is omitted.

One of the simplest, and yet one of the most important, items in the theory of the line integral is the substitution rule. In the development given above, the points q_i on the tangent lines are determined by an intrinsic notion (arc length) on M and have no connection with the variables u and v. Therefore, if u is continuous, v differentiable, and w an admissible coordinate, then

$$\int_{\mathbf{M}} u \, dv = \int_{\mathbf{M}} u \, D_w v \, dw. \tag{24}$$

Proof. The substitution $dv = D_w v \, dw$ does not change the sums (23); hence it does not change their limit (22).

At this point reexamine the definition of

$$\int_a^b f \circ x \, dx \tag{25}$$

as given in Section 1-3. Clearly, (25) is a very special case of a line integral. However, as shown in Section 1-3, this special case is equal to the integral of a function; hence it may be evaluated by the fundamental theorem of calculus. In general, to evaluate the line integral (22) reduce it to a form similar to integral (25).

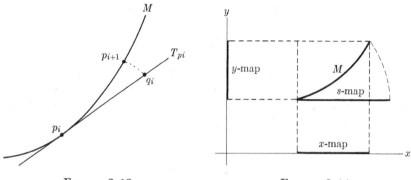

FIGURE 2–13   FIGURE 2–14

The machinery for this reduction is set up in the following manner. By definition, an admissible coordinate w on the whole of M maps M onto a real number interval, hence in a natural way onto a straight line interval. This straight line interval will be called the w-map of M. Note Fig. 2–14; the s-map of a curve is obtained by unrolling it onto a straight line; the x- and y-maps are obtained by projection.

Let M' be the w-map of M, where w is an arbitrary admissible coordinate over M. If p is a point on M, let p' be the corresponding point of M'. Variables on M are transferred to M' by the following rule. If u is any variable on M, define u' on M' by

$$u'(p') = u(p).$$

It is easily checked that this preserves all connecting functions. That is, if

$$u = f \circ v$$

on M, then

$$u' = f \circ v'$$

on M'. The variable w' is the arc-length variable on the straight line segment M'; therefore it is an admissible coordinate. Since functional relations are preserved, each primed variable has the same classification (continuous, differentiable, or admissible coordinate) as the corresponding unprimed variable. The important thing is that integrals are also invariant under this mapping process; that is, let M' be the w-map of M. Let u be continuous and v differentiable on M. Then,

$$\int_{\mathbf{M}} u \, dv = \int_{\mathbf{M}'} u' \, dv'. \tag{26}$$

The proof is omitted.

Now, the following procedure may be used to evaluate

$$\int_M u \, dv.$$

If v maps M one to one onto a line interval, use (26) to transfer the integral to the v-map of M. If v does not give a one-to-one mapping, use the substitution theorem (24) to introduce a variable w that does. Then, transfer to the w-map. Since w is one-to-one on M, there is a function f such that $uD_w v = f \circ w$; so on the w-map one has

$$\int_{M'} f \circ w' \, dw'. \tag{27}$$

Now, M' is a line segment on which w' measures distance; so (27) is precisely the same thing as (25). Change the notation to that of (25) and evaluate the integral in the usual way.

There is one important exception to the rule that the mapping used in this process must be one to one. If M is a *simple closed curve* (a one-to-one, continuous image of a circle) there is no single coordinate on the whole of M. However, in this case it is legitimate to generate the mapping by a variable that is double valued at only one point. For example, one can use the angle variable θ to transfer an integral around the unit circle to an integral from 0 to 2π even though θ assumes the values 0 and 2π at the same point on the circle. A one-point overlap like this does not affect an integral.

There are a few points concerning notation to be discussed. The basic notation for a line integral is

$$\int_M u \, dv, \tag{28}$$

in which the symbol M indicates the domain of integration. Frequently, it is convenient to introduce limits of integration in a notation reminiscent of that for the integral of a function. This will be done in the following manner. One may write

$$\int_{w=a}^{w=b} u \, dv \tag{29}$$

indicating that the curve on which the integral is based is specified in the context and that the integral is over the portion of the curve bounded by points at which $w = a$ and $w = b$. Note that the variable characterizing the end points of the domain of integration may or may not appear specifically in the remainder of the form. If $w = v$ in (29), the notation will be abbreviated to read

$$\int_a^b u \, dv. \tag{30}$$

It must be noted that the form (28) describes precisely the notion that has been defined here; integrals (29) and (30) may be introduced only when they are unambiguous. The principal sources of ambiguity are the following: (a) Values of w in (29) may not characterize points of M uniquely. (b) Even if they do, on a simple closed curve there are two ways of getting from one point to another.

Finally, here is a word about the substitution rule. Comparison of the substitution rule of this section with that for the integral of a function reveals that, in the integral of a function, the domain of integration seems to change with a substitution while in a line integral it does not. The situation is simply this. A substitution in the form (28) above does not change the domain M. A substitution in (30) still does not change the underlying curve, but it does change its description because of the notation convention that the a and b in (30) are linked to v. The following statement is the substitution rule for (30): Substitute for v and dv any variables that are equal to them. Change the numerical limits of integration so that, in terms of the new variable whose differential appears, they describe the same points on the underlying manifold as did the original limits.

EXAMPLES

1. Evaluate

$$\int_0^1 \sqrt{1 - x^2}\, dx.$$

The secret to the solution is the substitution

$$x = \sin u,$$

$$dx = \cos u\, du.$$

Points at which $x = 0$ and $x = 1$ are characterized, respectively, by $u = 0$ and $u = \pi/2$; so

$$\int_0^1 \sqrt{1 - x^2}\, dx = \int_0^{\pi/2} \sqrt{1 - \sin^2 u}\, \cos u\, du = \int_0^{\pi/2} \cos^2 u\, du$$

$$= \int_0^{\pi/2} \frac{1}{2}(1 + \cos 2u)\, du = \frac{u}{2} + \frac{1}{4}\sin 2u\Big|_0^{\pi/2} = \frac{\pi}{4}.$$

2. Evaluate

$$\int_M (x\, dy - y\, dx),$$

where M is the arc of the parabola on which $y = x^2$ from $(0, 0)$ to $(1, 1)$.

On the tangent bundle for M,

$$dy = 2x\,dx;$$

so (substitution rule)

$$\int_M (x\,dy - y\,dx) = \int_M (2x^2\,dx - x^2\,dx) = \int_M x^2\,dx.$$

Now, by formula (24), transfer to the x-map, and the result is

$$\int_0^1 x^2\,dx = \tfrac{1}{3}x^3\Big|_0^1 = \tfrac{1}{3}.$$

Strictly speaking, x should have been replaced by x' after the transfer. As a general rule, this will not be done. This is another instance in which precision in notation is helpful in explaining an idea, but is commonly abandoned in practice.

3. Evaluate

$$\int_M \frac{x\,dy - y\,dx}{x^2 + y^2},$$

where M is the unit circle traversed counterclockwise. Introduce the polar coordinate variable θ. On M,

$$x = \cos\theta, \qquad dx = -\sin\theta\,d\theta,$$
$$y = \sin\theta, \qquad dy = \cos\theta\,d\theta,$$
$$x^2 + y^2 = 1;$$

so,

$$\int_M \frac{x\,dy - y\,dx}{x^2 + y^2} = \int_M \cos^2\theta\,d\theta + \sin^2\theta\,d\theta = \int_M d\theta.$$

Transfer to the θ-map:

$$\int_M d\theta = \int_0^{2\pi} d\theta = 2\pi.$$

EXERCISES

1. In each of the following make the indicated substitution and evaluate the integral.

(a) $\displaystyle\int_0^2 x^3\sqrt{4 - x^2}\,dx;\ x = 2\sin u$

(b) $\displaystyle\int_0^2 x^3\sqrt{4 - x^2}\,dx;\ u = 4 - x^2$

(c) $\displaystyle\int_3^5 \frac{\sqrt{25-x^2}\,dx}{x^2}$; $x = 5\,\text{sech}\,u$

(d) $\displaystyle\int_2^4 \frac{\sqrt{x^2-4}\,dx}{x}$; $x = 2\sec u$

(e) $\displaystyle\int_0^2 \frac{x^5\,dx}{(1+x^3)^{3/2}}$; $u^2 = 1 + x^3$

(f) $\displaystyle\int_0^1 x\sqrt{1-x}\,dx$; $u^2 = 1 - x$

2. Evaluate each of the following line integrals. In each case x and y are the rectangular coordinates; r and θ are the polar coordinates; s is the arc-length variable on the manifold in question.

(a) $\int_M (y\,dx + x\,dy)$, where M is the parabolic arc on which $y = x^2$ from $(0, 0)$ to $(2, 4)$

(b) $\int_M x\,ds$, where M is the parabolic arc in a

(c) $\int_M y \sin x\,ds$, where M is the arc on which $y = \cos x$ from $(0, 1)$ to $(\pi/2, 0)$

(d) $\int_M r^2\,d\theta$, where M is the circle on which $r = \cos\theta$, traversed counterclockwise

(e) $\int_M x\,d\theta$, where M is the circle in d

(f) $\int_M y\,d\theta$, where M is the circle in d

(g) $\int_M (xy\,dx + x^2\,dy)$, where M is the broken line from $(0, 1)$ to $(2, 1)$ to $(2, 3)$

(h) $\int_M (xy\,dx + x^2\,dy)$, where M is the straight line from $(0, 1)$ to $(2, 3)$

3. Let M be a simple closed curve in the plane, and let x and y be the usual rectangular coordinates restricted to M. To avoid complications, assume that M is strictly convex; each tangent line touches it in only one point. Let the positive direction on M be counterclockwise. Show that each of the following line integrals gives the area of the plane figure enclosed by M.

(a) $\displaystyle\int_M x\,dy$ (b) $\displaystyle\int_M -y\,dx$ (c) $\displaystyle\frac{1}{2}\int_M (x\,dy - y\,dx)$

4. Let the M of Exercise 3 be the unit circle, and let θ be the arc-length variable measured counterclockwise from the point $(1, 0)$. Clearly, $x = \cos\theta$ and $y = \sin\theta$ on M. Use the substitution rule to convert each of the integrals in Exercise 3 into an integral in terms of θ. Evaluate each of these directly, and check against the result stated in Exercise 3.

5. Use the following device to solve Exercise 2(a) above. Define u on M by $u = xy$. Compute du, substitute, and transfer to the u-map.

6. From the considerations in Exercise 5 show that if M is any simple closed curve in the plane, then

$$\int_M (y \, dx + x \, dy) = 0.$$

2–6 Comments on basic concepts. The equation

$$y = f \circ \mathrm{x} \tag{31}$$

has the structure of a sentence in which y, f, and x are nouns and the equal sign is a verb. A sentence may be used to define one of its words, but, more generally, the words are supposed predefined, and the sentence conveys additional information. Obviously, the information conveyed depends on the definitions ascribed to the words. Read aloud, "The Internal Revenue Service taxes the doctor's patients (patience)."

In print the meaning is clear here, but in a typical calculus problem there is an ambiguity in the meaning of (31) that is not cleared up even by the printed form. In analytic geometry x and y are regarded as variables, each defined over the entire plane; and (31) is thought of as defining a locus in the plane. In the calculus of variables, x and y are presumed defined only on a manifold M in such a way that (31) is an identity—for every p in M, $y(p) = f[x(p)]$. In the broader picture, both points of view emerge. One starts with the equation (31), takes the analytic geometry point of view and comes up with a locus, lets that locus be the manifold M, and then reinterprets the symbols x and y so that the conditional nature of (31) disappears. With a little practice this shift in the interpretation of (31) becomes purely automatic, and in many discussions of calculus it is ignored altogether. This is unfortunate, to say the least, because unless the meaning of (31) is clearly understood, calculus is little more than a systematic set of marks on the page.

The calculus of variables is set up in such a way that the derivation of useful formulas becomes routine algebraic manipulation. Often this "formal calculus" is presented as nothing but a set of tricks which happen to give correct answers but which have no valid logical foundation and therefore do not prove anything. This is not a fair evaluation of the subject, and it has been the aim of this chapter to indicate how the calculus of variables may be established on a firm foundation so that its extremely helpful techniques may be used with confidence.

In many presentations of calculus, a differential is defined by something analogous to the fundamental theorem; that is, one differential, say dx, is defined arbitrarily, and from there on, $du = D_x u \, dx$ serves as a definition of du. Now, in such a scheme, it is pertinent to ask, Would it change anything if one started with dt instead of dx? The way in which differentials are often defined causes an embarrassing affirmative answer to this question, and hence the question is generally ignored. In the development of differentials given here, the question is not pertinent because each differential is defined separately. They are interdependent, but that is a consequence of the definition, not a part of it.

On the other hand, one usually finds du by relating u to x and then (fundamental theorem) finding du from dx. An analogous situation exists with regard to derivatives of functions. A function f determines f' with no help from other functions; but to find f' one relates f to I, sin, and ln and gets appropriate formulas. Practical methods for finding things must not be confused with definitions.

Finally, a few words about notation. The history of calculus notation is, in general, a story of confusion. The simultaneous development of calculus in England and Germany produced two different sets of notation, each with its ardent devotees and a considerable body of literature. About a hundred years after Newton and Leibnitz, various French mathematicians tried to simplify the situation by substituting completely new notation, but all they managed to do was to found a third school of thought on the subject. In current literature one can find three different symbols for a derivative:

$$f',\qquad \text{after Newton,}$$

$$\frac{dy}{dx},\qquad \text{after Leibnitz,}$$

$$D_x y,\qquad \text{after the French.}$$

Here all three notations have been introduced, but with different shades of meaning. The symbol $'$ is attached to *functions* to indicate differentiation. If x and y are *variables*, then $D_x y$ is the derivative of y with respect to x; and if $y = f \circ x$, then $D_x y = f' \circ x$. The *differentials dx and dy* are defined separately, and dy/dx means $dy \div dx$; however, there is a theorem that $dy/dx = D_x y$.

The student will find in the literature many variations of these basic forms, usually without the benefit of any careful definitions. Frequently, the trouble comes from writing $y = f \circ x$ and then confusing the symbols y and f. This leads to the notation y', df/dx, $D_x f$. If x, y, and f mean what they usually do, it seems doubtful that any one of these three symbols can be defined satisfactorily. At any rate, they have not been defined here and will not be used in this book. On the other hand, such notation does appear elsewhere in the literature, and it is to be hoped that the present study will help the student to translate such inexact notation into something meaningful on the basis of careful definitions.

CHAPTER 3

VECTORS IN THE PLANE

3–1 Vectors. Many physical quantities (displacements, forces, velocities, etc.) have two determining features, direction and magnitude. Thus, it is useful to isolate these two notions to define a purely mathematical idea that can be used to represent first one then another of these physical quantities. An ordered pair (direction and magnitude) is commonly called a *vector*. The current discussion will be limited to vectors in the plane. Other vector spaces will be considered later.

An arrow is generally used to describe a vector. The orientation of the shaft and position of the arrowhead indicate a direction, and the length of the shaft indicates magnitude. The only thing that is misleading about this picture is that one must draw the arrow *somewhere*, and position is not a part of a vector. Thus, two arrows pointing in the same direction and having the same length must be regarded as describing the same vector. Given a plane with coordinate axes, one could agree to draw all vectors as arrows emanating from the origin. For many purposes, this is a good plan, but it is not always the natural thing to do. For example, shortly there will be a discussion of vectors tangent to a curve. There is an obvious place to draw tangent vectors; but the important thing is to compare them, and this is best done by translating them to a common origin (see Fig. 3–1).

In modern algebra a *vector space* is a set of elements (called vectors) in which the operations of addition, subtraction, and multiplication by real numbers are defined. The sum or difference of two vectors is a uniquely

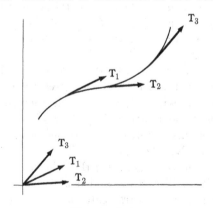

FIGURE 3–1

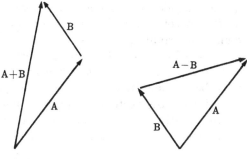

FIGURE 3–2

defined vector, and the product of a number and a vector is a uniquely determined vector. Finally, the expected rules of algebra hold for these operations. Addition of vectors is *commutative*,

$$\mathbf{A} + \mathbf{B} = \mathbf{B} + \mathbf{A},$$

and *associative*,

$$\mathbf{A} + (\mathbf{B} + \mathbf{C}) = (\mathbf{A} + \mathbf{B}) + \mathbf{C}.$$

Addition and subtraction are *inverse operations*,

$$(\mathbf{A} + \mathbf{B}) - \mathbf{B} = \mathbf{A}.$$

Multiplication by real numbers is *commutative*,

$$a\mathbf{A} = \mathbf{A}a,$$

associative,

$$a(b\mathbf{A}) = (ab)\mathbf{A},$$

and *distributive over addition* (of both numbers and vectors),

$$a(\mathbf{A} + \mathbf{B}) = a\mathbf{A} + a\mathbf{B}, \qquad (a + b)\mathbf{A} = a\mathbf{A} + b\mathbf{A}.$$

In the above examples, and in all subsequent discussions of vectors, the bold-faced symbols denote vectors, and the italicized symbols denote *scalars* (a vector-analysis term meaning real numbers). The above summary of basic operations in vector algebra is not a postulational description of a vector space. Rather it is an informal list of properties designed to answer the frequently heard question, Can I do this with vectors?

The directed magnitudes in the plane form an example of a vector space when addition and subtraction are defined as shown in Fig. 3–2, and multiplication by a scalar a consists of multiplying the magnitude by $|a|$ and taking the same or opposite direction according as a is positive or negative.

Vectors in the plane are independent of any coordinate system, and this is one of the salient features of vector analysis. It furnishes coordinate-free descriptions of many important quantities. However, if a coordinate system is introduced, it is important to relate the vectors to it, and this may be done by introducing *unit coordinate vectors* frequently called *basis vectors*. Let x and y be rectangular coordinates in the plane. Standard notation for the associated basis vectors is $\mathbf{i}$ and $\mathbf{j}$. These are vectors of unit magnitude in the positive x- and y-directions, respectively. Now, each vector $\mathbf{A}$ in the plane (Fig. 3–3) has a unique representation

$$\mathbf{A} = A_x\mathbf{i} + A_y\mathbf{j}.$$

The numbers A_x and A_y are called, respectively, the x- and y-*components* of $\mathbf{A}$. Thus, each vector is associated with an ordered pair of numbers:

$$\mathbf{A} \leftrightarrow (A_x, A_y).$$

However, the number-pair representation depends on the choice of coordinates. Exercises 6 and 7 below illustrate this dependence, and the general theory is discussed in Chapter 4.

Conspicuously absent among the vector algebra operations are multiplication and division of vectors. For vectors in three-space only, there is defined a vector product that acts *somewhat* like multiplication; this will be discussed later. *Division by vectors is not defined.* In *some* vector spaces, and in particular in the space of vectors in the plane, there is defined a notion of *scalar product*. For the plane vectors, scalar product may be defined as follows:

$$\mathbf{A} \cdot \mathbf{B} = |\mathbf{A}||\mathbf{B}| \cos \phi.$$

Here ϕ is the angle between the direction of $\mathbf{A}$ and $\mathbf{B}$. At this time (and subsequently), absolute-value signs around a vector symbol are used to denote the magnitude of the vector. Note that this is not really an operation within the vector space. As the terminology suggests, *the scalar product of two vectors is a scalar.*

Key properties of the scalar product are as follows:

$$\mathbf{A} \cdot \mathbf{B} = \mathbf{B} \cdot \mathbf{A},$$

$$\mathbf{A} \cdot \mathbf{A} = |\mathbf{A}|^2,$$

$$(a\mathbf{A}) \cdot \mathbf{B} = a(\mathbf{A} \cdot \mathbf{B}),$$

$$\mathbf{C} \cdot (\mathbf{A} + \mathbf{B}) = (\mathbf{C} \cdot \mathbf{A}) + (\mathbf{C} \cdot \mathbf{B}).$$

The first two properties are obvious from the definition. The third follows

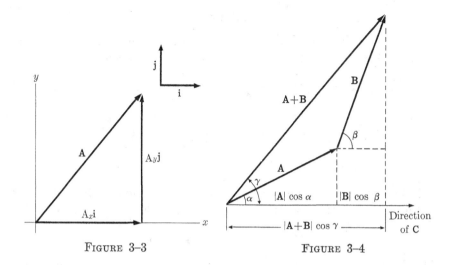

FIGURE 3-3   FIGURE 3-4

from the definition of multiplication of a vector by a scalar; that is, $|a\mathbf{A}| = |a||\mathbf{A}|$, and if $a < 0$, the angle between $a\mathbf{A}$ and $\mathbf{B}$ differs by π from the angle between $\mathbf{A}$ and $\mathbf{B}$.

With this third property established, it suffices (why?) to prove the distributive law for the case $|\mathbf{C}| = 1$. To this end, note that scalar multiplication by a unit vector gives the projection onto the direction of the unit vector. This being the case, the distributive law is apparent from Fig. 3-4.

It is worthwhile noting that components are given by scalar products. Specifically,

$$A_x = \mathbf{A} \cdot \mathbf{i}, \qquad A_y = \mathbf{A} \cdot \mathbf{j}.$$

Finally, there is an important formula for scalar products in terms of components. The following multiplication table for $\mathbf{i}$ and $\mathbf{j}$ is readily verified from the definition of scalar product.

	i	j
i	1	0
j	0	1

From this definition and the properties of scalar products listed above, it follows that

$$\begin{aligned}
\mathbf{A} \cdot \mathbf{B} &= (A_x\mathbf{i} + A_y\mathbf{j}) \cdot (B_x\mathbf{i} + B_y\mathbf{j}) \\
&= A_xB_x\mathbf{i} \cdot \mathbf{i} + A_xB_y\mathbf{i} \cdot \mathbf{j} + A_yB_x\mathbf{j} \cdot \mathbf{i} + A_yB_y\mathbf{j} \cdot \mathbf{j} \\
&= A_xB_x + A_yB_y.
\end{aligned}$$

EXAMPLES

1. Write in the form $A_x\mathbf{i} + A_y\mathbf{j}$ the vector $\mathbf{A}$ whose magnitude is 6 and which makes an angle $\pi/3$ with the positive x-axis. *Solution.*

$$A_x = \mathbf{A} \cdot \mathbf{i} = |\mathbf{A}|\,|\mathbf{i}|\,\cos\frac{\pi}{3} = (6)(1)(\tfrac{1}{2}) = 3,$$

$$A_y = \mathbf{A} \cdot \mathbf{j} = |\mathbf{A}|\,|\mathbf{j}|\,\cos\left(\frac{\pi}{2} - \frac{\pi}{3}\right) = (6)(1)\left(\frac{\sqrt{3}}{2}\right) = 3\sqrt{3},$$

$$\mathbf{A} = 3\mathbf{i} + 3\sqrt{3}\,\mathbf{j}.$$

2. Find the direction and magnitude of $3\mathbf{i} - 4\mathbf{j}$, and find the unit vector in this same direction. First, we get the magnitude:

$$3\mathbf{i} - 4\mathbf{j} = \sqrt{3^2 + 4^2} = 5.$$

The direction is clearly from the origin to $(3, -4)$, and the required unit vector is obtained by dividing the given vector by its magnitude:

$$\tfrac{3}{5}\mathbf{i} - \tfrac{4}{5}\mathbf{j}.$$

3. Let

$$\mathbf{A} = 2\mathbf{i} - 3\mathbf{j} \quad \text{and} \quad \mathbf{B} = 4\mathbf{i} + \mathbf{j}.$$

Find $\mathbf{A} + \mathbf{B}$, $\mathbf{A} \cdot \mathbf{B}$ and the component of $\mathbf{A}$ in the direction of $\mathbf{B}$. It follows from the basic rules of vector algebra listed above (which ones?) that vectors may be added a component at a time:

$$
\begin{aligned}
\mathbf{A} &= 2\mathbf{i} - 3\mathbf{j} \\
\mathbf{B} &= 4\mathbf{i} + \ \ \mathbf{j} \\
\hline
\mathbf{A} + \mathbf{B} &= 6\mathbf{i} - 2\mathbf{j}
\end{aligned}
$$

The formula for scalar multiplication in terms of components yields

$$\mathbf{A} \cdot \mathbf{B} = (2)(4) + (-3)(1) = 5.$$

The unit vector in the direction of $\mathbf{B}$ is

$$\frac{\mathbf{B}}{|\mathbf{B}|} = \frac{\mathbf{B}}{\sqrt{17}};$$

so the component of $\mathbf{A}$ in this direction is

$$\mathbf{A} \cdot \frac{\mathbf{B}}{\sqrt{17}} = \frac{5}{\sqrt{17}}.$$

Exercises

1. Let $\mathbf{A} = 2\mathbf{i} + \mathbf{j}$ and $\mathbf{B} = -\mathbf{i} + 3\mathbf{j}$. Find
   (a) $\mathbf{A} + \mathbf{B}$
   (b) $\mathbf{A} - \mathbf{B}$
   (c) $\mathbf{A} \cdot \mathbf{B}$
   (d) $|\mathbf{A}|$
   (e) $|\mathbf{B}|$
   (f) the component of $\mathbf{A}$ in the direction of $\mathbf{B}$
   (g) the component of $\mathbf{B}$ in the direction of $\mathbf{A}$

2. In each of the following exercises two operations are indicated. In each case insert parentheses to indicate the order in which the operations should be performed. Show in each case that there is only one way to insert parentheses to obtain a meaningful expression.

   (a) $\mathbf{A} \cdot \mathbf{BC}$ (b) $\mathbf{A} \cdot \mathbf{B} + \mathbf{C}$ (c) $\mathbf{A} \cdot \mathbf{B} + a$
   (d) $\mathbf{A}/\mathbf{B} \cdot \mathbf{C}$ (e) $\mathbf{AB} \cdot \mathbf{C}$ (f) $\mathbf{A}/a \cdot \mathbf{B}$
   (g) $\mathbf{A} + \mathbf{B} \cdot \mathbf{C}$ (h) $\mathbf{A}/a + \mathbf{B}$ (i) $a + \mathbf{A} \cdot \mathbf{B}$

3. Let $\mathbf{A}$ be the vector from the origin to p, and $\mathbf{B}$ the vector from the origin to q. What is the significance of $(\mathbf{A} + \mathbf{B})/2$?

4. For a triangle with sides a, b, and c with angle C opposite side c, the law of cosines reads

$$c^2 = a^2 + b^2 - 2ab \cos C.$$

Give a vector proof of this by letting $\mathbf{A}$ and $\mathbf{B}$ form sides of a triangle and expanding $(\mathbf{A} - \mathbf{B}) \cdot (\mathbf{A} - \mathbf{B})$.

5. How does scalar multiplication indicate perpendicularity of vectors?

6. Let $\mathbf{A} = 3\mathbf{i} - 4\mathbf{j}$ and $\mathbf{B} = 8\mathbf{i} + 6\mathbf{j}$.
   (a) Show that $\mathbf{A}$ and $\mathbf{B}$ are perpendicular.
   (b) Find unit vectors $\mathbf{u}$ and $\mathbf{v}$ in the directions of $\mathbf{A}$ and $\mathbf{B}$, respectively.
   (c) Express $2\mathbf{i} + 3\mathbf{j}$ in the form $a\mathbf{u} + b\mathbf{v}$.

7. Let $\mathbf{u} = \mathbf{i} \cos \alpha - \mathbf{j} \sin \alpha$, $\mathbf{v} = \mathbf{i} \sin \alpha + \mathbf{j} \cos \alpha$.
   (a) Show that $\mathbf{u}$ and $\mathbf{v}$ are perpendicular unit vectors. Make a sketch.
   (b) Find $\mathbf{i}$ and $\mathbf{j}$ in terms of $\mathbf{u}$ and $\mathbf{v}$.
   (c) Let $\mathbf{A} = A_x\mathbf{i} + A_y\mathbf{j} = A_u\mathbf{u} + A_v\mathbf{v}$. Find A_u and A_v in terms of α, A_x, and A_y.
   (d) The vector $\mathbf{A}$ has two number pair representations, (A_x, A_y) and (A_u, A_v). For what values of α are these identical?

3–2 Differential geometry of plane curves. Suppose M is a one-dimensional manifold set, and u and v are variables on M. Setting

$$\mathbf{w} = u\mathbf{i} + v\mathbf{j}$$

defines a *vector variable* $\mathbf{w}$ on M. This mapping $\mathbf{w}$ associates with each point of M a vector in the plane.

If t is an admissible coordinate on M, and the u and v above are differentiable, then one can define $D_t\mathbf{w}$, the derivative of $\mathbf{w}$ with respect to t:

$$D_t\mathbf{w}(p) = \lim_{q \to p} \frac{\mathbf{w}(q) - \mathbf{w}(p)}{t(q) - t(p)}.$$

This difference quotient may be written

$$\mathbf{i}\,\frac{u(q) - u(p)}{t(q) - t(p)} + \mathbf{j}\,\frac{v(q) - v(p)}{t(q) - t(p)};$$

so, clearly,

$$D_t\mathbf{w} = D_t u\mathbf{i} + D_t v\mathbf{j}. \tag{1}$$

Now, *define*

$$d\mathbf{w} = du\,\mathbf{i} + dv\,\mathbf{j}. \tag{2}$$

Although the definition of $d\mathbf{w}$ is given in terms of basis vectors $\mathbf{i}$ and $\mathbf{j}$, $d\mathbf{w}$ actually does not depend on the choice of basis vectors. This fact will be assumed throughout the remainder of this chapter, and a proof will appear in Chapter 4.

Comparison of (1) and (2) and use of the fundamental theorem on differentials for scalar variables shows that the fundamental theorem still goes:

$$d\mathbf{w} = D_t\mathbf{w}\,dt. \tag{3}$$

This establishes a calculus of variables for vector variables on a manifold. The calculus of variables is itself a powerful tool, and so is vector algebra. When the two are combined the results are fascinating indeed. The remainder of this chapter will present a few of the many geometric and physical problems that can be solved by simple manipulations with vector differentials.

Given rectangular-coordinate variables x and y, define the *position vector* $\mathbf{r}$ by

$$\mathbf{r} = x\mathbf{i} + y\mathbf{j}.$$

In other words, for each point p in the plane, $\mathbf{r}(p)$ is the vector from the origin to p. Note that parametric equations for a curve,

$$x = f \circ t, \qquad y = g \circ t,$$

may now be written in vector form:

$$\mathbf{r} = (f \circ t)\mathbf{i} + (g \circ t)\mathbf{j}.$$

If these variables are restricted to a one-dimensional manifold M, then on the tangent bundle for M,

$$d\mathbf{r} = dx\,\mathbf{i} + dy\,\mathbf{j} = \mathbf{T}\,ds, \tag{4}$$

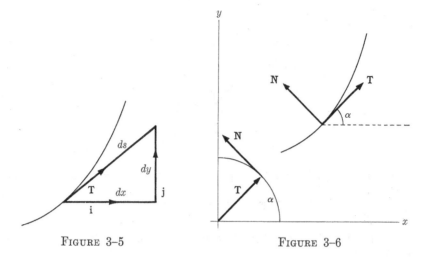

FIGURE 3–5   FIGURE 3–6

where $\mathbf{T}$ is a unit tangent vector to M (see Fig. 3–5). One must not confuse $d\mathbf{r}$ and ds here. Given p on M and q on the tangent line at p, $ds_p(q)$ is the numerical measure of directed distance from p to q; $d\mathbf{r}_p(q)$ is the vector from p to q.

Now, regarded as a position vector itself, $\mathbf{T}$ sweeps out an arc of the unit circle. Arc length on the unit circle is the angle variable, but the angle in question is α, the slope angle for M. The unit tangent to this circle is $\mathbf{N}$, the unit normal to M (Fig. 3–6). So (4) may be applied again with $\mathbf{r}$ replaced by $\mathbf{T}$, s by α, and $\mathbf{T}$ by $\mathbf{N}$:

$$d\mathbf{T} = \mathbf{N}\, d\alpha. \tag{5}$$

Note that use of α, together with the standard convention for positive angles, establishes the convention that the rotation angle from $\mathbf{T}$ to $\mathbf{N}$ is $+\pi/2$; that is, counterclockwise.

From (4) and (5) and the fundamental theorem (3), it now follows that

$$D_s\mathbf{r} = \frac{d\mathbf{r}}{ds} = \mathbf{T},$$

$$D_s^2\mathbf{r} = D_s\mathbf{T} = \frac{d\mathbf{T}}{ds} = \mathbf{N}\frac{d\alpha}{ds}.$$

The scalar $d\alpha/ds$ (rate of change of slope angle with respect to arc length) is called the *curvature* of M, which is commonly denoted by κ. Thus,

$$D_s^2\mathbf{r} = \kappa\mathbf{N}. \tag{6}$$

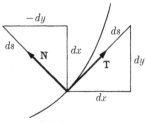

FIGURE 3–7

To find an expression for κ in terms of derivatives of x and y, note (Fig. 3–7) that

$$\mathbf{N} = -D_s y \mathbf{i} + D_s x \mathbf{j};$$

then take the scalar product of each side of (6) with $\mathbf{N}$, and recall that $\mathbf{N} \cdot \mathbf{N} = 1$:

$$\kappa = D_s^2 \mathbf{r} \cdot \mathbf{N} = D_s^2 y \, D_s x - D_s^2 x \, D_s y. \tag{7}$$

Parametric equations for the circle with its center at the origin and radius a are

$$x = a \cos \theta, \qquad y = a \sin \theta,$$

where θ is the angle variable. Thus, arc length is given by $s = a\theta$, and the equation may be written

$$x = a \cos \frac{s}{a}, \qquad y = a \sin \frac{s}{a}.$$

The derivatives are

$$D_s x = -\sin \frac{s}{a}, \qquad D_s y = \cos \frac{s}{a},$$

$$D_s^2 x = -\frac{1}{a} \cos \frac{s}{a}, \qquad D_s^2 y = -\frac{1}{a} \sin \frac{s}{a}.$$

Substitution in (7) yields

$$\kappa = \frac{1}{a} \sin^2 \frac{s}{a} + \frac{1}{a} \cos^2 \frac{s}{a} = \frac{1}{a}.$$

The curvature of a circle is the reciprocal of its radius. This explains the definition

$$R = \frac{1}{\kappa},$$

where R is called the *radius of curvature* of M.

If from a point p of M, one proceeds a directed distance $R(p)$ along the axis determined by $\mathbf{N}(p)$, the resulting point is called the *center of curvature* of M at p. The circle through p with its center at the center of curvature is called the *circle of curvature* for M at p. The position vector for the center of curvature is

$$\mathbf{r} + R\mathbf{N},$$

in components,

$$(x - R\,D_s y)\mathbf{i} + (y + R\,D_s x)\mathbf{j}.$$

Formula (7) for curvature is not too practical because derivatives with respect to arc length (particularly second derivatives) are somewhat cumbersome to compute. Another formula in terms of an arbitrary parameter t may be derived in the following manner. Return to (4) and divide by dt:

$$D_t \mathbf{r} = \frac{d\mathbf{r}}{dt} = \mathbf{T}\frac{ds}{dt} = \mathbf{T}\,D_t s.$$

Differentiate with respect to t:

$$D_t^2 \mathbf{r} = \mathbf{T}\,D_t^2 s + D_t \mathbf{T}\,D_t s.$$

To find $D_t\mathbf{T}$, return to (5):

$$d\mathbf{T} = \mathbf{N}\,d\alpha = \kappa\mathbf{N}\,ds,$$

$$D_t\mathbf{T} = \frac{d\mathbf{T}}{dt} = \kappa\mathbf{N}\frac{ds}{dt} = \kappa\mathbf{N}\,D_t s.$$

Therefore,

$$D_t^2\mathbf{r} = \mathbf{T}\,D_t^2 s + \kappa\mathbf{N}\,(D_t s)^2. \tag{8}$$

Take the scalar product with $\mathbf{N}$ and recall that $\mathbf{N}\cdot\mathbf{N} = 1$, $\mathbf{N}\cdot\mathbf{T} = 0$:

$$\kappa\left(\frac{ds}{dt}\right)^2 = D_t^2\mathbf{r}\cdot\mathbf{N}$$

$$= D_t^2 y\frac{dx}{ds} - D_t^2 x\frac{dy}{ds}$$

$$= D_t^2 y\frac{dx}{dt}\frac{dt}{ds} - D_t^2 x\frac{dy}{dt}\frac{dt}{ds}.$$

Thus,

$$\kappa\left(\frac{ds}{dt}\right)^3 = D_t^2 y\,D_t x - D_t^2 x\,D_t y,$$

and since $ds/dt = \sqrt{(D_t x)^2 + (D_t y)^2}$, then

$$\kappa = \frac{D_t^2 y\,D_t x - D_t^2 x\,D_t y}{[(D_t x)^2 + (D_t y)^2]^{3/2}}. \tag{9}$$

Finally, note that the description,

$$y = f \circ x,$$

of a curve may be regarded as a special case of a parametric description by writing it as

$$x = t, \qquad y = f \circ t.$$

It is easily verified that, for this special case, formula (9) for curvature reduces to

$$\kappa = \frac{D_x^2 y}{[1 + (D_x y)^2]^{3/2}}.$$

EXAMPLES

1. For the ellipse whose parametric equations are

$$x = 3 \cos t, \qquad y = 4 \sin t,$$

sketch the circle of curvature at the point at which $t = \pi/4$. First, compute the derivatives:

$$D_t x = -3 \sin t, \qquad D_t^2 x = -3 \cos t,$$

$$D_t y = 4 \cos t, \qquad D_t^2 y = -4 \sin t.$$

When $t = \pi/4$, the following variables have the indicated values:

x	y	$D_t x$	$D_t y$	$D_t^2 x$	$D_t^2 y$	$\dfrac{dx}{ds}$	$\dfrac{dy}{ds}$	R
$\dfrac{3}{\sqrt{2}}$	$\dfrac{4}{\sqrt{2}}$	$\dfrac{-3}{\sqrt{2}}$	$\dfrac{4}{\sqrt{2}}$	$\dfrac{-3}{\sqrt{2}}$	$\dfrac{-4}{\sqrt{2}}$	$\dfrac{-3}{5}$	$\dfrac{4}{5}$	$\dfrac{125}{24\sqrt{2}}$

The values of dx/ds and dy/ds indicate the positive tangential direction. The positive normal direction goes to the left from this. Since R is positive, proceed a distance $125/24\sqrt{2}$ in the positive normal direction. This gives the center of curvature, then draw the circle. A sketch is shown in Fig. 3–8.

2. Find the point of maximum curvature on the locus of $y = e^x$. For this curve, $\kappa = e^x/(1 + e^{2x})^{3/2}$; hence,

$$D_x \kappa = e^x (1 + e^{2x})^{-3/2} - 3e^{3x}(1 + e^{2x})^{-5/2}$$

$$= e^x (1 - 2e^{2x})(1 + e^{2x})^{-5/2}.$$

Setting $D_x \kappa = 0$ yields

$$1 - 2e^{2x} = 0, \qquad e^{2x} = \tfrac{1}{2}, \qquad x = \tfrac{1}{2} \ln \tfrac{1}{2} = -\ln \sqrt{2}.$$

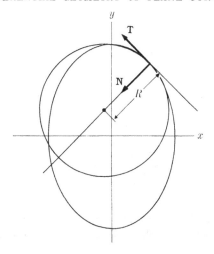

FIGURE 3–8

Since $2e^{2x}$ is increasing, $D_x\kappa$ changes from positive to negative at this point, therefore κ is a maximum. The point on the curve is $(-\ln\sqrt{2}, 1/\sqrt{2})$.

3. Find parametric equations for the locus of all centers of curvature of the curve defined by

$$x = t - \sin t, \qquad y = \cos t - 1.$$

This locus of centers of curvature is called the *evolute* of the given curve. *Solution.*

$$D_t x = 1 - \cos t, \qquad D_t y = -\sin t,$$

$$D_t^2 x = \sin t, \qquad D_t^2 y = \cos t,$$

$$ds^2 = (1 - 2\cos t + \cos^2 t + \sin^2 t)\, dt^2 = (2 - 2\cos t)\, dt^2,$$

$$R = \frac{(2 - 2\cos t)^{3/2}}{-\cos t\,(1 - \cos t) + \sin^2 t} = \frac{(2 - 2\cos t)^{3/2}}{1 - \cos t} = 2(2 - 2\cos t)^{1/2}.$$

Thus, parametric equations for the evolute are

$$X = x - R\frac{dy}{ds} = t - \sin t - 2(2 - 2\cos t)^{1/2}\frac{-\sin t}{(2 - 2\cos t)^{1/2}}$$

$$= t + \sin t,$$

$$Y = y + R\frac{dx}{ds} = \cos t - 1 + 2(2 - 2\cos t)^{1/2}\frac{1 - \cos t}{(2 - 2\cos t)^{1/2}}$$

$$= 1 - \cos t.$$

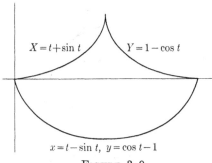

$$X = t + \sin t \qquad Y = 1 - \cos t$$

$$x = t - \sin t, \quad y = \cos t - 1$$

FIGURE 3–9

A sketch is shown in Fig. 3–9. The given curve is a cycloid, and so is the evolute. An interesting mechanical application of this fact is described in Exercise 7, Section 3–4.

4. Find the equation of the evolute of the parabola whose equation is $y = x^2$. *Solution.*

$$D_x y = 2x,$$

$$D_x^2 y = 2,$$

$$ds^2 = (1 + 4x^2)\, dx^2,$$

$$R = \frac{(1 + 4x^2)^{3/2}}{2}.$$

Thus, parametric equations for the evolute (with x as parameter) are

$$X = x - R\frac{dy}{ds} = x - \frac{(1 + 4x^2)^{3/2}}{2}\frac{2x}{(1 + 4x^2)^{1/2}} = -4x^3,$$

$$Y = y + R\frac{dx}{ds} = x^2 + \frac{(1 + 4x^2)^{3/2}}{2}\frac{1}{(1 + 4x^2)^{1/2}} = 3x^2 + \frac{1}{2}.$$

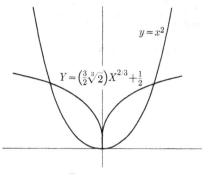

$$y = x^2$$

$$Y = \left(\tfrac{3}{2}\sqrt[3]{2}\right)X^{2/3} + \tfrac{1}{2}$$

FIGURE 3–10

To obtain the rectangular equation for this evolute, eliminate x between these two equations:

$$x = \left(\frac{-X}{4}\right)^{1/3},$$

$$Y = 3\left(\frac{-X}{4}\right)^{2/3} + \frac{1}{2} = \frac{3}{2\sqrt[3]{2}} X^{2/3} + \frac{1}{2}.$$

A sketch is shown in Fig. 3–10.

EXERCISES

1. Show that $\kappa^2 = |D_s^2 \mathbf{r}|^2$.

2. Derive formula (9) from the definition $\kappa = d\alpha/ds$ by writing

$$\alpha = \arctan (dy/dx)$$

and computing $d\alpha$ and ds in terms of dt.

3. Note the orientation conventions for $\mathbf{T}$ and $\mathbf{N}$ [remark following equation (5)] and discuss the geometric significance of the sign of κ.

4. Sketch the locus of each of the following and show the positive tangential and normal directions and the circle of curvature at the point indicated.

(a) $x = \sin t,\ y = \sin 2t \quad (t = \pi/4)$

(b) $x = \cos t,\ y = \cos 2t \quad (t = \pi/2)$

(c) $x = \cos t,\ y = \cos 2t \quad (t = 3\pi/2)$

(d) $x = 3 \cos t,\ y = 4 \sin t \quad (t = \pi/2)$

(e) $x = 3 \cos t,\ y = 4 \sin t \quad (t = 0)$

(f) $x = \sec t,\ y = \tan t \quad (t = \pi)$

(g) $x = t - \sin t,\ y = 1 - \cos t \quad (t = \pi)$

(h) $x = 2 \cos t - \cos 2t,\ y = 2 \sin t - \sin 2t \quad (t = \pi/2)$

(i) $x = \dfrac{3t}{1 + t^3},\ y = \dfrac{3t^2}{1 + t^3} \quad (t = 0)$

(j) $x = \dfrac{3t}{1 + t^3},\ y = \dfrac{3t^2}{1 + t^3} \quad (t = 1)$

5. Sketch the locus of each of the following and in each case find the curvature in terms of arc length measured from $t = 0$.

(a) $x = \displaystyle\int_0^t \frac{\cos u}{\sqrt{u}}\, du,\quad y = \int_0^t \frac{\sin u}{\sqrt{u}}\, du$

(b) $x = \displaystyle\int_0^t \cos u^2\, du,\quad y = \int_0^t \sin u^2\, du$

6. Find the radius and center of curvature for each of the following curves at the point indicated.

(a) $y = x^2$ $\quad(0, 0)$

(b) $y = x^3/3$ $\quad(2, 8/3)$

(c) $y = x^{3/2}$ $\quad(1, 1)$

(d) $y = \ln x$ $\quad(e, 1)$

(e) $y = \ln \cos x$ $\quad(0, 0)$

(f) $y = e^x$ $\quad(0, 1)$

(g) $y = \tan x$ $\quad(\pi/4, 1)$

(h) $y = \cosh x$ $\quad(0, 1)$

(i) $y = \sin x$ $\quad(\pi/2, 1)$

(j) $y = 2 \sin 2x$ $\quad(\pi/4, 2)$

7. Find the points of maximum and minimum curvature on the locus of $y = 3x - x^3$ and show that they are not the same as the maximum and minimum points on the curve.

8. Show that the locus of $y = x^3/3a^2$ has curvature that increases monotonically from 0 at $x = 0$ to a maximum value at $x = a/\sqrt[4]{5}$. Find the minimum value of the radius of curvature.

9. Let

$$y = \begin{cases} 0 & \text{for} \quad x \le 0 \\ \dfrac{x^3}{3} & \text{for} \quad 0 < x \le 1 \\ \dfrac{4}{3} - \sqrt{2 - x^2} & \text{for} \quad 1 < x \le \sqrt{2} \end{cases}$$

Show that the locus is continuous and has continuous slope and curvature. Thus, the arc of the locus of $y = x^3/3$ gives a transition from a straight line to a circle for which the curvature is continuous. For further discussion of such curves, see Section 3–4.

10. Find parametric equations for the evolute of the locus of each of the following.

(a) $y = x^3$

(b) $y = \ln \sec x$

(c) $y = e^x$

(d) $x = t - \sin t,\ y = 1 - \cos t$

(e) $x = \sec t,\ y = \tan t$

(f) $x = \cos^3 t,\ y = \sin^3 t$

(g) $x = \cos^4 t,\ y = \sin^4 t$

(h) $y = \cosh x$

(i) $x = \sin t,\ y = \cos 2t$

(j) $y = x^{3/2}$

11. The *involute* of a curve C is the curve for which C is the evolute. Show that the locus of

$$x = \cos t + t \sin t, \qquad y = \sin t - t \cos t$$

is the involute of a circle.

12. Find the rectangular equation for the evolute of the locus of each of the following.

(a) $x = \cos t,\ y = \sin t$

(b) $x = t,\ y = t^2$

(c) $x = 3 \cos t,\ y = 4 \sin t$

(d) $x = t - \tanh t,\ y = \operatorname{sech} t$

(e) $x = 2 \cos t + \cos 2t,\ y = 2 \sin t + \sin 2t$

3-3 Area. Let C be a simple closed curve in the plane directed counterclockwise. It is readily verified that each of the line integrals

$$\int_C -y\,dx, \qquad \int_C x\,dy$$

gives the area of the figure enclosed by C; therefore,

$$\tfrac{1}{2}\int_C (x\,dy - y\,dx) \tag{10}$$

also gives the area of the enclosed figure. This formula is particularly useful if C is described parametrically. If

$$x = f \circ t, \qquad y = g \circ t$$

on C, then direct substitution in (10) yields

$$\tfrac{1}{2}\int_C [(f \circ t)(g' \circ t) - (g \circ t)(f' \circ t)]\,dt$$

as a formula for the area of the enclosed figure.

Recall that the unit normal vector introduced in Section 3-2 may be written

$$\mathbf{N} = -\frac{dy}{ds}\mathbf{i} + \frac{dx}{ds}\mathbf{j},$$

and the position vector is $\mathbf{r} = x\mathbf{i} + y\mathbf{j}$. Thus, (10) may be written in vector form:

$$\tfrac{1}{2}\int_C -\mathbf{r} \cdot \mathbf{N}\,ds. \tag{11}$$

A vector formula is coordinate-free; hence, it may be used as a point of departure to describe a given quantity in various different coordinate

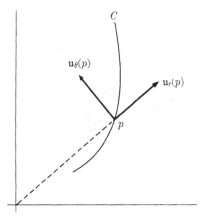

FIGURE 3-11

systems. As an illustration of this, a polar-coordinate formula for area will be derived from (11).

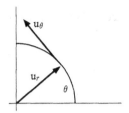

Let r and θ be the polar-coordinate variables restricted to C. At each point p of C, introduce unit vectors $\mathbf{u}_r(p)$ and $\mathbf{u}_\theta(p)$ in the directions of increasing r and θ, respectively (see Fig. 3–11). Note that

FIGURE 3–12

unlike the basis vectors $\mathbf{i}$ and $\mathbf{j}$, $\mathbf{u}_r$ and $\mathbf{u}_\theta$ vary from point to point. The position vector $\mathbf{r}$ may now be written

$$\mathbf{r} = r\mathbf{u}_r.$$

Since $d\mathbf{r} = \mathbf{T}\,ds$ (Section 3–2),

$$\mathbf{T}\,ds = d\mathbf{r} = dr\,\mathbf{u}_r + r\,d\mathbf{u}_r.$$

To find $d\mathbf{u}_r$, note (Fig. 3–12) that, as a position vector, $\mathbf{u}_r$ sweeps out the unit circle on which θ measures arc length and $\mathbf{u}_\theta$ is the unit tangent vector. Thus,

$$d\mathbf{u}_r = \mathbf{u}_\theta\,d\theta,$$

and

$$\mathbf{T}\,ds = dr\,\mathbf{u}_r + r\,d\theta\,\mathbf{u}_\theta; \tag{12}$$

hence (Fig. 3–13)

$$\mathbf{N}\,ds = -r\,d\theta\,\mathbf{u}_r + dr\,\mathbf{u}_\theta. \tag{13}$$

Since $\mathbf{u}_r \cdot \mathbf{u}_r = 1$, $\mathbf{u}_r \cdot \mathbf{u}_\theta = 0$,

$$-\mathbf{r} \cdot \mathbf{N}\,ds = -r\mathbf{u}_r \cdot (-r\,d\theta\,\mathbf{u}_r + dr\,\mathbf{u}_\theta) = r^2\,d\theta.$$

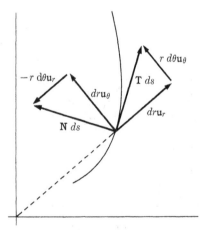

FIGURE 3–13

Substitution in (11) now yields

$$\tfrac{1}{2}\int_C r^2\, d\theta$$

as a formula for the area of the figure enclosed by C.

EXAMPLES

1. Find the area of the figure inside the cardioid whose equations are

$$x = 2\cos t - \cos 2t, \qquad y = 2\sin t - \sin 2t.$$

This curve is directed counterclockwise and is swept out as t runs from 0 to 2π. First, compute $x\,dy$ and $y\,dx$ in terms of t:

$$\begin{aligned}
x\,dy &= (2\cos t - \cos 2t)(2\cos t - 2\cos 2t)\,dt \\
&= (4\cos^2 t - 6\cos t \cos 2t + 2\cos^2 2t)\,dt \\
&= (4\cos^2 t - 6\cos t + 12\sin^2 t \cos t + 2\cos^2 2t)\,dt;
\end{aligned}$$

$$\begin{aligned}
y\,dx &= (2\sin t - \sin 2t)(-2\sin t + 2\sin 2t)\,dt \\
&= (-4\sin^2 t + 6\sin t \sin 2t - 2\sin^2 2t)\,dt \\
&= (-4\sin^2 t + 12\sin^2 t \cos t - 2\sin^2 2t)\,dt.
\end{aligned}$$

When the subtraction is performed, there is considerable simplification (recall that $\sin^2 + \cos^2 = 1$), and one obtains

$$x\,dy - y\,dx = (6 - 6\cos t)\,dt;$$

thus the required area is

$$\int_0^{2\pi} (3 - 3\cos t)\,dt = 6\pi.$$

2. Find the area of the figure enclosed by the locus of

$$x = \cos t, \qquad y = t\sin 2t \qquad (0 \le t \le 2\pi).$$

Note that here

$$-y\,dx = 2t\sin^2 t \cos t$$

is simpler than the symmetric form (contrast the situation in Example 1); so find the area by integrating this form. The required primitive may be found by integration by parts as follows:

$$\begin{aligned}
F(t) &= \int 2t\sin^2 t \cos t\, dt \\
&= \tfrac{2}{3}t\sin^3 t - \int \tfrac{2}{3}\sin^3 t\, dt = \tfrac{2}{3}t\sin^3 t - \tfrac{2}{3}(\tfrac{1}{3}\cos^3 t - \cos t).
\end{aligned}$$

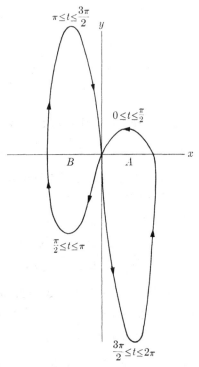

FIGURE 3–14

A sketch is shown in Fig. 3–14. From the sketch, it appears that

$$\alpha(A) = F(t)\Big|_0^{\pi/2} + F(t)\Big|_{3\pi/2}^{2\pi} = \left(\frac{\pi}{3} - \frac{4}{9}\right) + \left(\frac{4\pi}{3} + \frac{4}{9} + \pi\right) = \frac{8\pi}{3},$$

$$\alpha(B) = -F(t)\Big|_{\pi/2}^{3\pi/2} = \pi + \frac{\pi}{3} = \frac{4\pi}{3}.$$

Thus, the required total area is the sum of these two, or 4π.

3. Find the area of the figure inside the cardioid whose equation is

$$r = 1 - \cos\theta$$

and outside the circle whose equation is

$$r = 1.$$

A sketch is shown in Fig. 3–15. To follow the boundary of the figure in question counterclockwise, one follows the cardioid from $\theta = \pi/2$ to

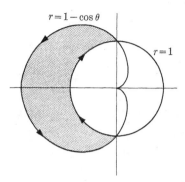

FIGURE 3–15

$\theta = 3\pi/2$ and then follows the circle from $\theta = 3\pi/2$ back to $\theta = \pi/2$. So,

$$\frac{1}{2}\int_C r^2\,d\theta = \frac{1}{2}\int_{\pi/2}^{3\pi/2} (1-\cos\theta)^2\,d\theta + \frac{1}{2}\int_{3\pi/2}^{\pi/2} 1^2\,d\theta$$

$$= \frac{1}{2}\int_{\pi/2}^{3\pi/2}\left[1 - 2\cos\theta + \frac{1}{2}(1+\cos 2\theta)\right]d\theta + \frac{1}{2}\left(\frac{\pi}{2} - \frac{3\pi}{2}\right)$$

$$= \frac{3\theta}{4} - \sin\theta + \frac{1}{8}\sin 2\theta\,\Big|_{\pi/2}^{3\pi/2} - \frac{\pi}{2}$$

$$= \frac{3\pi}{4} - (-2) - \frac{\pi}{2} = 2 + \frac{\pi}{4}.$$

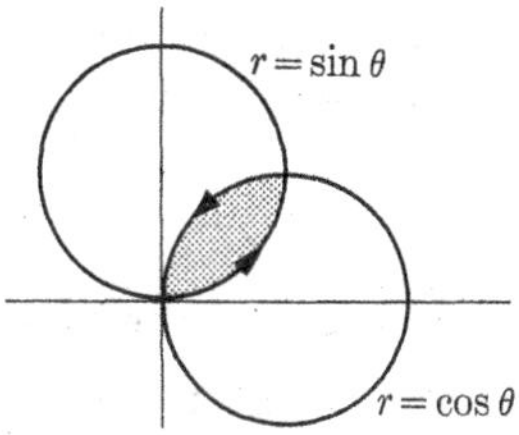

4. Find the area inside the loci of both

$$r = \sin\theta \qquad \text{and} \qquad r = \cos\theta.$$

FIGURE 3–16

To follow this boundary (see Fig. 3–16) one follows the locus of $r = \sin\theta$ from 0 to $\pi/4$ and the locus of $r = \cos\theta$ from $\pi/4$ to $\pi/2$:

$$\frac{1}{2}\int_C r^2\,d\theta = \frac{1}{2}\int_0^{\pi/4} \sin^2\theta\,d\theta + \frac{1}{2}\int_{\pi/4}^{\pi/2} \cos^2\theta\,d\theta$$

$$= \frac{1}{4}\int_0^{\pi/4}(1-\cos 2\theta)\,d\theta + \frac{1}{2}\int_{\pi/4}^{\pi/2}(1+\cos 2\theta)\,d\theta$$

$$= \frac{\theta}{4} - \frac{1}{8}\sin 2\theta\,\Big|_0^{\pi/4} + \frac{\theta}{4} + \frac{1}{8}\sin 2\theta\,\Big|_{\pi/4}^{\pi/2}$$

$$= \frac{\pi}{16} - \frac{1}{8} + \frac{\pi}{16} - \frac{1}{8} = \frac{\pi}{8} - \frac{1}{4}.$$

EXERCISES

1. Find the area of an ellipse (a) from the rectangular equation

$$\frac{x^2}{a^2} + \frac{y^2}{b^2} = 1,$$

and (b) from the parametric equations $x = a \cos t$, $y = b \sin t$.

2. Find the area enclosed by the locus of each of the following.
   (a) $x = 2 \cos^2 t$, $y = \sin 2t$
   (b) $x = \cos^2 t \sin t$, $y = \sin^2 t \cos t$
   (c) $x = \cos t \cos 2t$, $y = \sin t \cos 2t$
   (d) $x = 3t/(1 + t^3)$, $y = 3t^2/(1 + t^3)$
   (e) $x = t^2$, $y = t^3 - t$
   (f) $x = \sin 2t$, $y = \cos t$
   (g) $x = \sin t$, $y = t \sin 2t$ $(0 \le t \le 2\pi)$

3. Given the limacon whose parametric equations are

$$x = \cos t - \cos 2t, \qquad y = \sin t - \sin 2t,$$

   (a) find the total area enclosed;
   (b) find the area inside the little loop;
   (c) find the area between the loops;
   (d) explain what

$$\tfrac{1}{2} \int_{t=0}^{t=2\pi} (x \, dy - y \, dx)$$

   gives in terms of areas.

4. Find the area of each of the following figures:
   (a) Inside the circle on which $r = 1$ and outside the cardioid on which $r = 1 - \cos \theta$.
   (b) Between the loops of the limaçon on which $r = 1 - 2 \cos \theta$.
   (c) Inside both cardioids, that on which $r = 1 - \cos \theta$ and that on which $r = 1 - \sin \theta$.
   (d) Inside the circle on which $r = 2 \cos \theta$ and outside the circle on which $r = 1$.
   (e) In the first quadrant, inside the circle on which $r = 1$, and outside both circles, the loci of $r = \cos \theta$ and $r = \sin \theta$.

5. A sector of a circle of radius r and central angle α has area $\alpha/2\pi$ times that of the entire circle; so its area is

$$\frac{\alpha}{2\pi} \pi r^2 = \frac{1}{2} r^2 \alpha.$$

Using this result, show how the polar-coordinate area formula could be arrived at geometrically by approximating an area by a sum of areas of circular sectors.

3–4 Plane motion. When written in vector form, the mathematical description of the motion of a particle in the plane is quite analogous to that for rectilinear motion. The motion is described by the parametric equation

$$\mathbf{r} = (f \circ t)\mathbf{i} + (g \circ t)\mathbf{j}. \tag{14}$$

To begin, $\mathbf{r}$ and t should be variables on the time continuum. However, (14) defines a curve in the plane, and, in most problems, it seems more natural to regard $\mathbf{r}$ and t as variables on this curve. The transfer of these and other variables from the time continuum to the path of the particle is the familiar transfer procedure first introduced in Section 2–5. For the remainder of this discussion, the path will be presumed given, and the pertinent variables will be defined on it.

On this path, vector variables,

$$\mathbf{v} = D_t\mathbf{r}, \text{ velocity,}$$

$$\mathbf{a} = D_t\mathbf{v}, \text{ acceleration,}$$

and scalar variables,

$$s, \text{ arc length,}$$

$$v = D_t s, \text{ speed,}$$

are introduced. In this notation, (8) may be written

$$\mathbf{a} = D_t v \mathbf{T} + \kappa v^2 \mathbf{N}. \tag{15}$$

That is, the component of acceleration tangent to the path is the rate of change of speed, and the component of acceleration normal to the path is the curvature times the square of the speed. These components will be denoted by a_T and a_N, respectively.

The result $a_N = \kappa v^2$ has many applications. For example, consider a particle moving along a circular path. The force toward the center of the circle that keeps the particle on the circle is called the *centripetal force*. The equal and opposite force exerted by the particle is called the *centrifugal force*. The magnitude of these forces must be

$$|ma_N| = m|\kappa|v^2 = \frac{mv^2}{r},$$

where r is the radius of the circle.

In building curves in a railroad track, the engineers never join a circular section onto a straight one. The bulk of the curve may be circular, but this is connected to the straight section by a *transition curve* following some formula such as $y = x^3$, so that the curvature will be continuous (see Exercise 9, Section 3–2). Even though the curve looked smooth, a discontinuity in κ would cause a discontinuity in a_N and give the train a sudden jerk to one side. The magnitude of this jerk is proportional to the square of the velocity. Thus transition curves may not be necessary for toy trains, but they are essential for real ones. Observe, however, the unusual swaying of a cheap toy train as it goes into a curve.

Work done along a given portion C of the path is defined as

$$\int_C m\mathbf{a} \cdot \mathbf{T} \, ds,$$

where m is the mass of the particle. Take the scalar product of (15) with $\mathbf{T}$:

$$\mathbf{a} \cdot \mathbf{T} = D_t v = \frac{dv}{dt} = \frac{dv}{ds}\frac{ds}{dt} = v\frac{dv}{ds}.$$

Thus,

$$\int_C m\mathbf{a} \cdot \mathbf{T} \, ds = \int_C mv \, dv = \tfrac{1}{2}mv^2 \Big|_p^q, \tag{16}$$

where p and q are the end points of C. Work is equal to a change in the variable $mv^2/2$, and the variable is called *kinetic energy*.

Suppose, now, that the particle slides without friction down a prescribed path whose equation is

$$\mathbf{r} = (\phi \circ u)\mathbf{i} + (\psi \circ u)\mathbf{j}.$$

The significance (if any) of the parameter u is immaterial. There are two forces on the particle, the gravitational force

$$-mg\mathbf{j}$$

and a force

$$C\mathbf{N}$$

normal to the path that keeps the particle on the prescribed path. Since $\mathbf{N} \cdot \mathbf{T} = 0$,

$$m\mathbf{a} \cdot \mathbf{T} \, ds = (-mg\mathbf{j} + C\mathbf{N}) \cdot \mathbf{T} \, ds = -mg\mathbf{j} \cdot \mathbf{T} \, ds = -mg \, dy;$$

hence (when divided by m) (16) becomes

$$\tfrac{1}{2}v^2 \Big|_p^q = -g\int_C dy = -gy \Big|_p^q. \tag{17}$$

Let the particle start from rest at height y_0; that is, let $v(p) = 0$, $y(p) = y_0$. Then it follows from (17) that over the entire path,

$$v^2 = -2g(y - y_0),$$

$$\frac{ds}{dt} = \sqrt{2g(y_0 - y)}.$$

Therefore,

$$dt = \frac{ds}{\sqrt{2g(y_0 - y)}} = \sqrt{\frac{(D_u x)^2 + (D_u y)^2}{2g(y_0 - y)}}\, du.$$

So, if M is the entire path, the total time elapsed as the particle slides down M is given by

$$\int_M dt = \int_M \sqrt{\frac{(D_u x)^2 + (D_u y)^2}{2g(y_0 - y)}}\, du.$$

EXAMPLES

1. Suppose that a particle has constant acceleration $\mathbf{a}_0$, initial velocity $\mathbf{v}_0$, and initial position $\mathbf{r}_0$. By studying x- and y-components of $\mathbf{a}$, $\mathbf{v}$, and $\mathbf{r}$ separately, one finds that

$$\mathbf{v} = \mathbf{a}_0 t + \mathbf{v}_0,$$

$$\mathbf{r} = \tfrac{1}{2}\mathbf{a}_0 t^2 + \mathbf{v}_0 t + \mathbf{r}_0.$$

Apply this to a projectile that starts from the origin with speed v_0 at an angle θ with the x-axis and moves under the force of gravity alone. In this case,

$$\mathbf{r}_0 = 0, \qquad \mathbf{v}_0 = v_0 \cos\theta\, \mathbf{i} + v_0 \sin\theta\, \mathbf{j}, \qquad \mathbf{a}_0 = -g\mathbf{j};$$

so,

$$\mathbf{v} = v_0 \cos\theta\mathbf{i} + (-gt + v_0 \sin\theta)\mathbf{j},$$

$$\mathbf{r} = v_0 t \cos\theta\mathbf{i} + (-\tfrac{1}{2}gt^2 + v_0 t \sin\theta)\mathbf{j}. \tag{18}$$

The path of the projectile is a parabola for which (18) gives parametric equations in vector form.

A number of other questions may now be answered.

(a) What is the maximum height reached? To maximize y, set

$$\frac{dy}{dt} = v_y = v_0 \sin\theta - gt = 0,$$

$$t = \frac{v_0}{g}\sin\theta, \qquad y = \frac{v_0^2 \sin^2\theta}{2g}.$$

(b) When and where will the projectile strike the ground? Note that $y = 0$ when

$$t[v_0 \sin\theta - \tfrac{1}{2}gt] = 0,$$

$$t = 0, \qquad t = \frac{2v_0 \sin\theta}{g}.$$

The first value of t is for the initial state; the second is for the state in which the projectile strikes the ground. For the latter state, substitution in (18) yields

$$x = \frac{v_0^2 \sin 2\theta}{g}. \tag{19}$$

(c) What value of θ will give maximum horizontal range? Equation (19) gives the range x in terms of θ. To maximize this x, set

$$\frac{dx}{d\theta} = \frac{2v_0^2 \cos 2\theta}{g} = 0,$$

$$\theta = \frac{\pi}{4},$$

$$x = \frac{v_0^2}{g}.$$

(d) Find general expressions for $|\mathbf{v}|$, $|\mathbf{a}|$, a_T and a_N. By direct substitution,

$$|\mathbf{v}| = \sqrt{v_0^2 \cos^2 \theta + (v_0 \sin \theta - gt)^2}$$

$$= \sqrt{v_0^2 - 2v_0 gt \sin \theta + g^2 t^2},$$

$$|\mathbf{a}| = \sqrt{0 + (-g)^2} = g.$$

Note that $|\mathbf{a}| \neq d|\mathbf{v}|/dt$:

$$a_T = \frac{dv}{dt} = \frac{g^2 t - v_0 g \sin \theta}{\sqrt{v_0^2 - 2v_0 gt \sin \theta + g^2 t^2}}.$$

For the path given by (18), the curvature is

$$\kappa = \frac{D_t x \, D_t^2 y - D_t y \, D_t^2 x}{[(dx/dt)^2 + (dy/dt)^2]^{3/2}} = \frac{v_x a_y - v_y a_x}{v^3}$$

$$= \frac{-g v_0 \cos \theta - 0}{(v_0^2 - 2v_0 gt \sin \theta + g^2 t^2)^{3/2}}.$$

Therefore,

$$a_N = \kappa v^2 = \frac{-g v_0 \cos \theta}{\sqrt{v_0^2 - 2v_0 gt \sin \theta + g^2 t^2}}.$$

2. Suppose that a particle is moving from left to right along the curve on which

$$y = \sin x.$$

with a constant speed v_0. Find the x- and y-components of velocity and acceleration. First, make the following computations:

$$dy = \cos x \, dx, \qquad ds = \sqrt{1 + \cos^2 x} \, dx,$$

$$\cos \alpha = \frac{dx}{ds} = \frac{1}{\sqrt{1 + \cos^2 x}}, \qquad \sin \alpha = \frac{dy}{ds} = \frac{\cos x}{\sqrt{1 + \cos^2 x}},$$

$$D_x^2 y = -\sin x, \qquad \kappa = \frac{-\sin x}{(1 + \cos^2 x)^{3/2}}.$$

Now, it is given that $\mathbf{v} = v_0 \mathbf{T}$; so, $D_t v = 0$, and by (15), $\mathbf{a}$ reduces to $\kappa v_0^2 \mathbf{N}$. Therefore,

$$v_x = \mathbf{v} \cdot \mathbf{i} = v_0 \mathbf{T} \cdot \mathbf{i} = v_0 \cos \alpha = \frac{v_0}{\sqrt{1 + \cos^2 x}},$$

$$v_y = \mathbf{v} \cdot \mathbf{j} = v_0 \mathbf{T} \cdot \mathbf{j} = v_0 \sin \alpha = \frac{v_0 \cos x}{\sqrt{1 + \cos^2 x}},$$

$$a_x = \mathbf{a} \cdot \mathbf{i} = \kappa v_0^2 \mathbf{N} \cdot \mathbf{i} = -\kappa v_0^2 \sin \alpha = \frac{v_0^2 \sin x \cos x}{(1 + \cos^2 x)^2},$$

$$a_y = \mathbf{a} \cdot \mathbf{j} = \kappa v_0^2 \mathbf{N} \cdot \mathbf{j} = \kappa v_0^2 \cos \alpha = \frac{-(v_0^2 \sin x)}{(1 + \cos^2 x)^2}.$$

EXERCISES

1. For each of the following pairs of equations of motion find v_x, v_y, a_x, a_y, $|\mathbf{v}|$, $|\mathbf{a}|$, a_T, and a_N.

    (a) $x = \cos t,\ y = \sin t$ (b) $x = 3 \cos t,\ y = 4 \sin t$

    (c) $x = e^{-t} \cos t,\ y = e^{-t} \sin t$ (d) $x = e^t,\ y = 2e^{-t}$

    (e) $x = \operatorname{sech} t,\ y = \tanh t$ (f) $x = 40t,\ y = 30t - 16t^2$

    (g) $x = \cos t,\ y = \sin 2t$ (h) $x = t,\ y = \sqrt{1 - t^2}$

2. Use the results from projectiles in Example 1 to answer the following questions.

    (a) Given $h < (v_0^2 \sin^2 \theta)/(2g)$, what will be the value of x when the projectile reaches a height h on the way down? That is, how far away will the projectile fall if it is shot up over the edge of a cliff of height h?

    (b) What will happen if $h = (v_0^2 \sin^2 \theta)/(2g)$?

    (c) What will happen if $h > (v_0^2 \sin^2 \theta)/(2g)$?

    (d) Given a point on the ground $(k, 0)$ where $k < v_0^2/g$, show that there are two angles, θ_1 and θ_2, at which the projectile will hit this point.

    (e) Show that, for the angles θ_1 and θ_2 of part (d),

$$\theta_2 - \pi/4 = \pi/4 - \theta_1.$$

3. In each of the following exercises, assume that a particle is moving from left to right along the locus of the given equation with constant speed v. Find v_x, v_y, a_x, and a_y in terms of x.

(a) $y = x^2$ (b) $y = e^{-x}$

(c) $y = \cosh x$ (d) $y = \ln \sec x$

(e) $y = x^{2/3}$ (f) $y = x^{3/2}$

4. Suppose a particle moves along the locus of

$$y = f \circ x$$

from left to right. Find expressions for v_x, v_y, a_x, a_y (a) if it moves with constant speed v_0, and (b) if it moves with variable speed $v = \phi \circ x$.

5. A simple pendulum of length L involves a particle moving under the influence of gravity alone along the locus of

$$x = L \sin \theta, \qquad y = -L \cos \theta$$

(See Fig. 3–17). Suppose it swings from $\theta = -\alpha$ to $\theta = \alpha$ and back. The period T is the time of one complete oscillation.

(a) Show that

$$T = \sqrt{\frac{2L}{g}} \int_{-\alpha}^{\alpha} \frac{d\theta}{\sqrt{\cos \theta - \cos \alpha}}.$$

(b) Substitute

$$u = \frac{\sin (\theta/2)}{\sin (\alpha/2)},$$

and show that

$$T = 2\sqrt{\frac{L}{g}} \int_{-1}^{1} \frac{du}{\sqrt{(1 - u^2)[1 - u^2 \sin^2 (\alpha/2)]}}.$$

For $\alpha \neq 0$, this is an elliptic integral. Furthermore, the value of T clearly depends on α; that is, the period of a simple pendulum varies with the amplitude of oscillation.

(c) Assume α small enough so that the factor $1 - u^2 \sin^2 (\alpha/2)$ is close enough to unity to be neglected, and derive the usual formula for small amplitude oscillations of a simple pendulum

$$T = 2\pi\sqrt{L/g},$$

6. Suppose a particle moves under the force of gravity alone on the cycloid on which

$$x = a (\theta - \sin \theta), \qquad y = a (\cos \theta - 1)$$

(Fig. 3–18). Let it oscillate from $\theta = \alpha$ to $\theta = 2\pi - \alpha$, and let T be the period of the oscillation.

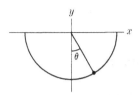

FIGURE 3–17

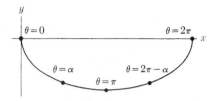

FIGURE 3–18

(a) Show that

$$\frac{T}{4} = \sqrt{\frac{a}{g}} \int_\alpha^\pi \sqrt{\frac{1 - \cos\theta}{\cos\alpha - \cos\theta}} \, d\theta.$$

(b) Substitute

$$u = \frac{\cos(\theta/2)}{\cos(\alpha/2)},$$

and show that

$$T = 4\pi\sqrt{a/g};$$

that is, T is independent of α. Contrast Exercise 6(b). Thus, for a pendulum with a period independent of the amplitude, one should constrain the pendulum bob to move on a cycloid.

7. Show that the following procedure will serve to construct a cycloidal pendulum as called for in Exercise 6(b). Take two cycloidal half-arches with a cusp pointing upward in the center. Take a string with a length equal to that of a half arch. Fasten the string at the cusp point and let the bob swing from the other end with the string winding around the cycloidal arches as the bob swings. [*Hint:* See Example 3, Section 3–2.]

8. Find the time required for a particle to slide under gravity alone down a straight line from $(0, 0)$ to $(a, -2a)$. Compare this time with the time spent between these two points sliding on the cycloid of Exercise 6. [*Note:* It can be shown by calculus of variations that the cycloid is the brachistochrone curve, i.e., the curve giving minimum time for a sliding particle between two given points.]

9. Derive the formulas

$$a_T = \frac{v_x a_x + v_y a_y}{|\mathbf{v}|}, \qquad a_N = \frac{v_x a_y - v_y a_x}{|\mathbf{v}|}.$$

CHAPTER 4

TOPICS IN LINEAR ALGEBRA

4–1 Matrix algebra. A rectangular array of symbols, such as

$$\begin{bmatrix} a_{11} & a_{12} & a_{13} \\ a_{21} & a_{22} & a_{23} \end{bmatrix},$$

is called a *matrix* (plural, *matrices*). The individual symbols may have first one meaning and then another, although in the present study, they will usually be either numbers or variables. The essential requirement is that one be able to add and multiply them in order to define the operations of matrix algebra.

The above matrix has two *rows* and three *columns* and is called a 2×3 matrix. When double subscript notation is used for matrix entries, the above pattern is always followed. The first subscript gives the row number and the second gives the column number. Often

$$[a_{ij}]$$

is used to denote the matrix with entry a_{ij} in the ith row and jth column. That is, a symbol for a general matrix entry enclosed in brackets will constitute a symbol for the matrix itself. When such notation is used, the dimensions of the matrix must be clear from the context. Frequently, a single capital letter is also used to denote a matrix; thus one can write

$$A = [a_{ij}]$$

and use these symbols interchangeably.

The following operations with matrices form the basis for their principal applications.

(i) *Multiplication.* For reasons that will appear very shortly, multiplication of matrices is defined in the following way. If $[a_{ij}]$ is $m \times n$ and $[b_{ij}]$ is $n \times s$, then the product $[a_{ij}]\,[b_{ij}]$ is the $m \times s$ matrix $[c_{ij}]$ defined by

$$c_{ik} = \sum_{j=1}^{n} a_{ij} b_{jk}.$$

In words, to get the (i, k) entry in the product AB, line up the ith row of A with the kth column of B; multiply corresponding entries and add.

Note that for multiplication of matrices to be defined, the dimensions must match, as shown in Fig. 4–1.

A more important point to be noted is that in computing AB one lines up rows of A with columns of B; this rule is not symmetric. If A and B

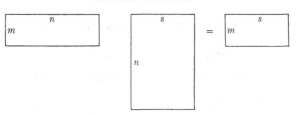

FIGURE 4-1

are both square matrices of the same dimension then both AB and BA are defined, but, in general, *these are not equal* (see Example 1 below). Perhaps this is the student's first encounter with *noncommutative multiplication*. Such operations are common in modern algebra, and other examples will appear later in this book.

On the other hand, matrix multiplication is *associative*. That is, if

$$A(BC)$$

is defined, so is

$$(AB)C,$$

and the two are equal. For this reason, it is unambiguous to write ABC.

The following considerations introduce the principal application of matrices, indicate why matrix multiplication is defined as it is, and, incidentally, suggest a simple proof of the associative law.

The pair of linear equations

$$u = a_{11}x + a_{12}y,$$
$$v = a_{21}x + a_{22}y,$$

$$(1)$$

define a mapping T_1 which carries ordered pairs (x, y) into ordered pairs (u, v); that is,

$$(u, v) = T_1(x, y). \tag{2}$$

A mapping of pairs into pairs is usually called a *transformation*, and one defined by a pair of linear equations is called a *linear transformation*. Now, suppose that there is another linear transformation T_2 such that

$$(z, w) = T_2(u, v) \tag{3}$$

means

$$z = b_{11}u + b_{12}v,$$
$$w = b_{21}u + b_{22}v.$$

$$(4)$$

By the definition of a composite mapping, given (2) and (3),

$$(z, w) = (T_2 \circ T_1)(x, y).$$

To find equations describing $T_2 \circ T_1$, substitute (1) into (4):

$$
\begin{aligned}
z &= b_{11}(a_{11}x + a_{12}y) + b_{12}(a_{21}x + a_{22}y) \\
&= (b_{11}a_{11} + b_{12}a_{21})x + (b_{11}a_{12} + b_{12}a_{22})y, \\
w &= b_{21}(a_{11}x + a_{12}y) + b_{22}(a_{21}x + a_{22}y) \\
&= (b_{12}a_{11} + b_{22}a_{21})x + (b_{21}a_{12} + b_{22}a_{22})y.
\end{aligned}
\tag{5}
$$

Now, each of these systems of linear equations may be written in matrix form. By the definition of matrix multiplication, (1) is the same as

$$
\begin{bmatrix} u \\ v \end{bmatrix} = \begin{bmatrix} a_{11} & a_{12} \\ a_{21} & a_{22} \end{bmatrix} \begin{bmatrix} x \\ y \end{bmatrix}.
\tag{1'}
$$

Similarly, Eq. (4) may be written

$$
\begin{bmatrix} z \\ w \end{bmatrix} = \begin{bmatrix} b_{11} & b_{12} \\ b_{21} & b_{22} \end{bmatrix} \begin{bmatrix} u \\ v \end{bmatrix}.
\tag{4'}
$$

Converting (5) in a similar way and noting again the definition of matrix multiplication yields

$$
\begin{aligned}
\begin{bmatrix} z \\ w \end{bmatrix} &= \begin{bmatrix} b_{11}a_{12} + b_{12}a_{21} & b_{11}a_{12} + b_{12}a_{22} \\ b_{12}a_{11} + b_{22}a_{21} & b_{21}a_{12} + b_{22}a_{22} \end{bmatrix} \begin{bmatrix} x \\ y \end{bmatrix} \\
&= \begin{bmatrix} b_{11} & b_{12} \\ b_{21} & b_{22} \end{bmatrix} \begin{bmatrix} a_{11} & a_{12} \\ a_{21} & a_{22} \end{bmatrix} \begin{bmatrix} x \\ y \end{bmatrix}.
\end{aligned}
\tag{5'}
$$

Thus, if ordered pairs are written as 2×1 matrices, a linear transformation is given by multiplication by a square matrix. Furthermore, comparison of (1'), (4'), and (5') shows that composition of linear transformations corresponds to multiplication of the associated matrices.

The associative law for matrix multiplication thus follows from the associativity of the $\circ$ operator.

(ii) *Addition.* Definition:

$$
[a_{ij}] + [b_{ij}] = [a_{ij} + b_{ij}].
$$

In words, matrices with identical dimensions are added by adding corresponding entries. Key properties (proofs left to the student):

$$
\begin{aligned}
A + B &= B + A, \\
A + (B + C) &= (A + B) + C, \\
A(B + C) &= AB + AC, \\
(A + B)C &= AC + BC.
\end{aligned}
$$

(iii) *Multiplication by scalars.* Definition:

$$[a_{ij}]c = c[a_{ij}] = [ca_{ij}].$$

In words, to multiply a matrix by a scalar, multiply each entry by the scalar. Key properties:

$$c(AB) = (cA)B = A(Bc),$$
$$c(A + B) = cA + cB.$$

(iv) *Transposition.* Definition:

$$[a_{ij}]' = [a_{ji}].$$

In words, the *transpose* of A, denoted by A' is obtained from A by converting rows into columns and vice versa. In the case of a square matrix, transposition may also be described as reflection through the main diagonal. Note that if A is $m \times n$, then A' is $n \times m$. The student should check the following properties:

$$(A + B)' = A' + B',$$
$$(AB)' = B'A'.$$

Note carefully the reversal of order when a product is transposed.

(v) *Inversion.* Consider the matrix $[\delta_{ij}]$, where δ_{ij} is the so-called "Kronecker δ" defined by

$$\delta_{ij} = \begin{cases} 1 & \text{for } i = j, \\ 0 & \text{for } i \neq j. \end{cases}$$

Thus $[\delta_{ij}]$ is a square matrix with 1's down the main diagonal and 0's elsewhere. The 2×2 and 3×3 cases are

$$\begin{bmatrix} 1 & 0 \\ 0 & 1 \end{bmatrix}, \qquad \begin{bmatrix} 1 & 0 & 0 \\ 0 & 1 & 0 \\ 0 & 0 & 1 \end{bmatrix}.$$

Now, it is easily verified that, for any matrix A,

$$A[\delta_{ij}] = A \qquad \text{and} \qquad [\delta_{ij}]A = A,$$

provided that the dimension of $[\delta_{ij}]$ is properly chosen. For this reason, $[\delta_{ij}]$ is called the *identity matrix*. It is often denoted by I, sometimes with a subscript to indicate dimension, I_n.

The *inverse* of a square matrix A is a matrix denoted by A^{-1}, such that

$$A^{-1}A = AA^{-1} = [\delta_{ij}].$$

Only a square matrix can have an inverse (see Exercise 4 below), but not all square matrices do. Conditions for the existence of an inverse will appear in Section 4–4.

Suppose A^{-1} exists: how does one find it? Consider the case in which A is $n \times n$. Let X and U each be $n \times 1$ and write a system of linear equations as in (1′):

$$U = AX.$$

If A^{-1} exists, this system is easily solved in matrix form:

$$A^{-1}U = A^{-1}AX = [\delta_{ij}]X = X.$$

Thus, finding the inverse of a square matrix is equivalent to solving a system of linear equations.

Note, finally, that

$$(B^{-1}A^{-1})(AB) = B^{-1}(A^{-1}A)B = B^{-1}[\delta_{ij}]B = B^{-1}B = [\delta_{ij}];$$

so that

$$(AB)^{-1} = B^{-1}A^{-1}.$$

We see that inversion reverses the order of factors, just as transposition does.

EXAMPLES

1. Let

$$A = \begin{bmatrix} 1 & 2 \\ 3 & 4 \end{bmatrix}, \quad B = \begin{bmatrix} 5 & 6 \\ 7 & 8 \end{bmatrix}.$$

Compute AB and BA.

$$AB = \begin{bmatrix} 1 \cdot 5 + 2 \cdot 7 & 1 \cdot 6 + 2 \cdot 8 \\ 3 \cdot 5 + 4 \cdot 7 & 3 \cdot 6 + 4 \cdot 8 \end{bmatrix} = \begin{bmatrix} 19 & 22 \\ 43 & 50 \end{bmatrix},$$

$$BA = \begin{bmatrix} 5 \cdot 1 + 6 \cdot 3 & 5 \cdot 2 + 6 \cdot 4 \\ 7 \cdot 1 + 8 \cdot 3 & 7 \cdot 2 + 8 \cdot 4 \end{bmatrix} = \begin{bmatrix} 23 & 34 \\ 31 & 46 \end{bmatrix}.$$

2. Let

$$A = \begin{bmatrix} 1 & -1 \\ 2 & 1 \end{bmatrix}.$$

Find A' and A^{-1}. Finding the transpose is easy:

$$A' = \begin{bmatrix} 1 & 2 \\ -1 & 1 \end{bmatrix}.$$

To find A^{-1}, write

$$u = x - y, \qquad v = 2x + y,$$

and solve. Adding the two equations, one gets

$$u + v = 3x;$$

subtracting twice the first equation from the second yields

$$-2u + v = 3y.$$

Thus,

$$x = \tfrac{1}{3} u + \tfrac{1}{3} v,$$
$$y = -\tfrac{2}{3} u + \tfrac{1}{3} v,$$

and it appears that

$$A^{-1} = \begin{bmatrix} \tfrac{1}{3} & \tfrac{1}{3} \\ -\tfrac{2}{3} & \tfrac{1}{3} \end{bmatrix}.$$

An alternative method would be to set

$$\begin{bmatrix} 1 & -1 \\ 2 & 1 \end{bmatrix} \begin{bmatrix} b_{11} & b_{12} \\ b_{21} & b_{22} \end{bmatrix} = \begin{bmatrix} 1 & 0 \\ 0 & 1 \end{bmatrix}$$

and solve for the four unknown entries. It is left to the student to carry out this operation and to see that it yields the same result as that obtained above. He may also check the answer; that is, he may verify directly that

$$\begin{bmatrix} 1 & -1 \\ 2 & 1 \end{bmatrix} \begin{bmatrix} \tfrac{1}{3} & \tfrac{1}{3} \\ -\tfrac{2}{3} & \tfrac{1}{3} \end{bmatrix} = \begin{bmatrix} \tfrac{1}{3} & \tfrac{1}{3} \\ -\tfrac{2}{3} & \tfrac{1}{3} \end{bmatrix} \begin{bmatrix} 1 & -1 \\ 2 & 1 \end{bmatrix} = \begin{bmatrix} 1 & 0 \\ 0 & 1 \end{bmatrix}.$$

EXERCISES

1. Let

$$A = \begin{bmatrix} 1 & -1 \\ 1 & 0 \end{bmatrix}, \qquad B = \begin{bmatrix} 1 & 2 & 0 \\ 0 & -1 & 2 \end{bmatrix},$$

$$C = \begin{bmatrix} 1 & -1 \\ 0 & 1 \\ 2 & 0 \end{bmatrix}, \qquad D = \begin{bmatrix} 1 & -1 & 0 \\ 0 & 1 & 2 \\ 2 & -1 & 1 \end{bmatrix}.$$

(a) Which of these products is defined? AB, BA, AC, CA, AD, DA, BC, CB, BD, DB, CD, DC.
(b) Compute each of the products mentioned in part (a) that is properly defined.
(c) Find A', B', C' and D'.
(d) Find A^{-1} and D^{-1}.

2. Write each of the following systems of equations in matrix form.

(a) $u = 2x - 3y$
  $v = 4x - y$

(b) $u = x + y$
  $v = -2y$

(c) $u = 2x - 3y + z$
  $v = x - 2y + 3z$
  $w = 4x - y + 5z$

(d) $u = x - y$
  $v = y - z$
  $w = z - x$

3. Let

$$X = \begin{bmatrix} x \\ y \end{bmatrix}, \qquad U = \begin{bmatrix} u \\ v \end{bmatrix}, \qquad A = \begin{bmatrix} 3 & -2 \\ 4 & 0 \end{bmatrix}.$$

Write the relation $U = AX$ as a system of linear equations.

4. Suppose

$$BA = [\delta_{ij}] \quad \text{and} \quad AC = [\delta_{ij}].$$

(a) Apply the associative law to the product BAC to show that $B = C$.
(b) Show from part (a) that only square matrices have inverses.
(c) Show from part (a) that if A^{-1} exists it is unique.

4–2 Linear transformations. Matrix equations may be used to describe important ideas in both analytic geometry and vector analysis, and in each case there are two points of view.

First, it is well to introduce a system of notation designed to describe these ideas. This notation will be followed throughout the remainder of this book.

In analytic geometry p and q are used to denote points, and x, y, u, v, etc. are used to denote variables on these points. In general, (x, y) will have the usual meaning, and in general discussions (u, v) will be another pair of variables. Variable pairs will be written as matrices:

$$X = \begin{bmatrix} x \\ y \end{bmatrix}, \qquad U = \begin{bmatrix} u \\ v \end{bmatrix}.$$

Now, when a single capital letter stands for a matrix of variables, that capital letter will be incorporated into appropriate notation to indicate operations on or with each of the individual variables. Specifically, the following will appear.

Application to a point.

$$U(p) \qquad \text{means} \qquad \begin{bmatrix} u(p) \\ v(p) \end{bmatrix}.$$

Differentiation.

$$D_t U \qquad \text{means} \qquad \begin{bmatrix} D_t u \\ D_t v \end{bmatrix}.$$

Differentials.

$$dU \qquad \text{means} \qquad \begin{bmatrix} du \\ dv \end{bmatrix}.$$

The components of a vector are analogous to the coordinates of a point. Indeed, if $\mathbf{w}$ is pictured as an arrow from the origin to p, then the components of $\mathbf{w}$ are exactly the coordinates of p. In vector analysis then x, y, u, v, etc. will be *component variables*. Each is a mapping that pairs vectors with their components. Specifically, $x(\mathbf{w})$ will be the **i**-component of $\mathbf{w}$ and $y(\mathbf{w})$ the **j**-component. That is,

$$\mathbf{w} = x(\mathbf{w})\mathbf{i} + y(\mathbf{w})\mathbf{j}.$$

Other pairs of basis vectors may be used and, when they are, the following convention will be followed. Basis vectors $\mathbf{b}_u$ and $\mathbf{b}_v$ will correspond to component variables u and v. That is,

$$\mathbf{w} = u(\mathbf{w})\mathbf{b}_u + v(\mathbf{w})\mathbf{b}_v.$$

Instead of $\mathbf{b}_x$ and $\mathbf{b}_y$ the customary **i** and **j** will still be used.

Now, look at the following four matrix equations, two from analytic geometry and two from vector analysis.

$$X(q) = AX(p) \tag{6}$$

$$U(p) = AX(p) \tag{7}$$

$$X(\mathbf{v}) = AX(\mathbf{w}) \tag{8}$$

$$U(\mathbf{w}) = AX(\mathbf{w}) \tag{9}$$

In each of these matrix equations, A is a square matrix of numbers; they describe transformations which are named as follows:

Linear point transformation (6). Note that p is carried into q. The equation gives the X-coordinates of q in terms of those of p.

Linear coordinate transformation (7). New coordinates U are given for each point p in terms of the original coordinates X. In general, the new coordinates will not be rectangular ones. The important case in which they are is discussed in Section 4–7. Any coordinates coming from rectangular ones by a linear coordinate transformation are called *affine coordinates*.

Linear transformation of the vector space into itself (8). The vector **w** is mapped onto the vector **v** with the mapping described by giving the X-components of **v** in terms of those of **w**.

Change of basis (9). The equation defines new component variables and thus introduces new basis vectors.

The matrices

$$\begin{bmatrix} 1 \\ 0 \end{bmatrix} \text{ and } \begin{bmatrix} 0 \\ 1 \end{bmatrix} \tag{10}$$

play an important role in the study of any of these four types of transformations. In terms of any affine coordinates, they give the unit points on the coordinate axes. In terms of any set of component variables, they give the basis vectors.

Thus, the following simple observation is the key to solving many specific problems. Check that

$$\begin{bmatrix} a_{11} & a_{12} \\ a_{21} & a_{22} \end{bmatrix}\begin{bmatrix} 1 \\ 0 \end{bmatrix} = \begin{bmatrix} a_{11} \\ a_{21} \end{bmatrix} \tag{11}$$

and

$$\begin{bmatrix} a_{11} & a_{12} \\ a_{21} & a_{22} \end{bmatrix}\begin{bmatrix} 0 \\ 1 \end{bmatrix} = \begin{bmatrix} a_{12} \\ a_{22} \end{bmatrix}. \tag{12}$$

The results of multiplying the key matrices in (10) by a square matrix A are displayed in the columns of A, itself.

EXAMPLES

1. Let X be the matrix of rectangular coordinate variables on the plane. Find the matrix equation in X for the point transformation that carries $(1, 0)$ into $(2, -3)$ and carries $(0, 1)$ into $(-1, 4)$. This comes directly from (11) and (12). *Answer.*

$$X(q) = \begin{bmatrix} 2 & -1 \\ -3 & 4 \end{bmatrix} X(p).$$

2. Let

$$A = \begin{bmatrix} 3 & 1 \\ 5 & 2 \end{bmatrix},$$

and let X be as in Example 1. For the coordinate transformation described by

$$U(p) = AX(p),$$

find the unit points on the U-axes and sketch. To find these points, one must have their X-coordinates. Now these are the points at which U assumes the values

$$\begin{bmatrix} 1 \\ 0 \end{bmatrix} \text{ and } \begin{bmatrix} 0 \\ 1 \end{bmatrix};$$

thus, solve the equation $U = AX$ for X:

$$A^{-1}U = X.$$

From (11) and (12) the desired coordinates appear in the columns of A^{-1}. For the matrix A given here, one finds that

$$A^{-1} = \begin{bmatrix} 2 & -1 \\ -5 & 3 \end{bmatrix}.$$

Therefore, the U unit points have X-coordinates $(2, -5)$ and $(-1, 3)$. The sketch is shown in Fig. 4–2. To illustrate the vagaries of an affine coordinate system, two points are plotted in Fig. 4–2 with their U-coordinates shown.

3. Find the coordinate transformation such that the new unit points are $(2, 3)$ and $(2, 4)$. The transformation is to be written in the form $U(p) = AX(p)$, and it appears from Example 2 that $(2, 3)$ and $(2, 4)$ must form the columns of A^{-1}; that is,

$$A^{-1} = \begin{bmatrix} 2 & 2 \\ 3 & 4 \end{bmatrix}.$$

On inverting this matrix, one finds that the required transformation is given by

$$U(p) = \begin{bmatrix} 2 & -1 \\ -\frac{3}{2} & 1 \end{bmatrix} X(p).$$

4. Let T be a linear transformation of the space of plane vectors into itself. It is customary to write $T\mathbf{w}$ rather than $T(\mathbf{w})$ for the vector into which T carries $\mathbf{w}$. Given that

$$T\mathbf{i} = 2\mathbf{i} - 3\mathbf{j}, \qquad T\mathbf{j} = -\mathbf{i} + 4\mathbf{j},$$

find the matrix equation for T in terms of X. This is the vector parallel

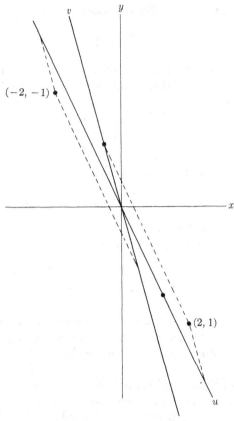

FIGURE 4-2

to the analytic geometry problem of Example 1. *Answer*.

$$X(\mathbf{v}) = \begin{bmatrix} 2 & -1 \\ -3 & 4 \end{bmatrix} X(\mathbf{w}).$$

5. Take the matrix A of Example 2 and consider the change of basis described by

$$U(\mathbf{w}) = AX(\mathbf{w}).$$

Find the basis vectors $\mathbf{b}_u$ and $\mathbf{b}_v$. This is an exact parallel to the problem of Example 2. *Answer*.

$$\mathbf{b}_u = 2\mathbf{i} - 5\mathbf{j}, \qquad \mathbf{b}_v = -\mathbf{i} + 3\mathbf{j}.$$

6. Let

$$\mathbf{b}_u = 2\mathbf{i} + 3\mathbf{j}, \qquad \mathbf{b}_v = 2\mathbf{i} + 4\mathbf{j};$$

find the matrix equation giving U in terms of X. This parallels Example 3. *Answer.*

$$U(\mathbf{w}) = \begin{bmatrix} 2 & -1 \\ -\frac{3}{2} & 1 \end{bmatrix} X(\mathbf{w}).$$

7. A scalar product is given in terms of **i**- and **j**-components by

$$\mathbf{v} \cdot \mathbf{w} = x(\mathbf{v})x(\mathbf{w}) + y(\mathbf{v})y(\mathbf{w}).$$

In matrix form, this may be written

$$\mathbf{v} \cdot \mathbf{w} = [x(\mathbf{v})\ y(\mathbf{v})] \begin{bmatrix} x(\mathbf{w}) \\ y(\mathbf{w}) \end{bmatrix} = [X(\mathbf{v})]'X(\mathbf{w}).$$

To find a formula for scalar products in terms of other component variables, proceed as follows. Let the change of basis be given by

$$U(\mathbf{w}) = AX(\mathbf{w});$$

then

$$X(\mathbf{w}) = A^{-1}U(\mathbf{w}),$$

and

$$\mathbf{v} \cdot \mathbf{w} = [A^{-1}U(\mathbf{v})]'[A^{-1}U(\mathbf{w})] = [U(\mathbf{v})]'(A^{-1})'A^{-1}U(\mathbf{w}).$$

For example, take the change of basis considered in Example 6. There

$$A^{-1} = \begin{bmatrix} 2 & 2 \\ 3 & 4 \end{bmatrix},$$

so

$$(A^{-1})'A^{-1} = \begin{bmatrix} 2 & 3 \\ 2 & 4 \end{bmatrix}\begin{bmatrix} 2 & 2 \\ 3 & 4 \end{bmatrix} = \begin{bmatrix} 13 & 16 \\ 16 & 20 \end{bmatrix},$$

and

$$\mathbf{v} \cdot \mathbf{w} = [u(\mathbf{v})\ v(\mathbf{v})] \begin{bmatrix} 13 & 16 \\ 16 & 20 \end{bmatrix}\begin{bmatrix} u(\mathbf{w}) \\ v(\mathbf{w}) \end{bmatrix}$$

$$= 13u(\mathbf{v})u(\mathbf{w}) + 16u(\mathbf{v})v(\mathbf{w}) + 16u(\mathbf{w})v(\mathbf{v}) + 20v(\mathbf{v})v(\mathbf{w}).$$

8. Let U and X be 2×1 matrices of variables, and let A be a 2×2 matrix of constants. Given $U = AX$, it follows by direct computation that

$$dU = A\ dX. \tag{13}$$

Now, the definition of differential of a vector variable given in Section 3–2 may be summarized in words by saying that the components of $d\mathbf{w}$ are the differentials of the components of $\mathbf{w}$. In terms of the matrix of component variables, X, this says

$$X(d\mathbf{w}) = d[X(\mathbf{w})]. \tag{14}$$

Suppose that a change of basis

$$U(\mathbf{w}) = AX(\mathbf{w})$$

is introduced. The same formula applies to the components of $d\mathbf{w}$; that is,

$$U(d\mathbf{w}) = AX(d\mathbf{w}).$$

Thus, accepting (14), it follows by (13) that

$$U(d\mathbf{w}) = AX(d\mathbf{w}) = A\,d[X(\mathbf{w})] = d[U(\mathbf{w})].$$

That is, the components of $d\mathbf{w}$ are formed in the same way no matter what basis is chosen. Thus, the definition given does not make $d\mathbf{w}$ dependent on the choice of basis.

EXERCISES

1. Let

$$A = \begin{bmatrix} 1/a & 0 \\ 0 & 1/b \end{bmatrix}.$$

Describe geometrically the point transformation defined by

$$X(q) = AX(p).$$

2. Let

$$A = \begin{bmatrix} \sqrt{3}/2 & 1/2 \\ -1/2 & \sqrt{3}/2 \end{bmatrix}.$$

Describe geometrically the point transformation defined by

$$X(q) = AX(p).$$

[*Hint:* Introduce polar coordinates and recall the addition formulas for sin and cos.]

3. Let

$$A = \begin{bmatrix} \cos\alpha & \sin\alpha \\ -\sin\alpha & \cos\alpha \end{bmatrix},$$

and let T be the linear transformation defined by $X(T\mathbf{w}) = AX(\mathbf{w})$.

(a) Describe geometrically the relation between **w** and $T\mathbf{w}$.

(b) Show that $AA' = [\delta_{ij}]$; then show that

$$T\mathbf{w} \cdot T\mathbf{w} = \mathbf{w} \cdot \mathbf{w}.$$

Explain how this proves that T preserves the lengths of vectors.

4. Consider the coordinate transformation defined by

$$U(p) = \begin{bmatrix} 1 & -2 \\ 1 & 2 \end{bmatrix} X(p).$$

(a) Sketch the new coordinate axes and locate the unit point on each.

(b) Given that

$$X(p) = \begin{bmatrix} 1 \\ -1 \end{bmatrix},$$

locate p on the sketch, and find $U(p)$.

(c) Given that

$$U(q) = \begin{bmatrix} 1 \\ -1 \end{bmatrix},$$

locate q on the sketch, and find $X(q)$.

5. Consider the change of basis defined by

$$U(\mathbf{w}) = \begin{bmatrix} 1 & -2 \\ 1 & 2 \end{bmatrix} X(\mathbf{w}).$$

(a) Find the basis vectors $\mathbf{b}_u$ and $\mathbf{b}_v$ in terms of **i** and **j**.

(b) Find the formula for $\mathbf{v} \cdot \mathbf{w}$ in terms of U-components.

(c) Find the U-components of $3\mathbf{i} + 4\mathbf{j}$, and use the formula derived in part (b) to verify that $|3\mathbf{i} + 4\mathbf{j}| = 5$.

6. Linear transformations of vector spaces may be characterized without reference to a basis. Let T be a mapping of vectors into vectors. T is called linear if

$$T(\alpha\mathbf{u} + \beta\mathbf{v}) = \alpha T\mathbf{u} + \beta T\mathbf{v}$$

for all vectors **u** and **v** and all scalars α and β. Let T be a linear transformation on the space of plane vectors, and let a basis be given; show that T then has a matrix representation. [*Hint:* In terms of the given basis

$$\begin{bmatrix} 1 \\ 0 \end{bmatrix} \rightarrow \begin{bmatrix} a \\ b \end{bmatrix}, \qquad \begin{bmatrix} 0 \\ 1 \end{bmatrix} \rightarrow \begin{bmatrix} c \\ d \end{bmatrix};$$

show that these conditions determine a 2×2 matrix; then show that an arbitrary vector transforms correctly.]

7. The matrix determined in Exercise 6 depends not only on T but also on the basis chosen. Suppose that T is defined by

$$X(T\mathbf{w}) = AX(\mathbf{w})$$

for a certain basis. Let

$$U(\mathbf{w}) = BX(\mathbf{w})$$

determine a change of basis, and show that, with respect to the new basis, the matrix of T is BAB^{-1}.

8. Let T be a linear transformation of the space of plane vectors such that

$$T\mathbf{i} = \mathbf{i} + \mathbf{j}, \qquad T\mathbf{j} = 2\mathbf{i} - \mathbf{j}.$$

Find the matrix representation for T in terms of $\mathbf{i}$- and $\mathbf{j}$-components.

9. Let

$$\mathbf{b}_u = \mathbf{i} - \mathbf{j}, \qquad \mathbf{b}_v = 3\mathbf{i} + \mathbf{j}.$$

With respect to $(\mathbf{b}_u, \mathbf{b}_v)$ as a basis, find the matrix representation for the transformation T of Exercise 8.

4–3 Determinants. Associated with each square matrix A of numbers is a number denoted by det A and called the *determinant* of A. Another notation in common use is $|A|$, although by association with the notion of absolute value, this gives the unfortunate impression that determinants must be positive, and such is not the case. On the other hand, when the matrix elements are displayed in a square array, the vertical bars are universally used to denote the determinant. Thus, if

$$A = \begin{bmatrix} a_{11} & a_{12} \\ a_{21} & a_{22} \end{bmatrix},$$

then

$$\det A = \begin{vmatrix} a_{11} & a_{12} \\ a_{21} & a_{22} \end{vmatrix}.$$

The definition of det A calls for the notions of *even* and *odd permutations*, and these may be described as follows. Given a finite set of distinct objects in some prescribed order, any rearrangement (change of order) of this set may be achieved through a succession of moves each consisting of a simple interchange of adjacent objects. For example,

$$1\ 2\ 3\ 4\ 5$$
$$1\ 3\ 2\ 4\ 5$$
$$1\ 3\ 4\ 2\ 5$$
$$1\ 4\ 3\ 2\ 5$$

indicates how 1 2 3 4 5 may be permuted in three elementary steps to

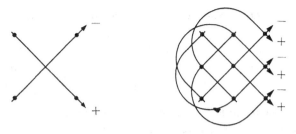

FIGURE 4–3

1 4 3 2 5. Now, it can be shown (proof omitted here) that if an odd number of successive interchanges of adjacent elements produces a given rearrangement of an ordered set, then any other succession of such interchanges, producing the same end result, will also involve an odd number of
steps. Thus, a *permutation* of an ordered set is another ordered set containing the same elements, and permutations are classified as odd or even,
according to the number of interchanges of adjacent elements required
to produce them.

Let A be an $n \times n$ matrix. There are $n!$ different sets

$$a_{1i} \cdot a_{2j} \cdots a_{nk},$$

each consisting of one element from each row of A and one from each
column. For each of these sets, take the product of the numbers involved
and append to each of these products a plus or minus sign according to
the following rule. Arrange the factors (as shown above) with row numbers
running $1, 2, \ldots, n$; append a plus sign if the column numbers then form
an even permutation of $1, 2, \ldots, n$ and a minus sign if they form an odd
permutation. Finally, take these products with signs appended, and add;
the result is, by definition, det A.

For the 2×2 and 3×3 cases, it is fairly easy to follow this definition
explicitly for the evaluation of determinants:

$$\begin{vmatrix} a_{11} & a_{12} \\ a_{21} & a_{22} \end{vmatrix} = a_{11}a_{22} - a_{12}a_{21},$$

$$\begin{vmatrix} a_{11} & a_{12} & a_{13} \\ a_{21} & a_{22} & a_{23} \\ a_{31} & a_{32} & a_{33} \end{vmatrix} = a_{11}a_{22}a_{33} + a_{12}a_{23}a_{31} + a_{13}a_{21}a_{32}$$
$$- a_{13}a_{22}a_{31} - a_{12}a_{21}a_{33} - a_{11}a_{23}a_{32}.$$

Figure 4–3 shows schematic diagrams for obtaining these products and
their signs.

In the above expansion of a third-order determinant, no two of the
entries a_{11}, a_{12}, a_{13} appear in the same product (why?); so, each of these

may be factored out of the two terms in which it does appear to yield

$$a_{11}(a_{22}a_{33} - a_{23}a_{32}) + a_{12}(a_{23}a_{31} - a_{21}a_{33}) + a_{13}(a_{21}a_{32} - a_{22}a_{31})$$

$$= a_{11} \begin{vmatrix} a_{22} & a_{23} \\ a_{32} & a_{33} \end{vmatrix} - a_{12} \begin{vmatrix} a_{21} & a_{23} \\ a_{31} & a_{33} \end{vmatrix} + a_{13} \begin{vmatrix} a_{21} & a_{22} \\ a_{31} & a_{32} \end{vmatrix}.$$

This exhibits the formula for *expansion by minors*, which can be generalized to any dimension. The general formula will be given here without proof, but the student should note that it follows the pattern set by the 3×3 case derived above.

Let A be an $n \times n$ matrix and a_{ij} one of its elements. The *minor* of a_{ij} is the $(n - 1) \times (n - 1)$ matrix obtained from A by deleting the ith row and the jth column; it will be denoted by A_{ij}. The *cofactor* of a_{ij}, to be denoted by α_{ij}, is defined by

$$\alpha_{ij} = (-1)^{i+j} \det A_{ij}.$$

Now, the pertinent theorem is that, for each i,

$$\det A = \sum_{j=1}^{n} a_{ij}\alpha_{ij}.$$

In words, take any row of A and multiply each element in this row by its cofactor; add these products to get det A.

If the ith row is chosen and this formula is applied, the process is called expansion of det A by minors on the ith row.

Note that if the words "row" and "column" are interchanged in the basic definition of a determinant, the overall result is unchanged. This leads to two interesting observations. First,

$$\det A' = \det A;$$

transposition of a matrix does not alter its determinant. Second, expansion by minors may equally well be based on an arbitrary column of the matrix.

Finally, it should be mentioned that determinants combine with matrix multiplication in a very pleasant manner; specifically for any square matrices A and B of the same dimension,

$$\det (AB) = (\det A)(\det B).$$

Proof of this is omitted and a proof by direct computation is probably not feasible. There is a hint of standard method of proof (not given in detail) in Section 4–4.

EXAMPLES

1. Compute

$$\begin{vmatrix} 1 & -2 & 3 \\ 0 & -1 & -2 \\ 3 & 1 & -1 \end{vmatrix}.$$

Answer. $1 + 12 + 0 - (-9) - (-2) - 0 = 24$. The expansion by minors on the first column is as follows:

$$1\begin{vmatrix} -1 & -2 \\ 1 & -1 \end{vmatrix} - 0\begin{vmatrix} -2 & 3 \\ 1 & -1 \end{vmatrix} + 3\begin{vmatrix} -2 & 3 \\ -1 & -2 \end{vmatrix}$$

$$= 1[1 - (-2)] - 0 + 3[4 - (-3)] = 3 + 21 = 24.$$

Note how the presence of a zero entry shortens the work in expanding by minors.

2. Compute

$$\begin{vmatrix} 3 & 0 & 0 & 0 \\ -1 & 2 & 0 & 0 \\ 1 & -1 & -2 & 0 \\ 2 & 5 & -3 & 1 \end{vmatrix}.$$

Expand by minors on the first row:

$$3\begin{vmatrix} 2 & 0 & 0 \\ -1 & -2 & 0 \\ 5 & -3 & 1 \end{vmatrix} + 0 + 0 + 0.$$

Again, expand by minors on the first row:

$$3(2\begin{vmatrix} -2 & 0 \\ -3 & 1 \end{vmatrix} + 0 + 0) = 3[2(-2 + 0)] = -12.$$

Note that the result is just the product of the main diagonal entries.

EXERCISES

1. Evaluate each of the following determinants.

(a) $\begin{vmatrix} 2 & -1 \\ 3 & 5 \end{vmatrix}$
(b) $\begin{vmatrix} -1 & 0 \\ 2 & 3 \end{vmatrix}$

(c) $\begin{vmatrix} \cos\alpha & -\sin\alpha \\ \sin\alpha & \cos\alpha \end{vmatrix}$

(d) $\begin{vmatrix} 2 & 2 & 1 \\ 1 & 3 & 2 \\ 3 & 1 & -1 \end{vmatrix}$

(e) $\begin{vmatrix} 2 & 0 & 2 \\ 0 & 3 & -3 \\ -3 & -2 & 0 \end{vmatrix}$

(f) $\begin{vmatrix} \cos\beta & -\sin\beta & 0 \\ \cos\alpha\sin\beta & \cos\alpha\cos\beta & -\sin\alpha \\ \sin\alpha\sin\beta & \sin\alpha\cos\beta & \cos\alpha \end{vmatrix}$

(g) $\begin{vmatrix} 1 & 3 & -1 & 2 \\ 2 & 1 & 3 & 1 \\ -1 & 2 & -1 & 3 \\ -2 & 1 & 2 & -3 \end{vmatrix}$

(h) $\begin{vmatrix} 4 & 3 & -2 & 7 \\ 8 & 1 & -4 & 6 \\ 6 & 2 & -3 & 11 \\ 10 & 4 & -5 & -8 \end{vmatrix}$

2. Let

$$A = \begin{bmatrix} 1 & -2 & 3 \\ 4 & 1 & -1 \\ 2 & 0 & 3 \end{bmatrix}, \qquad B = \begin{bmatrix} 1 & 0 & 2 \\ -1 & 2 & 3 \\ -3 & -1 & 4 \end{bmatrix};$$

show by direct computation that det (AB) = det A det B.

3. Given an ordered set of n objects, show that an interchange of two of them (not necessarily adjacent ones) produces an odd permutation.

4. A *cyclic* permutation of an ordered set is produced by placing the last element first and moving all others down one step. For example, 51234 is a cyclic permutation of 12345. Prove that a cyclic permutation of n factors is odd, when n is even, and even when n is odd.

5. Generalize Example 2. Let $[a_{ij}]$ be any square matrix with $a_{ij} = 0$ for $j > i$; then

$$\det(a_{ij}) = a_{11}a_{22}\ldots a_{nn}.$$

In words, if a matrix has all entries above the main diagonal zero, then its determinant is the product of the main diagonal entries.

4–4 Elementary operations. There are three basic *elementary row operations* on a square matrix:

    I. Interchanging two rows,

    II. multiplying each entry in some one row by a constant c, and

    III. replacing some row, say the ith, by the ith row plus c times the jth row $(i \neq j)$.

The present section deals with the effect of these operations on the determinant of a matrix.

Suppose A is a square matrix and let B_I, B_II, and B_III be, respectively, the matrices obtained by subjecting A to the elementary row operations I, II, and III. The effect on determinants is very quickly summarized as follows:

$$\det B_\mathrm{I} = -\det A; \tag{15}$$

$$\det B_\mathrm{II} = c \det A; \tag{16}$$

$$\det B_\mathrm{III} = \det A. \tag{17}$$

Operation I changes the sign of the determinant; II multiplies it by c; III does not change it.

Proofs of these rules are quite simple. To see (15), recall the definition of det A (Section 4–3). If in each product of elements the column numbers are arranged in order, an interchange of rows introduces an odd permutation in each set of row numbers and so changes the sign of each term in the expansion.

Expand det B_II by minors on the row that has been altered, and (16) follows at once.

To derive (17) requires an intermediate result that is of considerable importance in itself. Suppose A has two identical rows. Interchanging these rows yields the same matrix, hence the same determinant, but by (15) it changes the sign of the determinant. Thus, det $A = -$det A, so det $A = 0$. In brief, *a square matrix with two identical rows has a zero determinant.*

Now, if det B_III is expanded on the ith row, the result is

$$\det B_\mathrm{III} = \sum_k (a_{ik} + ca_{jk})\alpha_{ik}$$

$$= \sum_k a_{ik}\alpha_{ik} + c \sum_k a_{jk}\alpha_{ik}$$

$$= \det A + c \cdot 0.$$

Here α_{ik} is the cofactor (in A) of a_{ik}, and the last sum is zero because it gives the determinant of a matrix with rows i and j identical.

Since det $A = $ det A', there is an analogous theory for *elementary column operations.* Merely interchange the words "row" and "column" in the above discussion.

The rule for expansion of a determinant by minors and the fact that a square matrix with identical rows or columns has a zero determinant furnish a formula for the inversion of a matrix. As above, let

$$A = [a_{ij}],$$

and let α_{ij} be the cofactor of a_{ij}. The *adjoint* of the matrix A is defined as the transposed matrix of cofactors; that is,

$$\text{adj } A = [\alpha_{ij}]'.$$

Now, it follows at once that

$$A(\text{adj } A) = (\text{adj } A)A = \det A[\delta_{ij}]. \tag{18}$$

The (i, i)-entry in $A(\text{adj } A)$ is

$$\sum_{k=1}^{n} a_{ik}\alpha_{ik} = \det A,$$

and the (i, j)-entry is

$$\sum_{k=1}^{n} a_{ik}\alpha_{jk} = 0$$

because this multiplies row i by the cofactors of row j and effectively expands the determinant of a matrix with identical rows. A similar argument applies to the product $(\text{adj } A)A$.

So, it follows from (18) that if

$$\det A \neq 0,$$

then A has an inverse, and

$$A^{-1} = \frac{\text{adj } A}{\det A}.$$

EXAMPLES

1. Let

$$A = \begin{bmatrix} 1 & 0 & -1 \\ 2 & 3 & 2 \\ 4 & 2 & -3 \end{bmatrix},$$

and consider the following elementary row operations:

(i) Replace row 1 by row 1 plus one-half of row 2:

$$\begin{bmatrix} 2 & \frac{3}{2} & 0 \\ 2 & 3 & 2 \\ 4 & 2 & -3 \end{bmatrix}.$$

(ii) Replace row 2 by row 2 plus two-thirds of row 3:

$$\begin{bmatrix} 2 & \frac{3}{2} & 0 \\ \frac{14}{3} & \frac{13}{3} & 0 \\ 4 & 2 & -3 \end{bmatrix}.$$

(iii) Multiply row 1 by 26 and row 2 by 9:

$$\begin{bmatrix} 52 & 39 & 0 \\ 42 & 39 & 0 \\ 4 & 2 & -3 \end{bmatrix}.$$

(iv) Replace row 1 by row 1 minus row 2:

$$\begin{bmatrix} 10 & 0 & 0 \\ 30 & 39 & 0 \\ 4 & 2 & -3 \end{bmatrix}.$$

Call this last matrix B. Steps (i), (ii), and (iv) do not affect determinants; step (iii) multiplies the determinant by 9×26. By Exercise 5, Section 4–3,

$$\det B = 10 \times 39 \times (-3);$$

hence,

$$\det A = \frac{\det B}{9 \times 26} = - \frac{10 \times 39 \times 3}{9 \times 26} = -5.$$

2. Let

$$A = \begin{bmatrix} a_{11} & a_{12} \\ a_{21} & a_{22} \end{bmatrix};$$

find A^{-1}. For a 2×2 matrix the adjoint is sufficiently simple that it can be found by a mechanical rule. The matrix $[\alpha_{ij}]$ of cofactors is

$$\begin{bmatrix} a_{22} & -a_{21} \\ -a_{12} & a_{11} \end{bmatrix};$$

so,

$$\operatorname{adj} A = \begin{bmatrix} a_{22} & -a_{12} \\ -a_{21} & a_{11} \end{bmatrix}.$$

In words, interchange the main diagonal entries and change the signs of the other two entries. Finally,

$$A^{-1} = \frac{\begin{bmatrix} a_{22} & -a_{12} \\ -a_{21} & a_{11} \end{bmatrix}}{a_{11}a_{22} - a_{12}a_{21}}.$$

Using this rule, a 2×2 matrix may be inverted by inspection. Example:

$$\begin{bmatrix} 1 & 3 \\ -2 & 4 \end{bmatrix}^{-1} = \frac{1}{10}\begin{bmatrix} 4 & -3 \\ 2 & 1 \end{bmatrix}.$$

3. Let B be obtained from A by an elementary row operation and let E be obtained from $[\delta_{ij}]$ by the same operation. It is then not difficult to show that

$$B = EA.$$

Such a matrix E, obtained from the identity matrix by an elementary row operation, is called an *elementary matrix*. It is not difficult to verify by direct computation that if E_1 and E_2 are elementary matrices, then

$$\det (E_1E_2) = \det E_1 \det E_2. \tag{19}$$

The standard proof that $\det (AB) = \det A \det B$ may now be outlined as follows. Dispose of the zero cases separately; $\det (AB)$ is zero if and only if at least one of the other determinants is zero. If $\det A \neq 0$, then A has an inverse and

$$Y = AX$$

may be solved for X. Solving this system of linear equations by addition and subtraction is equivalent to reducing A to $[\delta_{ij}]$ by elementary row operations; therefore A must be a product of elementary matrices. The same goes for B; so the general result follows from the special case (19).

EXERCISES

1. Find

$$\begin{vmatrix} 1 & 3 & -2 \\ 2 & 1 & 4 \\ 3 & -2 & 1 \end{vmatrix}$$

(a) directly by the definition of a determinant,
(b) by expanding by minors on the first row,
(c) by using elementary row operations (as in Example 1) to reduce the matrix to a triangular one.

2. Let A be the matrix whose determinant appears in Exercise 1 and find
(a) adj A,
(b) the product A (adj A),
(c) the product (adj A)A.

3. Find the inverse of each of the following matrices.

(a) $\begin{bmatrix} 1 & 3 \\ -4 & 2 \end{bmatrix}$ 　　　　　(b) $\begin{bmatrix} -4 & 3 \\ -1 & 2 \end{bmatrix}$

4. *Cramer's Rule.* Let

$$X = \begin{bmatrix} x_1 \\ \vdots \\ x_n \end{bmatrix} \quad \text{and} \quad C = \begin{bmatrix} c_1 \\ \vdots \\ c_n \end{bmatrix},$$

and let A be a square matrix of numbers. The system of linear equations

$$AX = C \tag{20}$$

may be solved by the following rule. To find x_j, replace column j of A by C; compute the determinant of this modified matrix and divide by det A. Prove this rule in the following way. Solve (20) in matrix form:

$$X = A^{-1}C = \frac{(\text{adj } A)C}{\det A}. \tag{21}$$

Write out the $(j, 1)$ entry of the product $(\text{adj } A)C$, and note that this expands the modified matrix by minors on column j.

5. Use Cramer's Rule to solve each of the following systems of linear equations. [*Note:* In terms of the notation used above, X is

$$\begin{bmatrix} x \\ y \end{bmatrix} \quad \text{or} \quad \begin{bmatrix} x \\ y \\ z \end{bmatrix},$$

as the case may be; and C is a column matrix of constants. Give each solution in two forms: (i) Employ the rule as stated in words to find the value of each unknown as a quotient of determinants. (ii) Compute adj A and use (21) to display the entire solution in matrix form.]

(a) $x - 3y = 1$ 　　　　　(b) $x + 3y - 2z = 2$
　　$2x + y = -2$ 　　　　　　　　$2x + y + 4z = -1$
　　　　　　　　　　　　　　　　　$3x - 2y + z = 1$

6. Let $\mathbf{u} = u_1\mathbf{i} + u_2\mathbf{j}$, $\mathbf{v} = v_1\mathbf{i} + v_2\mathbf{j}$, and define

$$\det (\mathbf{u}, \mathbf{v}) = \begin{bmatrix} u_1 & v_1 \\ u_2 & v_2 \end{bmatrix}.$$

Use the results of this section to show that
　　(a) $\det (\mathbf{v}, \mathbf{u}) = -\det (\mathbf{u}, \mathbf{v})$,
　　(b) $\det (a\mathbf{u}, b\mathbf{v}) = ab \det (\mathbf{u}, \mathbf{v})$,
　　(c) $\det (\mathbf{u} + a\mathbf{v}, \mathbf{v}) = \det (\mathbf{u}, \mathbf{v} + b\mathbf{u}) = \det (\mathbf{u}, \mathbf{v})$.

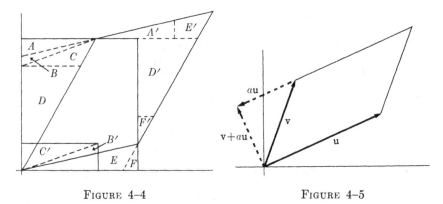

FIGURE 4–4   FIGURE 4–5

7. The determinant of a vector pair is defined in Exercise 6. Let T be a linear transformation of the space of plane vectors into itself, and let A be its matrix representation with respect to the basis $(\mathbf{i}, \mathbf{j})$ (see Exercise 6, Section 4–2). Show that det $(T\mathbf{u}, T\mathbf{v})$ = det A det $(\mathbf{u}, \mathbf{v})$.

8. Figure 4–4 shows a parallelogram formed by vectors $\mathbf{u}$ and $\mathbf{v}$. It also shows how to cut a jigsaw puzzle that can be put together either as this parallelogram or as the difference of two rectangles. Conclude from this that $|\det(\mathbf{u}, \mathbf{v})|$, as defined in Exercise 6, gives the area of the parallelogram.

9. Show that if $\mathbf{u} \cdot \mathbf{w} = 0$, then $|\mathbf{u}|\,|\mathbf{w}| = |\det(\mathbf{u}, \mathbf{w})|$. [*Hint:* Expand $|\mathbf{u}|^2\,|\mathbf{w}|^2$ and $|\det(\mathbf{u}, \mathbf{w})|^2$ in terms of components, and compare, recalling that $(\mathbf{u} \cdot \mathbf{w})^2 = 0$.

10. Figure 4–5 shows how the parallelogram of Exercise 8 may be regarded as having a base given by u and an altitude given by $\mathbf{v} + a\mathbf{u}$ for a suitable scalar a. Use this information, Exercises 9 and 6(c) to give another proof of the result stated in Exercise 8.

11. On the basis of Exercises 7 and 8 (or 10), discuss the effect of a linear point transformation of the plane on areas of parallelograms.

12. Show that det $(\mathbf{u}, \mathbf{v})$ is positive or negative according as the rotation from the direction of $\mathbf{u}$ to that of $\mathbf{v}$ (through the interior of their parallelogram) is counterclockwise or clockwise.

4–5 Vectors in three dimensions. The set of directed magnitudes in three-dimensional space may be made into a vector space. Addition and multiplication by scalars must be defined and this is done exactly as in the plane. An arrow representing $\mathbf{u} + \mathbf{v}$ is obtained by placing arrows representing $\mathbf{u}$ and $\mathbf{v}$, respectively, head to tail and constructing an arrow from start to finish. The product $a\mathbf{u}$ is defined by saying that

$$|a\mathbf{u}| = |a|\,|\mathbf{u}|$$

and that the direction of $a\mathbf{u}$ is that of $\mathbf{u}$ for $a > 0$ and exactly opposite to that of $\mathbf{u}$ for $a < 0$.

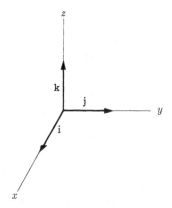

FIGURE 4–6

A basis in this three-dimensional vector space is furnished by a set of three mutually perpendicular unit vectors. If the three-dimensional space of points has a rectangular coordinate system in it, the associated unit vectors are usually denoted by **i**, **j**, **k**. Figure 4–6 shows a set of coordinate axes in three-space with the associated basis vectors for the vector space.

In Fig. 4–6 the basis vectors in the order **i**, **j**, **k** form a *right-handed* system. That is, their relative orientation is that of the thumb, forefinger, and middle finger of the right hand. If the fingers of the right hand are curled to indicate the direction of rotation from **i** to **j**, then the protruding right thumb will give the direction of **k**. A right-handed system may also be described by saying that when it is drawn on a piece of paper, the rotation from **i** to **j** to **k** is counterclockwise. For a reason that will appear very shortly, one should always use a right-handed basis in three-dimensional vector analysis. So, to preserve the usual correspondence between basis vectors and coordinate variables, one should always use a right-handed coordinate system in three-dimensional geometry.

The *scalar product* is defined as in the case of plane vectors:

$$\mathbf{u} \cdot \mathbf{v} = |\mathbf{u}|\,|\mathbf{v}|\cos\phi,$$

where ϕ is the angle between the directions of **u** and **v**. The algebra of scalar multiplication is just as before:

$$\mathbf{u} \cdot \mathbf{v} = \mathbf{v} \cdot \mathbf{u},$$

$$\mathbf{u} \cdot \mathbf{u} = |\mathbf{u}|^2,$$

$$(a\mathbf{u}) \cdot \mathbf{v} = a(\mathbf{u} \cdot \mathbf{v}),$$

$$\mathbf{u} \cdot (\mathbf{v} + \mathbf{w}) = (\mathbf{u} \cdot \mathbf{v}) + (\mathbf{u} \cdot \mathbf{w}).$$

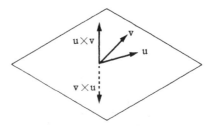

FIGURE 4–7

Proofs are omitted; however, the plane case was discussed in Section 3–1, and there is essentially nothing new here.

The following scalar multiplication table is easily verified

	i	j	k
i	1	0	0
j	0	1	0
k	0	0	1

From this and the basic properties listed above, it follows that

$$\mathbf{u} \cdot \mathbf{v} = (u_1\mathbf{i} + u_2\mathbf{j} + u_3\mathbf{k}) \cdot (v_1\mathbf{i} + v_2\mathbf{j} + v_3\mathbf{k})$$
$$= u_1v_1 + u_2v_2 + u_3v_3.$$

In terms of components with respect to $\mathbf{i}$, $\mathbf{j}$, and $\mathbf{k}$, the formula for scalar product follows the familiar pattern.

In three dimensions, and only in three dimensions, there is defined another product called the *vector product*. The vector product of $\mathbf{u}$ and $\mathbf{v}$ is denoted by $\mathbf{u} \times \mathbf{v}$ and is defined as follows. Magnitude:

$$|\mathbf{u} \times \mathbf{v}| = |\mathbf{u}|\,|\mathbf{v}|\,|\sin \phi|,$$

where ϕ is the angle between the $\mathbf{u}$- and $\mathbf{v}$-directions. Direction: $\mathbf{u} \times \mathbf{v}$ is perpendicular to the plane of $\mathbf{u}$ and $\mathbf{v}$, and of the two directions perpendicular to this plane, the one is chosen so that

$$\mathbf{u}, \qquad \mathbf{v}, \qquad \mathbf{u} \times \mathbf{v}$$

is a right-handed system.

Clearly (see Fig. 4–7),

$$\mathbf{u} \times \mathbf{v} = -\mathbf{v} \times \mathbf{u}. \tag{22}$$

Here is another example of noncommutative multiplication. However, unlike matrix algebra in which AB and BA may be completely different, here there is a fairly simple relation between $\mathbf{u} \times \mathbf{v}$ and $\mathbf{v} \times \mathbf{u}$.

Key properties of the vector product are as follows:

$$(a\mathbf{u}) \times \mathbf{v} = a(\mathbf{u} \times \mathbf{v}), \tag{23}$$

$$\mathbf{u} \times (\mathbf{v} + \mathbf{w}) = (\mathbf{u} \times \mathbf{v}) + (\mathbf{u} \times \mathbf{w}). \tag{24}$$

Proofs are omitted. Note, however, Exercise 3 below; another familiar property of multiplication fails in this system.

The following is the vector multiplication table for $\mathbf{i}$, $\mathbf{j}$, $\mathbf{k}$. Because of the noncommutative feature, it must be explained that the entry on the left is the first factor and that on top is the second factor.

	$\mathbf{i}$	$\mathbf{j}$	$\mathbf{k}$
$\mathbf{i}$	0	$\mathbf{k}$	$-\mathbf{j}$
$\mathbf{j}$	$-\mathbf{k}$	0	$\mathbf{i}$
$\mathbf{k}$	$\mathbf{j}$	$-\mathbf{i}$	0

Now, from this table and (22), (23), and (24) above, it is readily verified that

$$\mathbf{u} \times \mathbf{v} = (u_1\mathbf{i} + u_2\mathbf{j} + u_3\mathbf{k}) \times (v_1\mathbf{i} + v_2\mathbf{j} + v_3\mathbf{k})$$
$$= (u_2v_3 - u_3v_2)\mathbf{i} + (u_3v_1 - u_1v_3)\mathbf{j} + (u_1v_2 - u_2v_1)\mathbf{k}. \tag{25}$$

This is the standard formula for a vector product in terms of components.

The right-handed definition of vector multiplication is motivated by physical applications. Many right-hand rules of thumb in mechanics and electromagnetic theory are described simply by formulas involving vector products. With this convention dictated by physics and with the formula (25) established as standard, one must use a right-handed basis. Thus, in discussing vector analysis and analytic geometry side by side, it is obviously desirable to use right-handed coordinate systems. This will be done here, but the point is mentioned because many books on analytic geometry and calculus employ left-handed coordinate systems in three-space.

The formula for a vector product may be written in terms of a determinant as follows:

$$\mathbf{u} \times \mathbf{v} = \begin{vmatrix} \mathbf{i} & u_1 & v_1 \\ \mathbf{j} & u_2 & v_2 \\ \mathbf{k} & u_3 & v_3 \end{vmatrix}. \tag{26}$$

Expand this determinant by minors on the first column, and it reduces to (25).

By analogy with the idea introduced in Exercise 6, Section 4–4, one defines

$$\det (\mathbf{u}, \mathbf{v}, \mathbf{w}) = \begin{vmatrix} u_1 & v_1 & w_1 \\ u_2 & v_2 & w_2 \\ u_3 & v_3 & w_3 \end{vmatrix} .$$

It is readily verified that

$$\det (\mathbf{u}, \mathbf{v}, \mathbf{w}) = \mathbf{u} \cdot (\mathbf{v} \times \mathbf{w}),$$

and this determinant of three vectors is frequently referred to as the triple scalar product of the three vectors. (See Exercise 6 below.) However, this description of the determinant in terms of multiplication of vectors is limited to the three-dimensional case. Of more basic importance in calculus is the idea of a determinant of n vectors in n-space (see Section 4–6).

EXAMPLES

1. Find $(2\mathbf{i} - 3\mathbf{j} + \mathbf{k}) \times (\mathbf{i} + 2\mathbf{j} - 4\mathbf{k})$. Probably this is most quickly done from the determinant formula (26):

$$\begin{vmatrix} \mathbf{i} & 2 & 1 \\ \mathbf{j} & -3 & 2 \\ \mathbf{k} & 1 & -4 \end{vmatrix} = \mathbf{i}(12 - 2) - \mathbf{j}(-8 - 1) + \mathbf{k}(4 + 3) = 10\mathbf{i} + 9\mathbf{j} + 7\mathbf{k}.$$

2. Show that $\mathbf{i} + \mathbf{j} - \mathbf{k}$ and $\mathbf{i} + \mathbf{j} + 2\mathbf{k}$ are perpendicular. Find a third vector perpendicular to each of these and get unit vectors $\mathbf{u}$, $\mathbf{v}$, $\mathbf{w}$ in these mutually perpendicular directions. With respect to the basis $\mathbf{u}$, $\mathbf{v}$, $\mathbf{w}$, what are the components of $\mathbf{i} + \mathbf{j} + \mathbf{k}$?

First, check the scalar product of the given vectors;

$$(\mathbf{i} + \mathbf{j} - \mathbf{k}) \cdot (\mathbf{i} + \mathbf{j} + 2\mathbf{k}) = 1 + 1 - 2 = 0,$$

so they are perpendicular. The required third vector is given by the vector product:

$$(\mathbf{i} + \mathbf{j} - \mathbf{k}) \times (\mathbf{i} + \mathbf{j} + 2\mathbf{k}) = \begin{vmatrix} \mathbf{i} & 1 & 1 \\ \mathbf{j} & 1 & 1 \\ \mathbf{k} & -1 & 2 \end{vmatrix} = 3\mathbf{i} - 3\mathbf{j}.$$

To obtain unit vectors, divide each of the preceding by its magnitude:

$$\mathbf{u} = \frac{\mathbf{i} + \mathbf{j} - \mathbf{k}}{\sqrt{3}}, \qquad \mathbf{v} = \frac{\mathbf{i} + \mathbf{j} + 2\mathbf{k}}{\sqrt{6}}, \qquad \mathbf{w} = \frac{\mathbf{i} - \mathbf{j}}{\sqrt{2}}.$$

Since **w** is a positive multiple of **u** × **v**, the system **u**, **v**, **w** is right-handed. Components of **i** + **j** + **k** are scalar products of this vector with the basis vectors:

$$(\mathbf{i}+\mathbf{j}+\mathbf{k})\cdot\mathbf{u} = \frac{1+1-1}{\sqrt3} = \frac{1}{\sqrt3},$$

$$(\mathbf{i}+\mathbf{j}+\mathbf{k})\cdot\mathbf{v} = \frac{1+1+2}{\sqrt6} = \frac{4}{\sqrt6},$$

$$(\mathbf{i}+\mathbf{j}+\mathbf{k})\cdot\mathbf{w} = \frac{1-1+0}{\sqrt2} = 0.$$

EXERCISES

1. Let **u** = 2**i** − 3**j** + **k**, **v** = **i** + 4**j** + 5**k**, **w** = **i** + **j** + **k**. Find each of the following products.

(a) **u** · **v**   (b) **u** × **v**

(c) **v** × **u**   (d) **u** · (**v** + **w**)

(e) **u** · (**v** × **w**)   (f) **u** × (**v** + **w**)

2. Find the right-handed basis in which the first two vectors are positive multiples of **i** + **j** + **k** and 3**i** − 2**j** − **k**, respectively. With respect to this basis, find the components of 2**i** − **j** − **k**.

3. Show that
$$(\mathbf{i}\times\mathbf{j})\times(\mathbf{i}+\mathbf{j}) = -\mathbf{i}+\mathbf{j},$$
while
$$\mathbf{i}\times[\mathbf{j}\times(\mathbf{i}+\mathbf{j})] = \mathbf{j}.$$

Thus, vector multiplication is not associative.

4 Show that **u** × (**v** × **w**) is parallel to the plane of **v** and **w**, while (**u** × **v**) × **w** is parallel to that of **u** and **v**. Under what special circumstances will these be equal?

5. Explain why there is no ambiguity in writing **u** · **v** × **w** without parentheses.

6. Use the following steps to prove that **u** · **v** × **w** = **u** × **v** · **w**. (This is frequently stated as a rule. The dot and cross may be interchanged in a scalar triple product.)

(a) Use the theory of elementary row operations (Section 4–4) to show that
$$\det(\mathbf{u},\mathbf{v},\mathbf{w}) = \det(\mathbf{w},\mathbf{u},\mathbf{v}).$$
(b) Verify that
$$\det(\mathbf{u},\mathbf{v},\mathbf{w}) = \mathbf{u}\cdot\mathbf{v}\times\mathbf{w}.$$

(c) Get the desired result from parts (a) and (b) and the commutativity of scalar multiplication.

7. Show from the definitions of scalar and vector products that **u** · **v** × **w** is positive or negative according as **u**, **v**, **w** is a right- or left-handed system.

4–6 Orthonormal bases. We have given concrete examples, involving directed magnitudes, of vector spaces of dimensions two and three. For higher dimensions, the geometric picture fails, and an n-dimensional vector space is best described merely in terms of the vector algebra itself.

If three vectors in three-space are not coplanar, then no one of them is a linear combination of the other two. That is, for no a and b does $\mathbf{u} = a\mathbf{v} + b\mathbf{w}$. However, given four vectors in three space, there must be a relation of the form

$$\mathbf{u} = a\mathbf{v} + b\mathbf{w} + c\mathbf{x};$$

one of them must be a linear combination of the other three. (These statements can be proved, but here the student is asked only to convince himself intuitively that they are correct.)

Now, a finite set of vectors is called *linearly independent* if no one of them is a linear combination of the others. If a vector space contains a set of n linearly independent vectors but has the property that no set of $n + 1$ vectors can be linearly independent, then the space is called n-dimensional.

Suppose that $\mathbf{u}_1, \mathbf{u}_2, \ldots, \mathbf{u}_n$ is a set of n linearly independent vectors in an n-dimensional space. Then, given $\mathbf{v}$ in the space,

$$\mathbf{v} = \sum_{i=1}^{n} a_i \mathbf{u}_i.$$

Furthermore, the coefficients a_i are unique, because if

$$\sum_{i=1}^{n} a_i \mathbf{u}_i = \sum_{i=1}^{n} b_i \mathbf{u}_i \tag{27}$$

and if $a_j \neq b_j$, then equation (27) can be solved for $\mathbf{u}_j$ in terms of the others.

A *basis* in a vector space is defined as a set of vectors such that any vector in the space has a unique representation as a linear combination of the vectors in the basis. Thus, the discussion in the preceding paragraph shows that in an n-dimensional vector space any set of n linearly independent vectors forms a basis.

A scalar product can also be defined by its algebraic properties (with no reference to lengths, cosines, etc.). Specifically, it is required that

$$\mathbf{u} \cdot \mathbf{v} = \mathbf{v} \cdot \mathbf{u},$$

$$\mathbf{u} \cdot (a\mathbf{v} + b\mathbf{w}) = a(\mathbf{u} \cdot \mathbf{v}) + b(\mathbf{u} \cdot \mathbf{w}),$$

$$\mathbf{u} \cdot \mathbf{u} \geq 0.$$

Such an operation is definable in every n-dimensional vector space (see Exercise 4 below for outline of proof).

Given a vector space with a scalar product, one defines *length* of a vector by

$$|\mathbf{u}|^2 = \mathbf{u} \cdot \mathbf{u}.$$

Furthermore, $\mathbf{u}$ and $\mathbf{v}$ are defined to be *orthogonal* (perpendicular) provided

$$\mathbf{u} \cdot \mathbf{v} = 0.$$

A basis is called *orthonormal*, provided that any two vectors in it are orthogonal and each of them has unit length. More succinctly, $\mathbf{u}_1, \mathbf{u}_2, \ldots, \mathbf{u}_n$ is an orthonormal basis provided that $\mathbf{u}_i \cdot \mathbf{u}_j = \delta_{ij}$. A set of component variables with respect to an orthonormal basis will be called an orthonormal set of component variables.

The following procedure, known as the *Schmidt orthogonalization process*, may be used to construct an orthonormal basis from any given basis. Informally it is as follows. Suppose the given basis is $\mathbf{u}_1, \mathbf{u}_2, \ldots, \mathbf{u}_n$. Set

$$\mathbf{z}_1 = \mathbf{u}_1.$$

Next, add to $\mathbf{u}_2$ an appropriate multiple of $\mathbf{z}_1$ so that the result is orthogonal to $\mathbf{z}_1$:

$$\mathbf{z}_2 = \mathbf{u}_2 + a\mathbf{z}_1.$$

Get $\mathbf{z}_3$ by adding to $\mathbf{u}_3$ an appropriate combination of $\mathbf{z}_1$ and $\mathbf{z}_2$ so that the result is orthogonal to both $\mathbf{z}_1$ and $\mathbf{z}_2$:

$$\mathbf{z}_3 = \mathbf{u}_3 + b\mathbf{z}_1 + c\mathbf{z}_2.$$

Continue in this way to get a set $\mathbf{z}_1, \mathbf{z}_2, \ldots, \mathbf{z}_n$, each vector of which is orthogonal to all the others. Finally, unit lengths can be obtained by dividing each of these vectors by its length.

Figure 4–8 shows the orthogonalization process in two and three dimensions. In these cases, it is clear, geometrically, that appropriate multipliers do exist. However, simple algebraic computation gives a formula for these multipliers in general. If

$$\mathbf{z}_k = \mathbf{u}_k + \sum_{i=1}^{k-1} a_i \mathbf{z}_i$$

is to be orthogonal to $\mathbf{z}_j$, then $\mathbf{z}_k \cdot \mathbf{z}_j = 0$; and since already $\mathbf{z}_i \cdot \mathbf{z}_j = 0$ for $i \leq k - 1, j \leq k - 1, i \neq j$, this yields

$$0 = \mathbf{z}_k \cdot \mathbf{z}_j = \mathbf{u}_k \cdot \mathbf{z}_j + 0 + \cdots + a_j \mathbf{z}_j \cdot \mathbf{z}_j + \cdots + 0,$$

$$a_j = -\frac{\mathbf{u}_k \cdot \mathbf{z}_j}{\mathbf{z}_j \cdot \mathbf{z}_j}.$$

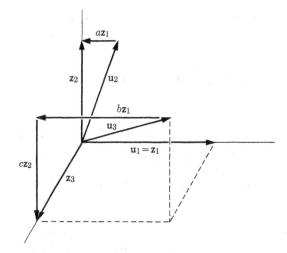

FIGURE 4-8

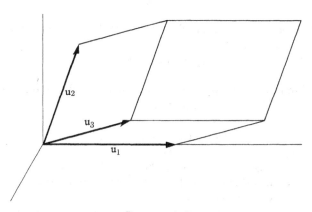

FIGURE 4-9

The Schmidt orthogonalization process was employed in Exercise 10, Section 4-4, to show that the determinant of two vectors in two-space gives the area of a parallelogram. The general process applies equally well in n dimensions to give a determinant formula for generalized volume.

The volume of a parallelepiped in three-space may be found in the following way. Take the length of some one edge; multiply by the altitude of some face containing this edge; then multiply by the altitude of the solid perpendicular to this face.

Now, Fig. 4-9 shows the same three vectors that appeared in Fig. 4-8, this time with a parallelepiped constructed on them. Comparison of the two figures shows that the lengths of z_2 and z_3 are precisely the altitudes needed to find the volume of the parallelepiped. The volume of the parallelepiped is $|z_1|\,|z_2|\,|z_3|$. So, given a linearly independent set $u_1, u_2, \ldots, u_n$ in n-dimensional space, define the *volume of their parallelepiped* to be

$$|z_1|\,|z_2|\,\ldots\,|z_n|,$$

where the z_i are the orthogonal vectors generated by the u_i through the Schmidt process.

Let $v_1, v_2, \ldots, v_n$ be the orthonormal basis obtained by dividing each z_i by its length, and let x_i be the v_i-component variable. Define

$$\det(u_1, u_2, \ldots, u_n) = \det[x_i(u_j)].$$

Now,

$$x_i(z_j) = |z_i|\,\delta_{ij},$$

so

$$\det(z_1, z_2, \ldots, z_n) = \begin{vmatrix} |z_1| & & & 0 \\ & |z_2| & & \\ & & \ddots & \\ 0 & & & |z_n| \end{vmatrix},$$

and this determinant is the volume of the parallelepiped. However,

$$\det(u_1, u_2, \ldots, u_n) = \det(z_1, z_2, \ldots, z_n)$$

because an examination of the Schmidt process shows that

$$[x_i(z_j)]$$

comes from

$$[x_i(u_j)]$$

by a succession of elementary column operations that do not change the value of a determinant.

Thus, the determinant of a set of vectors gives the volume of the parallelepiped they span provided that "determinant of a set of vectors" means determinant of the matrix of their components with respect to the orthonormal basis that they themselves generate through the Schmidt process. This theory will be completed in Section 4–7, where it will be shown that a change of basis to another orthonormal one changes, at most, the sign of $\det(u_1, u_2, \ldots, u_n)$.

The following purely mechanical point is worthy of note. The components of a single vector have been consistently given as a column matrix $(n \times 1)$. In the matrix $[x_i(u_j)]$ used above to define $\det(u_1, u_2, \ldots, u_n)$, the components of u_j appear as the jth column. Matrices of components of vectors will appear in many different connections henceforth, and this rule will be consistently invoked. That is, each column will be generated by a fixed vector and each row will be generated by a fixed component variable. Often there is no good reason for insisting on this. For example, where only a determinant is involved, to reverse the rule only transposes the matrix and changes nothing. However, if matrix multiplication is involved, things must be set up correctly; and it turns out that if vectors always come in columns, and components come in rows, then pertinent formulas assume the simplest and most natural form in terms of matrices.

EXAMPLES

1. Let

$$\mathbf{u} = \mathbf{i} + \mathbf{j}, \qquad \mathbf{v} = \mathbf{i} + 2\mathbf{j}.$$

Use the Schmidt process to orthogonalize this system. Set $\mathbf{z}_1 = \mathbf{u}$. Then

$$\mathbf{z}_2 = \mathbf{v} + a\mathbf{z}_1 = \mathbf{i} + 2\mathbf{j} + a(\mathbf{i} + \mathbf{j}),$$

where a is determined by the condition

$$0 = \mathbf{z}_1 \cdot \mathbf{z}_2 = (\mathbf{i} + \mathbf{j}) \cdot [(1 + a)\mathbf{i} + (2 + a)\mathbf{j}] = 1 + a + 2 + a;$$
$$2a + 3 = 0; \qquad a = -\tfrac{3}{2}.$$

Thus,

$$\mathbf{z}_2 = \mathbf{i} + 2\mathbf{j} - \tfrac{3}{2}(\mathbf{i} + \mathbf{j}) = -\tfrac{1}{2}\mathbf{i} + \tfrac{1}{2}\mathbf{j}.$$

See Fig. 4–10 for a sketch.

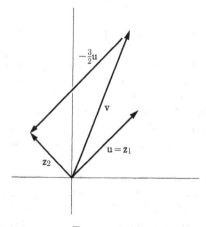

FIGURE 4–10

2. Orthogonalize the system

$$\mathbf{u} = \mathbf{i} + \mathbf{j} + \mathbf{k}, \qquad \mathbf{v} = \mathbf{i} + 2\mathbf{j} + \mathbf{k}, \qquad \mathbf{w} = \mathbf{i} - \mathbf{j} - \mathbf{k}.$$

Set

$$\mathbf{z}_1 = \mathbf{u}, \text{ and } \mathbf{z}_2 = \mathbf{v} + a\mathbf{z}_1 = (1 + a)\mathbf{i} + (2 + a)\mathbf{j} + (1 + a)\mathbf{k};$$

then the condition $\mathbf{z}_1 \cdot \mathbf{z}_2 = 0$ yields

$$(1 + a) + (2 + a) + (1 + a) = 0, \qquad a = -\tfrac{4}{3}.$$

Hence,

$$\mathbf{z}_2 = -\tfrac{1}{3}\mathbf{i} + \tfrac{2}{3}\mathbf{j} - \tfrac{1}{3}\mathbf{k}.$$

Next, set

$$\mathbf{z}_3 = \mathbf{w} + b\mathbf{z}_1 + c\mathbf{z}_2$$

$$= \left(1 + b - \frac{c}{3}\right)\mathbf{i} + \left(-1 + b + \frac{2c}{3}\right)\mathbf{j} + \left(-1 + b - \frac{c}{3}\right)\mathbf{k}.$$

Impose the orthogonality conditions,

$$\mathbf{z}_1 \cdot \mathbf{z}_3 = 0: \quad \left(1 + b - \frac{c}{3}\right) + \left(-1 + b + \frac{2c}{3}\right) + \left(-1 + b - \frac{c}{3}\right) = 0,$$

$$-1 + 3b = 0;$$

$$\mathbf{z}_2 \cdot \mathbf{z}_3 = 0: \quad \left(-\frac{1}{3} - \frac{b}{3} + \frac{c}{9}\right) + \left(-\frac{2}{3} + \frac{2b}{3} + \frac{4c}{9}\right) + \left(\frac{1}{3} - \frac{b}{3} + \frac{c}{9}\right) = 0,$$

$$-\frac{2}{3} + \frac{6c}{9} = 0.$$

Thus $b = \tfrac{1}{3}$, $c = 1$, and $\mathbf{z}_3 = \mathbf{i} - \mathbf{k}$.

3. For the three mutually orthogonal vectors $\mathbf{z}_1$, $\mathbf{z}_2$, and $\mathbf{z}_3$ of Example 2,

$$|\mathbf{z}_1|\,|\mathbf{z}_2|\,|\mathbf{z}_3| = \sqrt{3} \cdot \frac{\sqrt{6}}{3} \cdot \sqrt{2} = 2,$$

so this should be the volume of the parallelepiped determined by $\mathbf{u}$, $\mathbf{v}$, and $\mathbf{w}$. The determinant formula yields

$$\begin{vmatrix} 1 & 1 & 1 \\ 1 & 2 & -1 \\ 1 & 1 & -1 \end{vmatrix} = -2 - 1 + 1 - 2 + 1 + 1 = -2,$$

so $|\det (\mathbf{u}, \mathbf{v}, \mathbf{w})| = 2$ as expected. The negative value of $\det (\mathbf{u}, \mathbf{v}, \mathbf{w})$ indicates that $\mathbf{u}$, $\mathbf{v}$, $\mathbf{w}$, is a left-handed system. (See Exercise 7, Section 4–5.)

EXERCISES

1. Use the Schmidt process to orthogonalize each of the following systems. For each of the two-dimensional cases, draw a sketch.

(a) $\mathbf{u} = \mathbf{i} + 2\mathbf{j}, \quad \mathbf{v} = -\mathbf{i} + 2\mathbf{j}$

(b) $\mathbf{u} = 2\mathbf{i} + \mathbf{j}, \quad \mathbf{v} = 3\mathbf{i} - \mathbf{j}$

(c) $\mathbf{u} = \mathbf{i}, \quad \mathbf{v} = \mathbf{i} + \mathbf{j}$

(d) $\mathbf{u} = 2\mathbf{i} + \mathbf{j} + \mathbf{k}, \quad \mathbf{v} = 2\mathbf{i} - \mathbf{j} + \mathbf{k}, \quad \mathbf{w} = 2\mathbf{i} + \mathbf{j} - \mathbf{k}$

(e) $\mathbf{u} = \mathbf{i} + \mathbf{j} + \mathbf{k}, \quad \mathbf{v} = \mathbf{i} + \mathbf{j} - \mathbf{k}, \quad \mathbf{w} = \mathbf{i} - \mathbf{j} + \mathbf{k}$

2. In each part of Exercise 1 compute the determinant of the given vectors, the determinant of the orthogonal set, and the product of the lengths of the orthogonal vectors. How must these three numbers be related? Why?

3. Let T be a linear transformation on a vector space; set $\mathbf{v} = T\mathbf{u}$, and write this in matrix form, $X(\mathbf{v}) = AX(\mathbf{u})$, where X is the matrix of component variables with respect to some orthonormal basis.

(a) Show that if $\mathbf{v}_i = T\mathbf{u}_i$, then

$$\det (\mathbf{v}_1, \mathbf{v}_2, \ldots, \mathbf{v}_n) = \det A \det (\mathbf{u}_1, \mathbf{u}_2, \ldots, \mathbf{u}_n).$$

(b) In the light of part (a), discuss the effect of a linear transformation on volumes of parallelepipeds.

(c) The transformation T is called *orientation preserving* if $\det A > 0$ and *orientation reversing* if $\det A < 0$. Check this in dimensions two and three. Specifically, ordered pairs of vectors in the plane may be characterized as counterclockwise or clockwise, and ordered triples in three-space may be characterized as either right- or left-handed. Show that these characteristics are preserved by a transformation with a positive determinant and reversed by one with a negative determinant.

(d) Let $\mathbf{u}_1, \mathbf{u}_2, \ldots, \mathbf{u}_n$ be carried into $\mathbf{z}_1, \mathbf{z}_2, \ldots, \mathbf{z}_n$ by the Schmidt orthogonalization process. Let T be a linear transformation such that $T\mathbf{u}_i = \mathbf{z}_i$. Show that T is uniquely determined and is orientation preserving.

4. Let $\mathbf{u}_1, \mathbf{u}_2, \ldots, \mathbf{u}_n$ be a linearly independent set in an n-dimensional vector space. As noted in the text, this set forms a basis, and each $\mathbf{v}$ in the space has a unique representation

$$\mathbf{v} = \sum_{i=1}^{n} v_i \mathbf{u}_i.$$

If, also,

$$\mathbf{w} = \sum_{i=1}^{n} w_i \mathbf{u}_i,$$

define a scalar product by

$$\mathbf{v} \cdot \mathbf{w} = \sum_{i=1}^{n} v_i w_i.$$

(a) Show that this scalar product has all the properties listed in the text.
(b) Show that, with the above definition of scalar product, the basis, $\mathbf{u}_1, \mathbf{u}_2, \ldots, \mathbf{u}_n$ is automatically orthonormal.
(c) Consider an example of this in two dimensions. Let $\mathbf{u}_1 = 2\mathbf{i} + \mathbf{j}$, $\mathbf{u}_2 = \mathbf{i} + 2\mathbf{j}$. Define scalar products, as above, in terms of $\mathbf{u}_1$ and $\mathbf{u}_2$ components. Show that by such a definition neither $\mathbf{i}$ nor $\mathbf{j}$ is a unit vector, and that they are not orthogonal.

5. In three-space the scalar triple product $\mathbf{u} \cdot \mathbf{v} \times \mathbf{w}$ is zero if and only if the vectors $\mathbf{u}$, $\mathbf{v}$, and $\mathbf{w}$ are coplanar. Prove this in two ways:
(a) by considering perpendicularity of vectors,
(b) by considering volumes of parallelepipeds.

4-7 Rotations. A transformation T on a vector space is called an *isometry* (Greek: *isos*, equal + *metron*, measure) if it preserves the lengths of vectors, that is, if

$$|T\mathbf{w}| = |\mathbf{w}|$$

for every $\mathbf{w}$. Clearly, a rotation of a two- or three-dimensional vector space is a linear isometry. It is not quite true that every linear isometry is a rotation, but the exceptions are easily characterized, and the way to study the matrix description of rotations is to determine the matrix description of linear isometries.

The answer is as follows. Let X be a matrix of orthonormal component variables and let T be given by

$$X(T\mathbf{w}) = AX(\mathbf{w});$$

then T is an isometry if and only if

$$A'A = [\delta_{ij}]. \tag{28}$$

A matrix A satisfying (28) is called an *orthogonal* matrix.

To prove that an orthogonal matrix gives an isometry, suppose that A satisfies (28); then,

$$|T\mathbf{w}|^2 = [X(T\mathbf{w})]'X(T\mathbf{w}) = [AX(\mathbf{w})]'[AX(\mathbf{w})]$$
$$= [X(\mathbf{w})]'A'AX(\mathbf{w}) = [X(\mathbf{w})]'X(\mathbf{w}) = |\mathbf{w}|^2.$$

To prove the converse, that if the transformation is an isometry the matrix is orthogonal, note first that an isometry preserves scalar products. This follows from the identity

$$|\mathbf{v} - \mathbf{w}|^2 = |\mathbf{v}|^2 + |\mathbf{w}|^2 - 2\mathbf{v} \cdot \mathbf{w}.$$

A linear isometry preserves three terms; therefore it must preserve the fourth. This being the case, a linear isometry carries an orthonormal set of vectors into another orthonormal set. Let $\mathbf{b}_1, \ldots, \mathbf{b}_n$ be the basis vectors for the component matrix X; then the X-components of the vectors $T\mathbf{b}_1, \ldots, T\mathbf{b}_n$ are displayed in the columns of A. So, the columns of A give orthonormal vectors, and (28) follows at once.

Since $\det A' = \det A$, it follows from the definition (28) that if A is orthogonal, then

$$(\det A)^2 = \det A' \det A = \det (A'A) = \det [\delta_{ij}] = 1;$$

$$\det A = \pm 1.$$

This is the clue to identifying the rotations. Their matrices have determinant $+1$.

To see this, turn to the case $n = 2$.

Suppose that A is a 2×2 matrix and orthogonal:

$$\begin{bmatrix} a_{11} & a_{21} \\ a_{12} & a_{22} \end{bmatrix} \begin{bmatrix} a_{11} & a_{12} \\ a_{21} & a_{22} \end{bmatrix} = \begin{bmatrix} 1 & 0 \\ 0 & 1 \end{bmatrix},$$

$$a_{11}^2 + a_{21}^2 = 1, \tag{29}$$

$$a_{12}^2 + a_{22}^2 = 1, \tag{30}$$

$$a_{11}a_{12} + a_{21}a_{22} = 0. \tag{31}$$

It follows from (29) that there is an angle α such that

$$a_{11} = \cos \alpha, \qquad a_{21} = \sin \alpha;$$

thus, by (31)

$$\tan \alpha = \frac{a_{21}}{a_{11}} = \frac{-a_{12}}{a_{22}},$$

and it follows from (30) that

$$a_{12} = -\sin \alpha, \qquad a_{22} = \cos \alpha$$

or

$$a_{12} = \sin \alpha, \qquad a_{22} = -\cos \alpha.$$

Thus, A is one or the other of the two matrices

$$\begin{bmatrix} \cos \alpha & -\sin \alpha \\ \sin \alpha & \cos \alpha \end{bmatrix}, \quad \begin{bmatrix} \cos \alpha & \sin \alpha \\ \sin \alpha & -\cos \alpha \end{bmatrix}.$$

Suppose that A is the first of these, and consider the point transformation of the plane onto itself defined by

$$X(q) = AX(p).$$

Now, introduce the polar-coordinate variables r and θ; then

$$X(p) = \begin{bmatrix} r(p) \cos \theta(p) \\ r(p) \sin \theta(p) \end{bmatrix},$$

and

$$X(q) = AX(p) = \begin{bmatrix} \cos \alpha & -\sin \alpha \\ \sin \alpha & \cos \alpha \end{bmatrix} \begin{bmatrix} r(p) \cos \theta(p) \\ r(p) \sin \theta(p) \end{bmatrix}$$

$$= \begin{bmatrix} r(p) \cos \alpha \cos \theta(p) - r(p) \sin \alpha \sin \theta(p) \\ r(p) \sin \alpha \cos \theta(p) + r(p) \cos \alpha \sin \theta(p) \end{bmatrix}$$

$$= \begin{bmatrix} r(p) \cos [\alpha + \theta(p)] \\ r(p) \sin [\alpha + \theta(p)] \end{bmatrix}.$$

Thus,

$$r(q) = r(p), \qquad \theta(q) = \alpha + \theta(p);$$

p is carried into q by a rotation about the origin through an angle α.

Next, consider the point transformation defined by

$$X(q) = BX(p),$$

where

$$B = \begin{bmatrix} 1 & 0 \\ 0 & -1 \end{bmatrix}.$$

Here,

$$\begin{bmatrix} x(q) \\ y(q) \end{bmatrix} = \begin{bmatrix} 1 & 0 \\ 0 & -1 \end{bmatrix} \begin{bmatrix} x(p) \\ y(p) \end{bmatrix} = \begin{bmatrix} x(p) \\ -y(p) \end{bmatrix};$$

p is carried into q by reflection through the x-axis. Now,

$$\begin{bmatrix} \cos \alpha & -\sin \alpha \\ \sin \alpha & \cos \alpha \end{bmatrix} \begin{bmatrix} 1 & 0 \\ 0 & -1 \end{bmatrix} = \begin{bmatrix} \cos \alpha & \sin \alpha \\ \sin \alpha & -\cos \alpha \end{bmatrix},$$

and the second of the possible orthogonal matrices represents a reflection followed by a rotation.

Observe, now, that the rotation matrix has determinant 1, while the reflection-rotation matrix has determinant -1. This leads to the *definition* that a *rotation* in n-space is a linear transformation determined by an orthogonal matrix with positive determinant.   In Exercise 4 below, a proof that this is consistent with the usual geometric notion of rotation in three-space is outlined.

The significance of an orthogonal matrix in a change of basis formula is that orthogonal matrices and only orthogonal matrices carry orthonormal components into orthonormal components.   The outline of the proof is as follows.   Write

$$U(\mathbf{w}) \;=\; A X(\mathbf{w}),$$

and assume that X and U are both orthonormal.   The U-components of the basis vectors for X appear in the columns of A.   Now, scalar products are computed by the usual formula in terms of U-components; so the condition that the X basis be orthonormal reads $A'A = [\delta_{ij}]$.   On the other hand, if $\mathbf{b}_i$ and $\mathbf{b}_j$ are vectors from the U-basis, then automatically

$$[U(\mathbf{b}_i)]' U(\mathbf{b}_j) \;=\; \delta_{ij};$$

however, if A is orthogonal and X is orthonormal, then

$$\begin{aligned}
\delta_{ij} &= [U(\mathbf{b}_i)]' U(\mathbf{b}_j) = [AX(\mathbf{b}_i)]' AX(\mathbf{b}_j) \\
&= [X(\mathbf{b}_i)]' A'A X(\mathbf{b}_j) = [X(\mathbf{b}_i)]' X(\mathbf{b}_j) = \mathbf{b}_i \cdot \mathbf{b}_j,
\end{aligned}$$

and this is the definition of an orthonormal basis.

The discussion of volumes in Section 4–6 may now be completed.   Let $\mathbf{w}_1, \mathbf{w}_2, \ldots, \mathbf{w}_n$ be a set of n vectors.   Let X and U be two matrices of orthonormal component variables, and let A be the orthogonal matrix describing the change of basis.   The expression $\det (\mathbf{w}_1, \mathbf{w}_2, \ldots, \mathbf{w}_n)$ is either

$$\det [x_i(\mathbf{w}_j)] \qquad \text{or} \qquad \det [u_i(\mathbf{w}_j)],$$

and it does not matter which.   To see this, note that the change of basis equation applied to $\mathbf{w}_j$ reads

$$U(\mathbf{w}_j) \;=\; A X(\mathbf{w}_j),$$

and this is just the computation for the jth column to establish that

$$[u_i(\mathbf{w}_j)] \;=\; A[x_i(\mathbf{w}_j)].$$

Thus, since $\det A = \pm 1$, $\det [u_i(\mathbf{w}_j)] = \pm \det [x_i(\mathbf{w}_j)]$ and

$$|\det (\mathbf{w}_1, \mathbf{w}_2, \ldots, \mathbf{w}_n)|$$

is the same for all orthonormal bases and gives the volume of the parallelepiped spanned by $\mathbf{w}_1, \mathbf{w}_2, \ldots, \mathbf{w}_n$.

Examples

1. The rotation matrix

$$R = \begin{bmatrix} 1/2 & -\sqrt{3}/2 \\ \sqrt{3}/2 & 1/2 \end{bmatrix}$$

describes a rotation of the plane through an angle $\pi/3$. Note that

$$\begin{bmatrix} 1/2 & -\sqrt{3}/2 \\ \sqrt{3}/2 & 1/2 \end{bmatrix}\begin{bmatrix} 1 \\ 0 \end{bmatrix} = \begin{bmatrix} 1/2 \\ \sqrt{3}/2 \end{bmatrix},$$

$$\begin{bmatrix} 1/2 & -\sqrt{3}/2 \\ \sqrt{3}/2 & 1/2 \end{bmatrix}\begin{bmatrix} 0 \\ 1 \end{bmatrix} = \begin{bmatrix} -\sqrt{3}/2 \\ 1/2 \end{bmatrix},$$

$$\begin{bmatrix} 1/2 & -\sqrt{3}/2 \\ \sqrt{3}/2 & 1/2 \end{bmatrix}\begin{bmatrix} 2 \\ 2 \end{bmatrix} = \begin{bmatrix} 1 - \sqrt{3} \\ \sqrt{3} + 1 \end{bmatrix}.$$

Figure 4–11 shows these three point transformations.

2. Let

$$R = \begin{bmatrix} 1 & 0 & 0 \\ 0 & \cos\alpha & -\sin\alpha \\ 0 & \sin\alpha & \cos\alpha \end{bmatrix},$$

then

$$R\begin{bmatrix} 1 \\ 0 \\ 0 \end{bmatrix} = \begin{bmatrix} 1 \\ 0 \\ 0 \end{bmatrix}, \qquad R\begin{bmatrix} 0 \\ 1 \\ 0 \end{bmatrix} = \begin{bmatrix} 0 \\ \cos\alpha \\ \sin\alpha \end{bmatrix}, \qquad R\begin{bmatrix} 0 \\ 0 \\ 1 \end{bmatrix} = \begin{bmatrix} 0 \\ -\sin\alpha \\ \cos\alpha \end{bmatrix}.$$

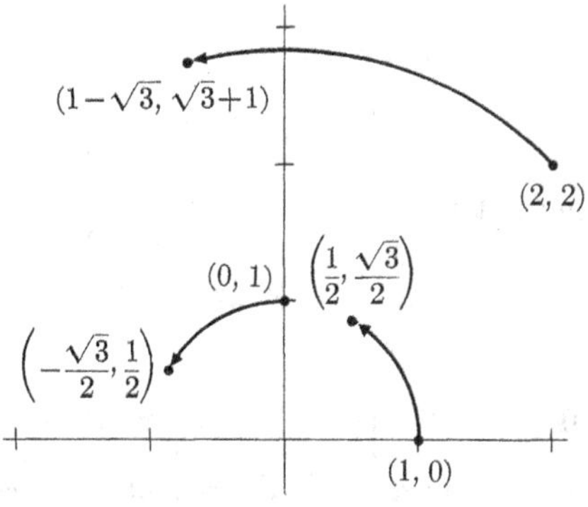

FIGURE 4–11

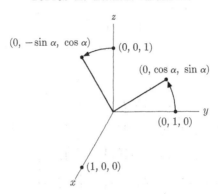

FIGURE 4–12

That is, R defines a linear transformation T such that

$$T\mathbf{i} = \mathbf{i}, \qquad T\mathbf{j} = \mathbf{j}\cos\alpha + \mathbf{k}\sin\alpha, \qquad T\mathbf{k} = -\mathbf{j}\sin\alpha + \mathbf{k}\cos\alpha.$$

These results are illustrated in Fig. 4–12; and since T is an isometry, it must consist of a rotation of the entire three-dimensional space about the x-axis.

EXERCISES

1. Let T be a rotation of the plane through an angle $\pi/6$.
   (a) Find the matrix of T.
   (b) Use matrices to find the transforms of $(1, 0)$, $(0, 1)$, $(-2, -1)$, and $(3, -2)$.
   (c) Sketch the points in part (b) and their transforms.

2. Let R_α be the matrix for the rotation of the plane through an angle α.
   (a) Show by direct computation that

$$R'_\alpha = R_\alpha^{-1} = R_{-\alpha}.$$

   (b) Why, for any orthogonal matrix A, is

$$A' = A^{-1}?$$

   (c) Why, in geometric terms, is

$$R_\alpha^{-1} = R_{-\alpha}?$$

3. Let

$$A_\alpha = \begin{bmatrix} 1 & 0 & 0 \\ 0 & \cos\alpha & -\sin\alpha \\ 0 & \sin\alpha & \cos\alpha \end{bmatrix}, \qquad B_\beta = \begin{bmatrix} \cos\beta & -\sin\beta & 0 \\ \sin\beta & \cos\beta & 0 \\ 0 & 0 & 1 \end{bmatrix}.$$

(a) Compute the matrix products $A_\alpha B_\beta$ and $B_\beta A_\alpha$.

(b) Describe geometrically the transformations defined by these products.

(c) Show, by computing the products, that

$$A_{\pi/2}B_{\pi/2} \neq B_{\pi/2}A_{\pi/2}.$$

(d) Experiment with a book, and show that the composite rotations described in part (c) are unequal.

4. The following steps may be used to prove that an orientation preserving, linear isometry on three-space is a rotation. Let T be such a transformation.

(a) Set $T\mathbf{i} = \mathbf{i}'$, $T\mathbf{j} = \mathbf{j}'$, $T\mathbf{k} = \mathbf{k}'$; then $\mathbf{i}'$, $\mathbf{j}'$, $\mathbf{k}'$ is a right-handed orthonormal basis. Why?

(b) Show that det $(\mathbf{i} - \mathbf{i}', \mathbf{j} - \mathbf{j}', \mathbf{k} - \mathbf{k}') = 0$. [*Hint:* Write this as a scalar triple product; expand to a sum of eight scalar triple products, and use Exercise 6, Section 4–5.

(c) Apply Exercise 5, Section 4–6; there is a vector $\mathbf{u}$ such that

$$\mathbf{u} \cdot (\mathbf{i} - \mathbf{i}') = \mathbf{u} \cdot (\mathbf{j} - \mathbf{j}') = \mathbf{u} \cdot (\mathbf{k} - \mathbf{k}') = 0.$$

Write $\mathbf{u} = (\mathbf{u} \cdot \mathbf{i})\mathbf{i} + (\mathbf{u} \cdot \mathbf{j})\mathbf{j} + (\mathbf{u} \cdot \mathbf{k})\mathbf{k}$; apply T, and show that $T\mathbf{u} = \mathbf{u}$.

(d) T carries each plane perpendicular to $\mathbf{u}$ into itself (why?). Therefore, it rotates each such plane and is, overall, a rotation of three-space about the line through the origin in the direction of $\mathbf{u}$.

5. More generally, it is true that any orientation-preserving isometry of two-space or three-space that leaves the origin fixed must be a rotation. To complete the proof, one needs to show that such a transformation must be linear. The following steps will accomplish this.

(a) An isometry preserves collinearity of points. [*Hint:* Describe the condition that three points be collinear in terms of distances alone.]

(b) An isometry preserves mid-points of line segments.

(c) The diagonals of a parallelogram have a common mid-point. Conversely, if two line segments have a common mid-point, their end points are the vertices of a parallelogram.

(d) An isometry preserves parallelograms; hence it preserves addition of vectors. That is, if T is an origin-preserving isometry, then

$$T(\mathbf{u} + \mathbf{v}) = T\mathbf{u} + T\mathbf{v}.$$

(e) Multiplication of vectors by scalars may be defined in terms of collinearity and ratios of distances. An isometry preserves collinearity and ratios of distances, so $T(a\mathbf{u}) = aT\mathbf{u}$.

(f) Results (d) and (e) together, make T linear.

6. Let $\mathbf{u}_1$, $\mathbf{u}_2$, ..., $\mathbf{u}_n$ be a set of n vectors in n-space, and let $[x_i(\mathbf{u}_j)]$ be the matrix of their components with respect to some orthonormal basis. Show that $[x_i(\mathbf{u}_j)]$ is an orthogonal matrix if and only if $\mathbf{u}_1$, $\mathbf{u}_2$, ..., $\mathbf{u}_n$ is itself an orthonormal basis. [*Hint:* Prove this directly from (28).]

7. Let T be a linear transformation described by

$$X(T\mathbf{w}) = AX(\mathbf{w}),$$

where X is orthonormal. Show that

$$\det (T\mathbf{w}_1, T\mathbf{w}_2, \ldots, T\mathbf{w}_n) = \det A \det (\mathbf{w}_1, \mathbf{w}_2, \ldots, \mathbf{w}_n).$$

Describe this result geometrically. What effect does a linear transformation have on volumes?

8. Prove the equivalence of the following statements.
   (a) A is orthogonal.
   (b) $A' = A^{-1}$.
   (c) A^{-1} is orthogonal.
   (d) A' is orthogonal.
   (e) The columns of A give components of orthonormal vectors with respect to an orthonormal basis.
   (f) The rows of A give components of orthonormal vectors with respect to an orthonormal basis.

4–8 Quadratic and bilinear forms. Let $x_1, x_2, \ldots, x_n$ and $y_1, y_2, \ldots, y_n$ be two sets of variables and form matrices from them as follows:

$$X = \begin{bmatrix} x_1 \\ x_2 \\ \vdots \\ x_n \end{bmatrix}, \qquad Y = \begin{bmatrix} y_1 \\ y_2 \\ \vdots \\ y_n \end{bmatrix}.$$

Let A be an $n \times n$ matrix of numbers; then $Y'AX$ is 1×1 (a scalar), and indeed

$$Y'AX = \sum_{i=1}^{n} \sum_{j=1}^{n} a_{ij} y_i x_j.$$

This latter is called a *bilinear form* in the variables x_i and y_j, and the procedure described here yields a matrix representation for bilinear forms. If $Y = X$, the above equation reduces to

$$X'AX = \sum_{i=1}^{n} \sum_{j=1}^{n} a_{ij} x_i x_j;$$

and the result is called a *quadratic form* in the variables x_i.

Consider the following examples of quadratic and bilinear forms.

(i) Let x and y be rectangular coordinates on the plane, and let

$$X = \begin{bmatrix} x \\ y \end{bmatrix}.$$

Let

$$A = \begin{bmatrix} 1/a^2 & 0 \\ 0 & 1/b^2 \end{bmatrix};$$

then

$$X'AX = \frac{x^2}{a^2} + \frac{y^2}{b^2},$$

and $X'AX = 1$ is the equation of an ellipse.

(ii) Let X be as in (i) and let

$$B = \begin{bmatrix} 1/a^2 & 0 \\ 0 & -1/b^2 \end{bmatrix};$$

then $X'BX = 1$ is the equation of a hyperbola.

(iii) Let x and y be rectangular coordinates on a one-dimensional manifold, and define

$$X = \begin{bmatrix} x \\ y \end{bmatrix}, \quad dX = \begin{bmatrix} dx \\ dy \end{bmatrix}, \quad A = \begin{bmatrix} 0 & 1 \\ -1 & 0 \end{bmatrix}, \quad B = \begin{bmatrix} 1 & 0 \\ 0 & 1 \end{bmatrix}.$$

Then,

$$(dX)'B\, dX = ds^2,$$

and

$$X'A\, dX = x\, dy - y\, dx$$

is the bilinear form whose integral gives area.

(iv) Take X, A, and B as in (iii); let t be a coordinate on the manifold, and define

$$D_t X - \begin{bmatrix} D_t x \\ D_t y \end{bmatrix}, \quad D_t^2 X - \begin{bmatrix} D_t^2 x \\ D_t^2 y \end{bmatrix}.$$

Then, curvature is given by

$$\kappa = \frac{(D_t X)'A\, D_t^2 X}{[(D_t X)'B\, D_t X]^{3/2}}.$$

(v) Let

$$X = \begin{bmatrix} x \\ y \\ 1 \end{bmatrix}, \quad C = \begin{bmatrix} a & b & d \\ b & c & e \\ d & e & f \end{bmatrix};$$

then

$$X'CX = ax^2 + 2bxy + cy^2 + 2dx + 2ey + f$$

is the general second-degree polynomial in x and y.

With X defined as in (v), let

$$R = \begin{bmatrix} \cos\alpha & -\sin\alpha & 0 \\ \sin\alpha & \cos\alpha & 0 \\ 0 & 0 & 1 \end{bmatrix};$$

then

$$\begin{bmatrix} x(q) \\ y(q) \\ 1 \end{bmatrix} = R \begin{bmatrix} x(p) \\ y(p) \\ 1 \end{bmatrix}$$

describes a rotation through an angle α carrying p into q. Now, if $X'CX = 0$ at q, then at p, $(RX)'C(RX) = X'(R'CR)X = 0$. Thus, if the rotation carries a locus M onto M' and if

$$M' \text{ has the equation } X'CX = 0,$$

then

$$M \text{ has the equation } X'R'CRX = 0.$$

If in (v) above, $b = 0$ (the polynomial has no xy-term), then $X'CX = 0$ is recognizable as one of the equations of a conic section, in standard form. However, for any matrix C, as in (v), $X'CX = 0$ is the equation of a conic, because there is always a rotation matrix R such that

$$R'CR = \begin{bmatrix} a' & 0 & d' \\ 0 & c' & e' \\ d' & e' & f' \end{bmatrix}.$$

To determine the angle α in terms of the entries in C, it is necessary to compute only enough of the matrix product,

$$\begin{bmatrix} \cos\alpha & \sin\alpha & 0 \\ -\sin\alpha & \cos\alpha & 0 \\ 0 & 0 & 1 \end{bmatrix} \begin{bmatrix} a & b & d \\ b & c & e \\ d & e & f \end{bmatrix} \begin{bmatrix} \cos\alpha & -\sin\alpha & 0 \\ \sin\alpha & \cos\alpha & 0 \\ 0 & 0 & 1 \end{bmatrix},$$

to get the entry for the second row, first column:

$$(-a\sin\alpha + b\cos\alpha)\cos\alpha + (-b\sin\alpha + c\cos\alpha)\sin\alpha = 0,$$

$$-\tfrac{1}{2}a\sin 2\alpha + b\cos 2\alpha + \tfrac{1}{2}c\sin 2\alpha = 0,$$

$$\tan 2\alpha = \frac{2b}{a-c}.$$

This describes the required rotation if $a - c \neq 0$; if $a - c = 0$, then $\cos 2\alpha = 0$ reduces the second row, first column, entry to 0.

EXAMPLES

1. Sketch the locus of

$$3x^2 + 4\sqrt{3}\,xy - y^2 - 7 = 0.$$

First, determine the rotation that will eliminate the xy-term:

$$\tan 2\alpha = \frac{4\sqrt{3}}{3 - (-1)} = \sqrt{3},$$

$$2\alpha = \frac{\pi}{3}, \qquad \alpha = \frac{\pi}{6},$$

$$\cos \alpha = \frac{\sqrt{3}}{2}, \qquad \sin \alpha = \frac{1}{2}.$$

The matrix of the new form is

$$\begin{bmatrix} \sqrt{3}/2 & 1/2 & 0 \\ -1/2 & \sqrt{3}/2 & 0 \\ 0 & 0 & 1 \end{bmatrix} \begin{bmatrix} 3 & 2\sqrt{3} & 0 \\ 2\sqrt{3} & -1 & 0 \\ 0 & 0 & -7 \end{bmatrix} \begin{bmatrix} \sqrt{3}/2 & -1/2 & 0 \\ 1/2 & \sqrt{3}/2 & 0 \\ 0 & 0 & 1 \end{bmatrix}$$

$$= \begin{bmatrix} 5\sqrt{3}/2 & 5/2 & 0 \\ 3/2 & -3\sqrt{3}/2 & 0 \\ 0 & 0 & -7 \end{bmatrix} \begin{bmatrix} \sqrt{3}/2 & -1/2 & 0 \\ 1/2 & \sqrt{3}/2 & 0 \\ 0 & 0 & 1 \end{bmatrix} = \begin{bmatrix} 5 & 0 & 0 \\ 0 & -3 & 0 \\ 0 & 0 & -7 \end{bmatrix}.$$

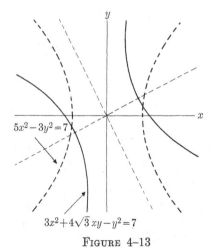

$5x^2 - 3y^2 = 7$

$3x^2 + 4\sqrt{3}\,xy - y^2 = 7$

FIGURE 4–13

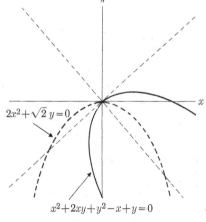

$2x^2 + \sqrt{2}\,y = 0$

$x^2 + 2xy + y^2 - x + y = 0$

FIGURE 4–14

Thus, the required locus comes from the locus of

$$5x^2 - 3y^2 - 7 = 0$$

by a rotation through $\pi/6$. The sketch is shown in Fig. 4–13.

2. Sketch the locus of

$$x^2 + 2xy + y^2 - x + y = 0.$$

Here, $a = c$, hence, $\cos 2\alpha = 0$ or $\alpha = \pi/4$ gives the required rotation. The matrix for the new form is

$$\begin{bmatrix} 1/\sqrt{2} & 1/\sqrt{2} & 0 \\ -1/\sqrt{2} & 1/\sqrt{2} & 0 \\ 0 & 0 & 1 \end{bmatrix} \begin{bmatrix} 1 & 1 & -1/2 \\ 1 & 1 & 1/2 \\ -1/2 & 1/2 & 0 \end{bmatrix} \begin{bmatrix} 1/\sqrt{2} & -1/\sqrt{2} & 0 \\ 1/\sqrt{2} & 1/\sqrt{2} & 0 \\ 0 & 0 & 1 \end{bmatrix}$$

$$= \begin{bmatrix} \sqrt{2} & \sqrt{2} & 0 \\ 0 & 0 & 1/\sqrt{2} \\ -1/2 & 1/2 & 0 \end{bmatrix} \begin{bmatrix} 1/\sqrt{2} & -1/\sqrt{2} & 0 \\ 1/\sqrt{2} & 1/\sqrt{2} & 0 \\ 0 & 0 & 1 \end{bmatrix} = \begin{bmatrix} 2 & 0 & 0 \\ 0 & 0 & 1/\sqrt{2} \\ 0 & 1/\sqrt{2} & 0 \end{bmatrix}.$$

The new equation is

$$2x^2 + \sqrt{2}\, y = 0,$$

and the required figure comes from the locus of this equation by a rotation through $\pi/4$. A sketch is shown in Fig. 4–14.

3. The number $ac - b^2$ is called the *discriminant* of the *homogeneous quadratic form* $ax^2 + 2bxy + cy^2$. In matrix notation, the quadratic form is

$$[x \ \ y] \begin{bmatrix} a & b \\ b & c \end{bmatrix} \begin{bmatrix} x \\ y \end{bmatrix} = X'CX;$$

and in these terms the discriminant is

$$\det C.$$

If R is the matrix of a rotation, then

$$\det (R'CR) = \det R' \det C \det R = \det C$$

because a rotation matrix has determinant 1. Thus, if M is mapped onto M' by a rotation and if two quadratic forms in x and y, one on M and the other on M' are equal at corresponding points, then their discriminants are equal. This is what is meant by the statement that the discriminant of a homogeneous quadratic form is *invariant* under rotations of the plane.

4. Note the following standard equations of conics:

$$\frac{x^2}{a^2} + \frac{y^2}{b^2} - 1 = 0,$$

$$\frac{x^2}{a^2} - \frac{y^2}{b^2} - 1 = 0,$$

$$x^2 - 4ay = 0.$$

The homogeneous quadratic forms involved have discriminants as follows:

$$\text{Ellipse,} \qquad \left(\frac{1}{a^2}\right)\left(\frac{1}{b^2}\right) - 0 > 0.$$

$$\text{Hyperbola,} \qquad \left(\frac{1}{a^2}\right)\left(-\frac{1}{b^2}\right) - 0 < 0.$$

$$\text{Parabola,} \qquad 1 \cdot 0 - 0 = 0.$$

Since this discriminant is invariant under rotations, the locus of

$$ax^2 + 2bxy + cy^2 + 2\,dx + 2ey + f = 0$$

can be typed merely by computing $ac - b^2$.

EXERCISES

1. Each of the following is the equation of a conic. In each case use the results of Example 4 to tell, by inspection, what type of conic the locus is.
   (a) $3x^2 - 2\sqrt{3}xy + y^2 - x = 0$
   (b) $x^2 + xy + y^2 - 3 = 0$
   (c) $x^2 - 3xy + y^2 - 5 = 0$
   (d) $3x^2 + 2xy + 3y^2 - 19 = 0$
   (e) $x^2 + 2xy + y^2 + y - 1 = 0$
   (f) $2x^2 + 2\sqrt{3}xy - 4 = 0$
   (g) $2xy - 1 = 0$

2. For each of the equations in Exercise 1, find a rotation to eliminate the xy-term and sketch the locus.

3. Let X and C be as in item (v) of the text. The discriminant of the form $X'CX$ is defined as $\det C$.
   (a) Show that $\det C$ is invariant under rotations of the plane.
   (b) Consider the following special cases of $X'CX = 0$:

$$ax^2 + 2ey + f = 0, \qquad ax^2 + by^2 + f = 0.$$

Show that their loci reduce to lines or points if and only if $\det C = 0$.

(c) Explain how to tell from the coefficients in a general quadratic equation whether the locus is a degenerate conic.

4. Let X, dX, D_tX, D_t^2X, A, and B be the matrices introduced in items (iii) and (iv) of the text.

(a) Show that if R is the matrix of a rotation, then $R'AR = A$ and $R'BR = B$.

(b) Conclude from this that arc length, area, and curvature are invariant under rotations.

5. Suppose that two curves intersect, and let D_xy_1 be the slope of one and D_xy_2 the slope of the other at the point of intersection. Introduce matrices

$$\begin{bmatrix} 1 \\ D_xy_1 \end{bmatrix} \quad \text{and} \quad \begin{bmatrix} 1 \\ D_xy_2 \end{bmatrix},$$

and use the theory of bilinear and quadratic forms to show that the angle between the two curves is invariant under rotations.

6. Suppose that A is a square matrix and R is an orthogonal matrix such that

$$R'AR = D,$$

where D is a diagonal matrix,

$$D = \begin{bmatrix} d_1 & & & 0 \\ & d_2 & & \\ & & \ddots & \\ 0 & & & d_n \end{bmatrix}.$$

(Note that elimination of cross-product terms in a quadratic form by rotation consists of finding such an R.)

(a) Show that $AR = RD$.

(b) Let R_j be the $n \times 1$ matrix consisting of the jth column of R; show that for each j,
$$AR_j = d_jR_j.$$

(c) Let X be the matrix of component variables. If, for some scalar λ and some vector $\mathbf{u}$,
$$AX(\mathbf{u}) = \lambda X(\mathbf{u}),$$

then $\mathbf{u}$ is called a *characteristic vector* for A and λ is called a *characteristic root* for A. Show how part (b) displays n mutually orthogonal characteristic vectors for A and their corresponding characteristic roots.

(d) Let A be a symmetric matrix; that is, let $A' = A$. Show that characteristic vectors for A corresponding to distinct characteristic roots must be orthogonal. [*Hint:* Let $AX(\mathbf{u}) = \lambda_1X(\mathbf{u})$, $AX(\mathbf{v}) = \lambda_2X(\mathbf{v})$; set $X(\mathbf{w}) = AX(\mathbf{v})$, and show that

$$\mathbf{w} \cdot \mathbf{u} - \mathbf{u} \cdot \mathbf{w} = (\lambda_1 - \lambda_2)\mathbf{u} \cdot \mathbf{v}.$$

4–9 Analytic geometry: lines and planes. The purpose of this section is to present certain standard forms for the equations of lines and planes in three-space in terms of rectangular coordinates x, y, and z. Before looking at details, it is well to note the following generalities.

(i) Roughly speaking, each equation in coordinate variables restricts the dimension of the locus by one. So, for a well-behaved system of simultaneous equations, the dimension of the locus plus the number of equations equals the dimension of the space. A precise description of well-behaved systems is beside the point here. It is sufficient to say that lines and planes fit this model; *thus a line has two equations, and a plane has one.*

(ii) Given a point p_0 and a vector $\mathbf{u}$ in three-space, there is a unique plane through p_0 perpendicular to $\mathbf{u}$, but there are many planes through p_0 parallel to $\mathbf{u}$. On the other hand, there is a unique line through p_0 parallel to $\mathbf{u}$, but there are many lines through p_0 perpendicular to $\mathbf{u}$. *To characterize orientation of a figure by a single direction, for a line make it parallel, for a plane, perpendicular.*

Let x, y, and z be rectangular coordinates in three-space, and let $\mathbf{r}$ be the vector position variable defined by

$$\mathbf{r} = x\mathbf{i} + y\mathbf{j} + z\mathbf{k}.$$

Let p_0 be a given point with coordinates (x_0, y_0, z_0), and let $\mathbf{u}$ be a given vector with components (a, b, c). Denote $\mathbf{r}(p_0)$ by $\mathbf{r}_0$. For an arbitrary point p, the direction from p_0 to p is that of the vector

$$\mathbf{r}(p) - \mathbf{r}_0.$$

Thus, the equation

$$\mathbf{u} \cdot [\mathbf{r}(p) - \mathbf{r}_0] = 0$$

asserts that the direction from p_0 to p is perpendicular to that of $\mathbf{u}$. This characterizes p as being on the plane through p_0 perpendicular to $\mathbf{u}$. Thus, the equation of this plane is

$$\mathbf{u} \cdot (\mathbf{r} - \mathbf{r}_0) = 0. \tag{32}$$

In terms of components, this equation assumes the form

$$a(x - x_0) + b(y - y_0) + c(z - z_0) = 0. \tag{33}$$

With the parentheses removed, equation (33) assumes the form

$$ax + by + cz + d = 0, \tag{34}$$

where a, b, and c are the same as in (33). Now, (34) is the general form for a linear equation in x, y, and z, so such a linear equation has a plane for a locus, and the coefficients of x, y, and z are components of a vector perpendicular to the locus.

Let $\mathbf{v} = v_1\mathbf{i} + v_2\mathbf{j} + v_3\mathbf{k}$ and $\mathbf{w} = w_1\mathbf{i} + w_2\mathbf{j} + w_3\mathbf{k}$ be nonparallel vectors. Through p_0 there is a unique plane parallel to both $\mathbf{v}$ and $\mathbf{w}$, and its equation is easily found as follows. The vector product $\mathbf{v} \times \mathbf{w}$ is perpendicular to both $\mathbf{v}$ and $\mathbf{w}$; hence one may set $\mathbf{u} = \mathbf{v} \times \mathbf{w}$ and use (32):

$$(\mathbf{r} - \mathbf{r}_0) \cdot \mathbf{v} \times \mathbf{w} = 0. \tag{35}$$

In terms of components, this is most easily written in determinant form:

$$\begin{vmatrix} x - x_0 & v_1 & w_1 \\ y - y_0 & v_2 & w_2 \\ z - z_0 & v_3 & w_3 \end{vmatrix} = 0. \tag{36}$$

The condition that p be on the line through p_0 parallel to $\mathbf{u}$ is that, for some scalar t,

$$\mathbf{r}(p) - \mathbf{r}_0 = t\mathbf{u}. \tag{37}$$

In component form, this reads

$$x = x_0 + at, \qquad y = y_0 + bt, \qquad z = z_0 + ct. \tag{38}$$

To derive the pair of rectangular equations for this line, solve each of the equations in (38) for t, and equate the three expressions each of which equals t:

$$\frac{x - x_0}{a} = \frac{y - y_0}{b} = \frac{z - z_0}{c}. \tag{39}$$

Given $\mathbf{v}$ and $\mathbf{w}$ as above, there is a unique line through p_0 perpendicular to both $\mathbf{v}$ and $\mathbf{w}$. To find its equations, set $\mathbf{u} = \mathbf{v} \times \mathbf{w}$ in (37):

$$\mathbf{r} - \mathbf{r}_0 = t\mathbf{v} \times \mathbf{w}. \tag{40}$$

The parametric equations in components are

$$x = x_0 + t\begin{vmatrix} v_2 & w_2 \\ v_3 & w_3 \end{vmatrix},$$
$$y = y_0 + t\begin{vmatrix} v_3 & w_3 \\ v_1 & w_1 \end{vmatrix}, \tag{41}$$
$$z = z_0 + t\begin{vmatrix} v_1 & w_1 \\ v_2 & w_2 \end{vmatrix}.$$

The pair of rectangular equations then becomes

$$\frac{x - x_0}{\begin{vmatrix} v_2 & w_2 \\ v_3 & w_3 \end{vmatrix}} = \frac{y - y_0}{\begin{vmatrix} v_3 & w_3 \\ v_1 & w_1 \end{vmatrix}} = \frac{z - z_0}{\begin{vmatrix} v_1 & w_1 \\ v_1 & w_2 \end{vmatrix}}. \tag{42}$$

EXAMPLES

1. Find the equation of the plane through the point with coordinates (2, 1, 4) perpendicular to each of the planes whose equations are

$$x + 5y + z = 3, \qquad 2x - 5y - 3z = 4.$$

Vectors perpendicular to the given planes are

$$\mathbf{i} + 5\mathbf{j} + \mathbf{k} \qquad \text{and} \qquad 2\mathbf{i} - 5\mathbf{j} - 3\mathbf{k},$$

respectively. The required plane will be parallel to each of these vectors, thus its equation is given by the form (36):

$$\begin{vmatrix} x - 2 & 1 & 2 \\ y - 1 & 5 & -5 \\ z - 4 & 1 & -3 \end{vmatrix} = 0,$$

$$-10(x - 2) + 5(y - 1) - 15(z - 4) = 0;$$
$$2x - y + 3x - 15 = 0.$$

2. Find the equations of the line through the point with coordinates (2, 1, 4) parallel to each of the planes whose equations are

$$x + 5y + z = 7, \qquad 2x - 5y - 3z = -1.$$

These planes are parallel to those in Example 1; so the required line is perpendicular to both $\mathbf{i} + 5\mathbf{j} + \mathbf{k}$ and $2\mathbf{i} - 5\mathbf{j} - 3\mathbf{k}$. The equations of the line are given by (42):

$$\frac{x - 2}{\begin{vmatrix} 5 & -5 \\ 1 & -3 \end{vmatrix}} = \frac{y - 1}{\begin{vmatrix} 1 & -3 \\ 1 & 2 \end{vmatrix}} = \frac{z - 4}{\begin{vmatrix} 1 & 2 \\ 5 & -5 \end{vmatrix}} ;$$

$$\frac{x - 2}{2} = \frac{y - 1}{-1} = \frac{z - 4}{3}.$$

The key consideration, in both Examples 1 and 2, is that

$$(\mathbf{i} + 5\mathbf{j} + \mathbf{k}) \times (2\mathbf{i} - 5\mathbf{j} - 3\mathbf{k}) = -5(2\mathbf{i} - \mathbf{j} + 3\mathbf{k}).$$

With the above equation in mind, the final results may be obtained from the forms (33) and (39).

3. Find the equations of the line through the point with coordinates (2, 3, −1) perpendicular to the plane whose equation is $3x - 4y + 2z = 5$.

The vector $3\mathbf{i} - 4\mathbf{j} + 2\mathbf{k}$ is perpendicular to the given plane; hence it is parallel to the required line. The equations are then given by (39):

$$\frac{x - 2}{3} = \frac{y - 3}{-4} = \frac{z + 1}{2}.$$

4. Find the equation of the plane through the point with coordinates $(2, 3, -1)$ perpendicular to the line whose equations are

$$\frac{x - 4}{3} = \frac{y + 2}{-4} = \frac{z - 6}{2}.$$

The key vector here is the same as that in Example 3, and the result is given by (33):

$$3(x - 2) - 4(y - 3) + 2(z + 1) = 0.$$

EXERCISES

1. Each of the following parts gives the equations of two planes and the coordinates of a point. Find the equation of the plane through the given point perpendicular to each of the given planes.

(a) $2x + 3y - z = 5$,　　$x - 2y + 2z = 3$,　　$(3, -5, 2)$
(b) $x - 3y - 7z = 0$,　　$2x + 3y - 4z = -6$,　　$(-1, 3, -5)$
(c) $3x + 5y + 6z = 2$,　　$4x - y - 2z = 9$,　　$(3, 0, -1)$
(d) $x - y - z = 0$,　　$x + y - 2z = 3$,　　$(0, 0, 1)$
(e) $3x + 2y + z = 5$,　　$x - 3y + 2z = 7$,　　$(-1, -3, -5)$

2. For each entry in Exercise 1, find the equations of the line through the given point parallel to each of the given planes.

3. For each entry in Exercise 1, write the equation of the plane through the given point parallel to the first of the given planes.

4. For each entry in Exercise 1, write the equations of the line through the given point perpendicular to the first of the given planes.

5. Each of the following parts gives the equation pairs for two lines and the coordinates of a point. In each case find the equation of the plane through the given point parallel to each of the given lines.

(a) $\dfrac{x - 1}{2} = \dfrac{y + 2}{3} = \dfrac{z - 3}{4}$,　$\dfrac{x + 4}{3} = \dfrac{y - 1}{7} = \dfrac{z + 3}{5}$,　$(3, 4, -2)$

(b) $\dfrac{x + 3}{2} = \dfrac{y - 7}{3} = \dfrac{z + 4}{-1}$,　$\dfrac{x}{3} = \dfrac{y + 2}{4} = \dfrac{z - 3}{-2}$,　$(3, 0, -2)$

(c) $\dfrac{x - 2}{4} = \dfrac{y}{2} = \dfrac{z + 1}{5}$,　$\dfrac{x + 8}{-3} = \dfrac{y - 3}{-1} = \dfrac{z}{3}$,　$(0, 0, 0)$

(d) $\dfrac{x}{2} = \dfrac{y}{5} = \dfrac{z}{3}$, $\dfrac{x+3}{3} = \dfrac{y-2}{3} = \dfrac{z}{2}$, $(3, 0, 1)$

(e) $\dfrac{x+2}{-1} = \dfrac{y-1}{-2} = \dfrac{z+5}{5}$, $\dfrac{x+4}{-2} = \dfrac{y-3}{-2} = \dfrac{z+1}{3}$, $(-1, 2, -5)$

6. For each of the entries in Exercise 5, find the equations of the line through the given point perpendicular to each of the given lines.

7. For each entry in Exercise 5, write the equation of the plane through the given point perpendicular to the first of the given lines.

8. For each entry in Exercise 5, write the equations of the line through the given point parallel to the first of the given lines.

CHAPTER 5

PARTIAL DERIVATIVES

5-1 Functions on n-tuples. When the notion of locus was introduced in Section 1–2, the form $y = f \circ x$ was used as the model for the analytic description of a locus. Often loci are not described in exactly this way. For example, the natural description of the unit circle is $x^2 + y^2 = 1$. This equation sets a mapping equal to a constant, but the structure of this mapping is the structure of the two coordinate variables followed by a mapping that carries number pairs into numbers. Mappings of this latter type are called *functions on pairs*.

In a composite mapping a function on pairs must follow an ordered pair of mappings, not a single one. That is, if ϕ is a function on pairs, $\phi \circ x$ is nonsense; $\phi \circ (x, y)$ is the meaningful form. Schematically, the mapping $\phi \circ (x, y)$ is shown as follows:

$$\text{point} \left\langle \begin{array}{l} x \to \text{number} \\ y \to \text{number} \end{array} \right\} \xrightarrow{\ \phi\ } \text{number}$$

From this there are obvious generalizations to functions on triples, quadruples, quintuples, and in general, n-tuples. If f is a function on n-tuples and $a_1, a_2, \ldots, a_n$ is an element of its domain, the symbol for function value is

$$f(a_1, a_2, \ldots, a_n). \tag{1}$$

Composition is defined for an ordered set, $x_1, x_2, \ldots, x_n$, of variables with a common domain followed by a function f on n-tuples. The natural symbol for the composite mapping is

$$f \circ (x_1, x_2, \ldots, x_n). \tag{2}$$

At this point, the process of changing from precise notation to notation in common use will be pushed one step further; that is,

$$f(x_1, x_2, \ldots, x_n) \tag{3}$$

will be used instead of (2). The similarity between (1) and (3) could cause confusion, but this is easily resolved if the meaning of the symbols in parentheses is made clear.

Let x and y be the rectangular coordinate variables on the plane, and let f be the function on pairs defined by

$$f(x, y) = x^2 - y^2. \tag{4}$$

One could equally well write

$$f(y, x) = y^2 - x^2. \tag{5}$$

This describes a different variable over the plane, but (4) and (5) involve the same function f. A formula for a function on pairs merely indicates how the value of the first entry contributes to the function value and how that of the second entry does. So far as the structure of the function is concerned the name and additional significance (if any) of an entry are immaterial.

Suppose that x, y, and z are the rectangular-coordinate variables on three-space. In general, the locus of

$$z = f(x, y) \tag{6}$$

is a surface in three-space. To draw a contour map of the locus of (6) is frequently helpful. A contour line for a surface is a curve in the plane over which the surface has constant height, and a contour map consists of a set of such curves. Thus, to sketch a contour map from (6), one must graph

$$f(x, y) = c$$

for various values of the constant c.

For many purposes a contour map is a more appropriate picture of the function than is the surface. In many (perhaps most) applications one considers an equation

$$u = f(x, y) \tag{7}$$

in which u is not to be interpreted as a coordinate in three-space at all, but as another variable *over the plane* related to x and y by (7).

For example, in a problem on heat dispersion, u might measure temperature. Then (7) would describe the way in which temperature varies from one point of the plane to another. In an electrostatics problem, u might measure electrostatic potential, and (7) would describe the dependence of potential on position in the plane. More often than not, a problem based on a function on pairs is a two-dimensional problem, not three.

Given (7), the locus of $f(x, y) = $ *constant* is called a *level curve* for u. Instead of talking about contour maps, then, it is more appropriate to say that a picture of the physical situation described by (7) is given by a set of level curves for u. If u is a temperature variable, these level curves are called *isothermal* curves. If u is a potential variable, they are called *equi-*

potential curves. A physicist would draw a temperature distribution by drawing some isothermals; he would draw a potential distribution by drawing some equipotentials.

If this point of view is borne in mind, there is no difficulty in seeing physical interpretations of

$$u = f(x, y, z). \tag{8}$$

Forget about sketching the locus of (8). The variable u in (8) is not a spatial coordinate. It gives numerical measurements of some physical quantity associated with the points of three-space. Again, temperature is a good example. The locus of $f(x, y, z) = constant$ is usually a surface; so given (8), one speaks of the *level surfaces* for u. In this case a sketch, even of the level loci, is probably not practical.

If temperature depends on time as well as position, one has a relation of the form

$$u = f(x, y, z, t). \tag{9}$$

Here the sketching problem is completely out of hand, and thus one might as well forget it. There is, however, no mystery about the physical significance of (9). To carry this line of discussion even further, note that the energy of an ideal gas depends on the velocity components of its individual molecules, three components for each molecule. Thus,

$$E = f(u_1, \ldots, u_n, \quad v_1, \ldots, v_n, \quad w_1, \ldots, w_n); \tag{10}$$

the pertinent relation involves $3n$ variables where n is the number of molecules. Despite the utter impossibility of graphing (10), it has a concrete and easily understood physical significance.

For functions on n-tuples the notion of continuity is a little more involved than for functions on numbers. If f is a function on pairs, one says that f is *continuous* at (a, b) provided that

$$\lim_{(x,y)\to(a,b)} f(x, y) = f(a, b). \tag{11}$$

Equation (11) looks familiar enough, but the notion of limit involved here is something new. It is what is called a *double limit*. By definition, (11) means that given $\epsilon > 0$, there is a $\delta > 0$ such that

$$|f(x, y) - f(a, b)| < \epsilon$$

for all x and y for which

$$|x - a| < \delta \quad \text{and} \quad |y - b| < \delta.$$

Informally, think of a "moving point" (x, y) "getting close to" (a, b). The complicating feature is that there are so many different paths that can

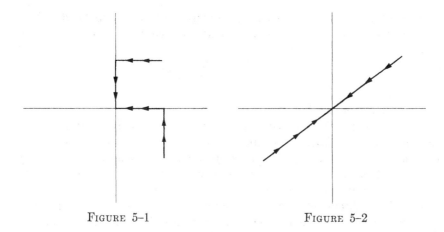

FIGURE 5-1   FIGURE 5-2

be followed by the moving point, and $f(x, y)$ must tend to the same limit along all these paths.

Consider, for example,

$$f(x, y) = \frac{xy}{x^2 + y^2}.$$

As $x \to 0$ for fixed $y \neq 0$, it is clear that $f(x, y) \to 0$. Similarly, $f(x, y) \to 0$ as $y \to 0$ for fixed $x \neq 0$. Thus, if $(x, y) \to (0, 0)$ along either of the paths in Fig. 5-1, $f(x, y) \to 0$. However, if $x = y$, then $f(x, y) = \frac{1}{2}$; thus $f(x, y) \to \frac{1}{2}$ as $(x, y) \to (0, 0)$ along the paths in Fig. 5-2. Therefore, the double limit of f does not exist at $(0, 0)$.

As with functions on numbers (see discussion in Section 1-1) the practical way to deal with limits and continuity is to learn what functions are continuous and then evaluate limits of these by substitution. The basic functions on pairs are given by

$$f(x, y) = x, \qquad g(x, y) = y,$$

$$\phi(x, y) = x + y, \qquad \psi(x, y) = xy,$$

$$\eta(x, y) = x/y.$$

Each of these is continuous wherever it is defined (proof omitted). Note that η is not defined where its second entry is zero. This information, together with the following, serves to unearth a large class of continuous functions.

THEOREM. On continuity of composite functions. Let F, G, and H be functions on pairs, each continuous wherever it is defined. Let f and g

be functions on numbers, each continuous wherever it is defined. Then, each of the following composite functions is continuous wherever it is defined:

$$F \circ (G, H), \qquad F \circ (f, g),$$

$$f \circ F, \qquad f \circ g.$$

The proof is omitted.

Combining these results with those in Section 1–1 shows that any function generated by basic algebraic operations between two entries, compounded with elementary functions on numbers, is continuous at every point at which the formula for function values does not indicate division by zero or the logarithm of zero.

The above discussion proves continuity for a large class of functions. Unfortunately, it does not cover cases in which the formula for function values yields an indeterminate form. These cases arise all too often in applications, and there is always the temptation to evaluate an indeterminate form in two variables by taking a limit in each variable separately by inspection or by L'Hôpital's rule. This is not sufficient for a study of continuity.

EXAMPLES

1. Let

$$f(x, y) = \begin{cases} \dfrac{\sin xy}{xy} & \text{for} \quad x \neq 0 \text{ and } y \neq 0, \\ 1 & \text{for} \quad x = 0 \text{ or } y = 0. \end{cases}$$

Show that f is continuous everywhere. Note that here

$$f(x, y) = g \circ F \circ (x, y),$$

where

$$F(x, y) = xy \text{ and } g(u) = \begin{cases} \dfrac{\sin u}{u} & \text{for} \quad u \neq 0, \\ 1 & \text{for} \quad u = 0. \end{cases}$$

Both F and g are continuous everywhere, hence the result follows by the theorem on continuity of composite functions.

2. Let

$$f(x, y) = \begin{cases} \dfrac{xy}{\sqrt{x^2 + y^2}} & \text{for} \quad x^2 + y^2 \neq 0, \\ 0 & \text{for} \quad x = 0 \text{ and } y = 0. \end{cases}$$

Show that f is continuous everywhere. Except at the origin, the continuity of f follows by the composite function theorem. To check the situation at the origin, note that

$$\left| \frac{xy}{\sqrt{x^2 + y^2}} \right| = \frac{1}{\sqrt{(1/y^2) + (1/x^2)}} .$$

Now, $\sqrt{(1/y^2) + (1/x^2)}$ may be made arbitrarily large by choosing $|x|$ or $|y|$ *or both* sufficiently small; hence

$$\lim_{(x,y)\to(0,0)} f(x, y) = 0,$$

as required for continuity. Note that the step of dividing by $|xy|$ is necessary in order to see the double limit. From the original form it is apparent only that if one argument is close to zero *and the other is not*, then the function value is close to zero. This is not sufficient to give a double limit.

EXERCISES

1. Give a detailed proof (using the theorem on continuity of composite functions) that each of the following functions f is continuous in the entire xy-plane.

(a) $f(x, y) = (x^2 - y^2)^{1/3}$ (b) $f(x, y) = xye^{x^2 + y^2}$

(c) $f(x, y) = \sin (x + y^2)$

(d) $f(x, y) = \begin{cases} \dfrac{\sin xy}{\sqrt{x^2 + y^2}} & \text{for} \quad x^2 + y^2 \neq 0 \\ 0 & \text{for} \quad x = y = 0 \end{cases}$

(e) $f(x, y) = \begin{cases} (x^2 + y^2) \ln (x^2 + y^2) & \text{for} \quad x^2 + y^2 \neq 0 \\ 0 & \text{for} \quad x = y = 0 \end{cases}$

(f) $f(x, y) = \begin{cases} y^2 \ln (x^2 + y^2) & \text{for} \quad x^2 + y^2 \neq 0 \\ 0 & \text{for} \quad x = y = 0 \end{cases}$

(g) $f(x, y) = \begin{cases} \dfrac{x^3 + y^3}{x^2 + y^2} & \text{for} \quad x^2 + y^2 \neq 0 \\ 0 & \text{for} \quad x = y = 0 \end{cases}$

(h) $f(x, y) = \begin{cases} \dfrac{1 - \cos (x - y)}{x - y} & \text{for} \quad x \neq y \\ 0 & \text{for} \quad x = y \end{cases}$

(i) $f(x, y) = \begin{cases} xy \ln (x^2 + y^2) & \text{for} \quad x^2 + y^2 \neq 0 \\ 0 & \text{for} \quad x = y = 0 \end{cases}$

2. Locate the points of essential discontinuity (points at which the double limit does not exist) for each of the following functions f.

(a) $f(x, y) = \dfrac{x^2 - y^2}{x^2 + y^2}$

(b) $f(x, y) = \dfrac{x}{x + y}$

(c) $f(x, y) = \dfrac{x + y}{x - y}$

(d) $f(x, y) = \dfrac{\sin(x + y)}{x}$

(e) $f(x, y) = \dfrac{1 - \cos(x + y)}{y^2}$

(f) $f(x, y) = |x|^{|1/y|}$

(g) $f(x, y) = \dfrac{(x + y)^2}{x^2 + y^2}$

(h) $f(x, y) = e^{-x/y}$

(i) $f(x, y) = e^{-|x/y|}$

3. Let
$$f(x, y) = \frac{(y^2 - x)^3 + x^2 y}{(y^2 - x)^2 + |y|^5}.$$

Show that $f(x, y) \to 0$ as $(x, y) \to (0, 0)$ along any straight line, but not along the parabola on which $x = y^2$.

4. Let
$$f(x, y) = \begin{cases} |x|^y & \text{for} \quad x \neq 0, \\ 1 & \text{for} \quad x = 0. \end{cases}$$

Show that $f(x, y) \to 1$ as $(x, y) \to (0, 0)$ along the locus of $y = mx^a$ $(a > 0)$. Show, however, that the double limit of $f(x, y)$ does not exist at $(0, 0)$.

5–2 Partial derivatives of functions. Let f be a function on pairs. The *partial derivative of f with respect to its first entry* is another function on pairs, denoted by f_1. Specifically,
$$f_1(a, b) = \lim_{h \to 0} \frac{f(a + h, b) - f(a, b)}{h}.$$

Informally, to generate f_1 reduce f to a function on numbers by regarding the first entry as the only operative one, and differentiate.

More generally, if f is a function on n-tuples, define
$$f_i(a_1, a_2, \ldots, a_n) = \lim_{h \to 0} \frac{f(a_1, \ldots, a_i + h, \ldots, a_n) - f(a_1, \ldots, a_n)}{h}.$$

The function f_i is the partial derivative of f with respect to its ith entry. Subscripts on symbols for functions on n-tuples will always indicate partial differentiation.

By taking a partial derivative of a partial derivative, one forms higher-order partial derivatives. There is a new feature here, though, in that one may change entries from one differentiation to the next. Thus, for a function on pairs, there are, technically, four second-order partial derivatives:

$$f_{11} = (f_1)_1, \qquad f_{12} = (f_1)_2,$$

$$f_{21} = (f_2)_1, \qquad f_{22} = (f_2)_2.$$

Effectively, there are only three, however. If the second-order partial derivatives of f are continuous, then (proof omitted)

$$f_{12} = f_{21}.$$

Computation of partial derivatives by formula is extremely simple. Given $f(a_1, a_2, \ldots, a_n)$ in terms of $a_1, a_2, \ldots, a_n$, temporarily regard all but one of these symbols as constants, and apply the appropriate derivative formulas.

With partial derivatives as coefficients, one gets a Taylor formula (polynomial approximation) for functions on n-tuples. Through quadratic terms, the Taylor expansion for a function on pairs is

$$f(a + h, b + k) = f(a, b) + f_1(a, b)h + f_2(a, b)k$$
$$+ \frac{1}{2!}[f_{11}(a, b)h^2 + 2f_{12}(a, b)hk + f_{22}(a, b)k^2] + \cdots$$

The cubic term would be

$$\frac{1}{3!}(f_{111}h^3 + 3f_{112}h^2k + 3f_{122}hk^2 + f_{222}k^3);$$

in general, the coefficients follow the binomial theorem formula.

In symbolic form, the Taylor expansion for a function on n-tuples is as follows. Let D_i be an operator denoting partial differentiation with respect to the ith entry. Then

$$f(a_1 + h_1, \ldots, a_n + h_n) \sim \sum_{k=0}^{\infty} \frac{1}{k!} \left(\sum_{i=1}^{n} h_i D_i \right)^k f(a_1, \ldots, a_n).$$

A remainder formula may be obtained or convergence of Taylor series may be studied; but the important consideration at present is that if f has continuous partial derivatives of order $k + 1$, then the coefficients in the kth-order term may be replaced by differentiable functions to obtain an exact equality. Specifically, for use in Section 5–4, note the following statement (proof omitted). If f has continuous third partial derivatives, then there exist functions g^{ij} $(i, j = 1, \ldots, n)$ with continuous first partial derivatives such that

$$f(a_1 + h_1, \ldots, a_n + h_n) = f(a_1, \ldots, a_n) + \sum_{i=1}^{n} f_i(a_1, \ldots, a_n)h_i$$

$$+ \sum_{i=1}^{n} \sum_{j=1}^{n} g^{ij}(h_1, \ldots, h_n)h_i h_j. \qquad (12)$$

For functions on numbers, the existence of a derivative answers many pertinent questions. For example, if f is differentiable at a, then f is continuous at a, and the graph of $y = f(x)$ has a unique tangent line at $[a, f(a)]$.

The existence of partial derivatives does not play an analogous role in the higher-dimensional theory. This is not too surprising because partial derivatives are defined by limits on very special paths, whereas the basic limit notion is that of the double limit. The analogy is restored if one replaces "is differentiable" by "has continuous partial derivatives." That is (proof omitted), if f has continuous partial derivatives at (a, b), then f is continuous at (a, b), and the locus of $z = f(x, y)$ has a unique tangent plane at $[a, b, f(a, b)]$. Note, too, that continuous third partial derivatives are called for in conjunction with (12) above.

Examples

1. Given that

$$f(x, y, z) = xy - xz + yz,$$

find $f_1(x, y, z)$, $f_2(x, y, z)$, and $f_3(x, y, z)$. In finding $f_1(x, y, z)$, one regards y and z as constants; so

$$f_1(x, y, z) = y - z.$$

Similarly,

$$f_2(x, y, z) = x + z, \qquad f_3(x, y, z) = y - x.$$

2. Given that

$$f(x, y) = \sin \frac{y}{x},$$

find $f_{12}(x, y)$ and $f_{21}(x, y)$. *Solution:*

$$f_1(x, y) = -\frac{y}{x^2} \cos \frac{y}{x}, \qquad f_{12}(x, y) = -\frac{1}{x^2} \cos \frac{y}{x} + \frac{y}{x^3} \sin \frac{y}{x},$$

$$f_2(x, y) = \frac{1}{x} \cos \frac{y}{x}, \qquad f_{21}(x, y) = -\frac{1}{x^2} \cos \frac{y}{x} + \frac{y}{x^3} \sin \frac{y}{x}.$$

As noted above, it is not mere coincidence that these are equal.

3. Let

$$f(x, y) = \begin{cases} \dfrac{xy}{x^2 + y^2} & \text{for} \quad x \neq 0 \text{ or } y \neq 0, \\ 0 & \text{for} \quad x = y = 0. \end{cases}$$

It was pointed out in Section 5–1 that f was not continuous at $(0, 0)$ because $f(x, 0) = 0$ and $f(0, y) = 0$ while $f(x, x) = \frac{1}{2}$. However, since $f(x, 0) = 0$ for all x,

$$f_1(0, 0) = 0,$$

and, similarly,

$$f_2(0, 0) = 0.$$

The partial derivatives exist, but the function is not continuous.

Exercises

1. For each of the following functions f, find all first-order partial derivatives.

(a) $f(x, y) = x^2 y - 2y^2 x$ (b) $f(x, y) = 2xe^{-y}$

(c) $f(r, \theta) = 2re^{-\theta}$ (d) $f(r, \theta) = e^{-r} \cos \theta$

(e) $f(u, v) = u/v$ (f) $f(v, u) = u/v$

(g) $f(r, \theta) = r \cos \theta$ (h) $f(r, \theta) = r \sin \theta$

(i) $f(x, y) = \sqrt{x^2 + y^2}$ (j) $f(x, y) = \arctan (y/x)$

(k) $f(x, y) = \ln (x^2 + y^2)$ (l) $f(x, y) = 1/\sqrt{x^2 + y^2}$

(m) $f(p, v) = pv^{1.4}$ (n) $f(x, t) = e^{-t} \cos At \sin Bx$

(o) $f(x, y, z) = \sqrt{x^2 + y^2 + z^2}$ (p) $f(x, y, z) = \dfrac{1}{\sqrt{x^2 + y^2 + z^2}}$

(q) $f(x, y, z) = \arctan (y/x)$ (r) $f(x, y, x) = \sqrt{x^2 + y^2}$

(s) $f(x, y, z) = \arccos \dfrac{z}{\sqrt{x^2 + y^2 + z^2}}$ (t) $f(\rho, \phi, \theta) = \rho \sin \phi \cos \theta$

(u) $f(\rho, \phi, \theta) = \rho \sin \phi \sin \theta$ (v) $f(\rho, \phi, \theta) = \rho \cos \phi$

(w) $f(r, \theta, z) = \sqrt{z^2 + r^2}$ (x) $f(r, \theta, z) = \arctan (z/r)$

(y) $f(\rho, \phi, \theta) = \rho \sin \phi$ (z) $f(r, \theta, z) = \arcsin \dfrac{r}{\sqrt{r^2 + z^2}}$

2. In general, $f_{12} = f_{21}$. Show that this is the case for each of the following functions f.

(a) $f(u, v) = u/v$ (b) $f(r, \theta) = r \cos \theta$

(c) $f(x, y) = xe^{-y}$ (d) $f(x, y) = x^3 y^2$

(e) $f(x, y) = \sqrt{x^2 + y^2}$ (f) $f(x, t) = e^{-t} \cos x$

3. Show that for each of the following functions f,

$$f_{11} + f_{22} = 0.$$

(a) $f(x, y) = \ln (x^2 + y^2)$ (b) $f(x, y) = x^2 - y^2$

(c) $f(x, y) = xy$ (d) $f(x, y) = e^x \cos y$

(e) $f(x, y) = \sin x \cosh y$ (f) $f(x, y) = \arctan (y/x)$

4. Let

$$f(x, y, z) = \frac{1}{\sqrt{x^2 + y^2 + z^2}}.$$

Show that

$$f_{11} + f_{22} + f_{33} = 0.$$

5. For each of the following functions f write out, through terms of the second degree, the Taylor expansion of f about $(0, 0)$.

(a) $f(x, y) = e^x \cos y$ (b) $f(x, y) = e^y \sin x$

(c) $f(x, y) = \sin (x + y)$ (d) $f(x, y) = \cos (x - y)$

6. Obtain each of the results in Exercise 1 by multiplication and/or addition of one-dimensional Taylor series.

7. Let f be a function on triples. Write out specifically, through terms of the second degree, the Taylor formula for $f(a + h, b + k, c + m)$.

8. Let

$$f(x, y) = \begin{cases} xy\dfrac{x^2 - y^2}{x^2 + y^2} & \text{for} \quad x^2 + y^2 \neq 0, \\ 0 & \text{for} \quad x = y = 0. \end{cases}$$

Show that $f_{12}(0, 0) = 1$ and $f_{21}(0, 0) = -1$.

5–3 Partial derivatives of variables. Each coordinate u on a one-dimensional manifold determines a derivative operator D_u as follows. Since a coordinate on a one-dimensional manifold is a one-to-one mapping, given w on the manifold, there is a function f such that $w = f(u)$. Then, if f is differentiable, $D_u w$ is defined to be $f'(u)$.

A technical definition of n-dimensional manifold will be given later (Section 5–6). The important feature is that a single coordinate does not characterize points uniquely. Rather, there is a set of n coordinates u_1, u_2, ..., u_n; and the postulate which makes the manifold n-dimensional is that the mapping,

$$p \to [u_1(p), u_2(p), \ldots, u_n(p)]$$

of points into n-tuples of numbers is one to one. For example, on the plane, x is not one to one and neither is y, but the mapping

$$p \to [x(p), y(p)]$$

is. The plane with rectangular coordinates is a basic example of a two-dimensional manifold.

So, given a coordinate set $u_1, u_2, \ldots, u_n$ on an n-dimensional manifold, and given another variable w on the manifold, there is a function f on n-tuples such that

$$w = f(u_1, u_2, \ldots, u_n).$$

The obvious analogy with the one-dimensional case is to call

$$f_i (u_1, u_2, \ldots, u_n)$$

the partial derivative of w with respect to u_i. Essentially, this will be done, but there are certain complications.

First, there is a matter of notation. The notation $D_{u_i} w$ appears occasionally for the partial derivative of w with respect to u_i, but there is a much more widely used notation that will be employed here. The derivative operator will be denoted by

$$\frac{\partial}{\partial u_i}, \tag{13}$$

and the result of applying this derivative operator to w will be written

$$\frac{\partial w}{\partial u_i}. \tag{14}$$

Note the distinction between (14) and the form

$$\frac{dy}{dx}. \tag{15}$$

The form (15) is a quotient. It separates as $dy \div dx$. The form (14) is not a quotient. It is an operator applied to a variable and it separates as

$$\frac{\partial}{\partial u_i} (w).$$

To explain the major complication, it is well to begin with the two-dimensional case. Suppose that

$$w = f(u, v);$$

then

$$\frac{\partial w}{\partial u} = f_1(u, v). \tag{16}$$

On the other hand, there might be a different coordinate set (u, t) also containing the variable u. In this case

$$w = g(u, t),$$

and

$$\frac{\partial w}{\partial u} = g_1(u, t). \tag{17}$$

In general (16) and (17) are not the same at all! The explanation is that the operator notation (13) is incomplete. On a manifold of dimension more than one, a derivative with respect to one coordinate depends not just on that coordinate but on all the others as well.

A notation commonly used to clarify this point is as follows: instead of (16) write

$$\left(\frac{\partial w}{\partial u}\right)_v = f_1(u, v), \tag{16'}$$

and for (17) write

$$\left(\frac{\partial w}{\partial u}\right)_t = g_1(u, t). \tag{17'}$$

The left side of (16') is read the *partial derivative of w with respect to u for constant v.* In this notation (13) is an abbreviation for

$$\left(\frac{\partial}{\partial u_i}\right)_{u_1 \ldots u_{i-1} u_{i+1} \ldots u_n}. \tag{13'}$$

For a dimension higher than two or three, the full-dress notation (13') becomes quite cumbersome. Therefore, in all cases, the abbreviated operator form (13) will be used whenever it is clear from the context what the other coordinates are. There are many cases of practical importance, however, in which the complete notation is required.

EXAMPLES

1. Let r and θ be polar coordinates in the plane. Recall that

$$x = r \cos \theta, \qquad y = r \sin \theta. \tag{18}$$

These equations may be solved for r and θ:

$$r = \sqrt{x^2 + y^2}, \qquad \theta = \arctan \frac{y}{x}. \tag{19}$$

From (18), it follows that

$$\left(\frac{\partial x}{\partial r}\right)_\theta = \cos \theta. \tag{20}$$

From (19), it follows that

$$\left(\frac{\partial r}{\partial x}\right)_y = \frac{x}{\sqrt{x^2 + y^2}} = \frac{r \cos \theta}{r} = \cos \theta. \tag{21}$$

Comparison of (20) and (21) here reveals no paradox, only coincidence.

However, this is because the complete notation (13′) is being used. In the abbreviated notation (13) these results would read

$$\frac{\partial x}{\partial r} = \cos \theta = \frac{\partial r}{\partial x},$$

and students often wonder why. Why not

$$\frac{\partial x}{\partial r} = 1\Big/\left(\frac{\partial r}{\partial x}\right)?$$

Any confusion here is due only to bad notation. From (18), it follows that $r = x \sec \theta$, so, compared with (20),

$$\left(\frac{\partial r}{\partial x}\right)_{\theta} = \sec \theta = 1\Big/\left(\frac{\partial x}{\partial r}\right)_{\theta},$$

as might be expected.

EXERCISES

1. From the relations $x = r \cos \theta$, $y = r \sin \theta$, and other appropriate relations derived from them, find the following partial derivatives.

(a) $\left(\dfrac{\partial x}{\partial r}\right)_{\theta}$ (b) $\left(\dfrac{\partial x}{\partial r}\right)_{y}$ (c) $\left(\dfrac{\partial x}{\partial \theta}\right)_{r}$

(d) $\left(\dfrac{\partial x}{\partial \theta}\right)_{y}$ (e) $\left(\dfrac{\partial x}{\partial y}\right)_{r}$ (f) $\left(\dfrac{\partial x}{\partial y}\right)_{\theta}$

(g) $\left(\dfrac{\partial y}{\partial r}\right)_{\theta}$ (h) $\left(\dfrac{\partial y}{\partial r}\right)_{x}$ (i) $\left(\dfrac{\partial y}{\partial \theta}\right)_{r}$

(j) $\left(\dfrac{\partial y}{\partial \theta}\right)_{x}$ (k) $\left(\dfrac{\partial y}{\partial x}\right)_{r}$ (l) $\left(\dfrac{\partial y}{\partial x}\right)_{\theta}$

(m) $\left(\dfrac{\partial r}{\partial x}\right)_{y}$ (n) $\left(\dfrac{\partial r}{\partial x}\right)_{\theta}$ (o) $\left(\dfrac{\partial r}{\partial y}\right)_{x}$

(p) $\left(\dfrac{\partial r}{\partial y}\right)_{\theta}$ (q) $\left(\dfrac{\partial r}{\partial \theta}\right)_{x}$ (r) $\left(\dfrac{\partial r}{\partial \theta}\right)_{y}$

(s) $\left(\dfrac{\partial \theta}{\partial x}\right)_{y}$ (t) $\left(\dfrac{\partial \theta}{\partial x}\right)_{r}$ (u) $\left(\dfrac{\partial \theta}{\partial y}\right)_{x}$

(v) $\left(\dfrac{\partial \theta}{\partial y}\right)_{r}$ (w) $\left(\dfrac{\partial \theta}{\partial r}\right)_{x}$ (x) $\left(\dfrac{\partial \theta}{\partial r}\right)_{y}$

2. Without referring to the answers for Exercise 1, group the twenty-four partial derivatives listed there into twelve pairs of reciprocals. Then, check them against the answers.

5–4 Differentials.   In brief outline, the algebraic definition of the differential given in Section 1–6 is stated in the following manner.

(i) A derivative operator D at a point p of the underlying manifold is defined by the postulates

$$D(au + bv)(p) = a\,Du(p) + b\,Dv(p),$$

$$D(uv)(p) = u(p)\,Dv(p) + v(p)\,Du(p).$$

(22)

(ii) Let u be a coordinate at p on a one-dimensional manifold, and let v be a variable on the manifold; then (by the definitions of coordinate and one-dimensional manifold), $v = f(u)$. Assume that f has three derivatives and write the polynomial formula for $f(u)$ in $u - u(p)$, terminating with the quadratic term. Apply D to this form, and

$$Dv(p) = f'\,[u(p)]\,Du(p) \tag{23}$$

because the postulates (22) lead to zero values for all other terms.

(iii) Introduce the operator D_u and note that it satisfies the postulates. Then rewrite (23):

$$Dv(p) = D_uv(p)\,Du(p). \tag{24}$$

Since this holds for all differentiable variables v, it says that at p,

$$D = Du(p)\,Du.$$

That is, the vector D is the component $Du(p)$ times the basis vector D_u. The set of all derivative operators at p is a one-dimensional vector space.

(iv) Identify this vector space with the tangent line T_p and thus establish the set of ordered pairs (p, D) as the domain for dv. Then, define

$$dv_p(D) = Dv(p). \tag{25}$$

Each v on the manifold thus generates dv on the tangent bundle.

(v) Substitute from (25) into (24) to get the fundamental theorem on differentials,

$$dv = D_uv\,du,$$

on the tangent bundle.

The theory in higher dimensions follows this pattern exactly. The only difference (see Section 5–3) is that instead of starting with $v = f(u)$, one starts with

$$v = f(u_1, u_2, \ldots, u_n). \tag{26}$$

The steps are now just those given above.

(i) Define derivative operators by postulate. Use (22) verbatim; not even the notation needs changing.

(ii) Assume that f in (26) has continuous third partial derivatives and apply (12):

$$v = v(p) + \sum_{i=1}^{n} f_i[u_1(p), \dots, u_n(p)][u_i - u_i(p)]$$

$$+ \sum_{i=1}^{n} \sum_{j=1}^{n} g^{ij}(u_1, \dots, u_n)[u_i - u_i(p)][u_j - u_j(p)].$$

Operate on the equation above with D; each term in the quadratic form vanishes, and the result is

$$Dv(p) = \sum_{i=1}^{n} f_i[u_1(p), \dots, u_n(p)] \, Du_i(p). \tag{23'}$$

(iii) Introduce the operators $\partial/\partial u_i$ (see Section 5–3) and rewrite (23'):

$$Dv(p) = \sum_{i=1}^{n} \frac{\partial v}{\partial u_i}(p) \, Du_i(p). \tag{24'}$$

Interpret (24') as giving the expansion

$$D = \sum_{i=1}^{n} Du_i(p) \frac{\partial}{\partial u_i}$$

of a vector D in terms of components $Du_i(p)$ and basis vectors $\partial/\partial u_i$. The set of derivative operators at p is an n-dimensional vector space.

(iv) Let this n-dimensional vector space be the tangent space T_p and define dv by (25) without even changing notation.

(v) Substitute from (25) into (24') to get the fundamental theorem on differentials,

$$dv = \sum_{i=1}^{n} \frac{\partial v}{\partial u_i} du_i,$$

on the tangent bundle.

The strength of this theorem lies in the fact that the choice of coordinates is purely arbitrary. That is, for a given variable v the *form* of the expression for dv is the same for all admissible sets of coordinates.

An important consideration in this connection is that of *submanifolds*. Suppose that $u_1, u_2, \dots, u_n$ is a coordinate set on M, and that $t_1, t_2, \dots, t_k$ $(k < n)$ is a set of variables on M. Suppose, further, that the parametric equations

$$u_i = \phi^i(t_1, t_2, \dots, t_k), \qquad i = 1, 2, \dots, n,$$

have a locus (subset of M) which is a k-dimensional manifold K with t_1,

$t_2, \ldots, t_k$ as coordinates. The tangent bundle for K is then a subset of that for M; thus on the smaller tangent bundle, one has, by the fundamental theorem, two expressions for a differential:

$$\sum_{i=1}^{n} \frac{\partial v}{\partial u_i} \, du_i = dv = \sum_{j=1}^{k} \frac{\partial v}{\partial t_j} \, dt_j.$$

This is particularly interesting in a case of $k = 1$. Let K be a one-dimensional submanifold defined by

$$u_i = \phi^i(t) \qquad i = 1, 2, \ldots, n, \tag{27}$$

then the theory of Section 2–2 is applicable on K and its tangent bundle, and

$$D_t v \, dt = dv = \sum_{i=1}^{n} \frac{\partial v}{\partial u_i} \, du_i$$

over the tangent bundle for K; thus

$$D_t v = \frac{dv}{dt} = \sum_{i=1}^{n} \frac{\partial v}{\partial u_i} \frac{du_i}{dt} \tag{28}$$

on the submanifold K.

The operation of dividing by dt in (28) recalls the remark (Section 2–2) that the extreme left member of (28) should be written

$$D_t v \, dt^0$$

instead of simply $D_t v$. As before, this constant factor will usually be dropped.

There is a mechanical point worth noting. Suppose that

$$v = f(u_1, u_2, \ldots, u_n) \tag{29}$$

on M, and suppose that K is defined by (27). If (28) is used to compute $D_t v$, the factors

$$\frac{\partial v}{\partial u_i}$$

are computed from (29) while the factors

$$\frac{du_i}{dt}$$

are computed from (27).

As in the one-dimensional case, the calculus of variables introduced here is designed to produce significant results largely by routine algebraic manipulation. The introduction of submanifolds enhances this study considerably, but it also furnishes possible sources of confusion. The essential clue to keeping things straight is to note that in the calculus of variables an equation seldom expresses equality of two numbers. More often it expresses equality of two variables and is interpreted to mean that these variables are identical over some domain. Therefore, to avoid confusion, *along with each equation describe its domain of validity.* Note the models set in Examples 1 and 2 below.

EXAMPLES

1. Let
$$z = e^{xy} \tag{30}$$

over the plane. Find dz/dx on the parabola on which

$$y = x^2. \tag{31}$$

First solution. Substitute (31) into (30):

$$z = e^{x^3}$$

on the parabola; hence

$$dz = 3x^2 e^{x^3} \, dx \tag{32}$$

on the tangent bundle to the parabola, and $dz/dx = 3x^2 e^{x^3}$ on the parabola.
Second solution. From (30),

$$\left(\frac{\partial z}{\partial x}\right)_y = ye^{xy}, \qquad \left(\frac{\partial z}{\partial y}\right)_x = xe^{xy}$$

on the plane, thus by the fundamental theorem,

$$dz = ye^{xy} \, dx + xe^{xy} \, dy \tag{33}$$

on the tangent bundle for the plane, and

$$\frac{dz}{dx} = ye^{xy} \, dx^0 + xe^{xy} \, \frac{dy}{dx} \tag{34}$$

on the tangent bundle to the plane. However, on the parabola $y = x^2$ and $dy/dx = 2x$, so on the parabola,

$$\frac{dz}{dx} = ye^{xy} + xe^{xy}(2x) = 3x^2 e^{x^3}.$$

There are several points worthy of note here.

(a) There are two manifolds involved: (30) holds on a two-dimensional manifold, and (31) holds on a one-dimensional submanifold of the two-dimensional manifold.

(b) Partial derivatives are defined only with respect to the two-dimensional manifold, therefore, they must be computed from (30). Quotients of differentials give rates of change only with respect to the one-dimensional manifold, therefore, (31) must be introduced for dz/dx to have its usual significance.

(c) Equations (32) and (33) each express the fundamental theorem, and the fact that they lead to equivalent results illustrates the fact that the fundamental theorem is independent of the choice of coordinates.

(d) There is a difference between (32) and (33) in that (32) holds only over the tangent bundle for the one-dimensional submanifold, while (33) holds over the entire tangent bundle for the two-dimensional manifold. The invariance of the fundamental theorem under coordinate changes here means that these are equivalent over the smaller domain. Indeed, (33) leads to the end result only with the substitution $dy = 2x\,dx$ which holds only on the smaller tangent bundle.

(e) Equation (33) could have been computed by the rules for differentials on one-dimensional manifolds without recourse to partial derivatives:

$$dz = d(e^{xy}) = e^{xy}\,d(xy) = e^{xy}(x\,dy + y\,dx).$$

(f) In (34) dx^0 is constant on each tangent plane, while dy/dx is not. Therefore dz/dx represents a variable on the manifold only after the restriction to the one-dimensional case.

2. Let

$$z = x^y$$

on the plane and find dz in terms of x, y, dx, dy. Here

$$\left(\frac{\partial z}{\partial x}\right)_y = yx^{y-1},$$

and writing $x^y = e^{y\,\ln x}$ yields

$$\left(\frac{\partial z}{\partial y}\right)_x = x^y \ln x.$$

Therefore, by the fundamental theorem,

$$dz = yx^{y-1}\,dx + x^y \ln x\,dy$$

on the tangent bundle for the plane. From this it follows that on any one-dimensional submanifold determined by $y = f(x)$,

$$D_x(x^y) = \frac{dz}{dx} = yx^{y-1} + x^y \ln x \, \frac{dy}{dx} \cdot$$

This confirms a result usually obtained by logarithmic differentiation. Compare comment (e) in Example 1 above.

3. Let

$$z = x^2 - y^2, \tag{35}$$

and introduce polar coordinates. Then,

$$z = r^2 \cos^2 \theta - r^2 \sin^2 \theta = r^2 \cos 2\theta. \tag{36}$$

From (35),

$$dz = 2x \, dx - 2y \, dy,$$

whereas from (36),

$$dz = 2r \cos 2\theta \, dr - 2r^2 \sin 2\theta \, d\theta.$$

However,

$$dx = \cos \theta \, dr - r \sin \theta \, d\theta, \qquad dy = \sin \theta \, dr + r \cos \theta \, d\theta,$$

hence

$$\begin{aligned} 2x \, dx - 2y \, dy &= 2r \cos \theta (\cos \theta \, dr - r \sin \theta \, d\theta) \\ &\quad - 2r \sin \theta (\sin \theta \, dr + r \cos \theta \, d\theta) \\ &= 2r(\cos^2 \theta - \sin^2 \theta) \, dr - 4r^2 \cos \theta \sin \theta \, d\theta \\ &= 2r \cos 2\theta \, dr - 2r^2 \sin 2\theta \, d\theta. \end{aligned}$$

The two expressions for dz are, indeed, equal. This is not mere coincidence; it is one of the principal consequences of the fundamental theorem.

EXERCISES

1. In each of the following cases w is given on a multidimensional manifold in terms of several variables; then each of these variables is given on a one-dimensional submanifold in terms of a single variable t. In each case, follow the model of Example 1 to find dw/dt on the submanifold in two ways.

(a) $w = x^2 + y^2 + z^2$; $\quad x = e^t \cos t, \; y = e^t \sin t, \; z = e^t$

(b) $w = xy/(x^2 + y^2)$; $\quad x = \cosh t, \; y = \sinh t$

(c) $w = e^{2x+2y} \cos 4z$; $\quad x = \ln t, \; y = \ln (t^2 + 1), \; z = t$

(d) $w = xyz$; $\quad x = \sin t, \; y = \cos t, \; z = t^2$

(e) $w = (1/z)e^{xy}$; $\quad x = t^2 y = 1 - t, \; z = t^3$

(f) $w = x^y$; $\quad x = \sin t, \; y = \tan t$

(g) $w = \log_x y$; $\quad x = t, \; y = \sin t$

2. Let $z = xy$ on the plane, and introduce polar coordinates. Among the four coordinate variables now introduced, there are six pairs, and z may be expressed in terms of any pair. From these six representations for z, compute six different forms for dz and show that they are all equal.

3. In each of the following cases, compute dz in two ways. First, use the fundamental theorem formula,

$$dz = \frac{\partial z}{\partial x}\, dx + \frac{\partial z}{\partial y}\, dy.$$

Then, regard x, y, and z as variables on a one-dimensional manifold and compute differentials by the rules developed in Chapter 2. Compare the results of both computations.

(a) $z = e^{-xy}$ (b) $z = \sqrt{x^2 - y^2}$

(c) $z = \ln (x^2 + y^2)$ (d) $z = \sin (x + y)$

(e) $z = \arctan \dfrac{y}{x}$ (f) $z = \cosh xy$

(g) $z = xe^{-y}$ (h) $z = \dfrac{x}{y}\sin \dfrac{y}{x}$

(i) $z = x^2 + xy + y^2$ (j) $z = x \ln y$

(k) $z = y^x$ (l) $z = e^{-x/y}$

(m) $z = x^2 \sin y^2$ (n) $z = \ln \dfrac{x - y}{x + y}$

(o) $z = e^x \sin y$ (p) $z = e^{x+y} \cos x \sin y$

4. Let

$$w = x^2 + y^2 + z^2$$

on three-space. Each of the following sets of equations introduces one, two, or three new coordinates on a submanifold of appropriate dimension. In each case substitute to get w in terms of the new coordinates; then compute dw in terms of dx, dy, and dz and also in terms of the new coordinate differentials. Show in each case that the two results are equal.

(a) $x = r \cos \theta,\quad y = r \sin \theta,\quad z = z$

(b) $x = \rho \sin \phi \cos \theta,\quad y = \rho \sin \phi \sin \theta,\quad z = \rho \cos \phi$

(c) $x = t + u + v,\quad y = t + u - v,\quad z = t - u - v$

(d) $x = \sin u \cos v,\quad y = \sin u \sin v,\quad z = \cos u$

(e) $x = u + v,\quad y = u - v,\quad z = u^2 + v^2$

(f) $x = tu,\quad y = uv,\quad z = tv$

(g) $x = \cos (u + v),\quad y = \sin (u + v),\quad z = \cos (u - v)$

(h) $x = \cos t,\quad y = \sin t,\quad z = t$

(i) $x = t^2,\quad y = 1 - t^3,\quad z = e^{-t}$

(j) $x = t^2 + u^2 + v^2,\quad y = t^2 + u^2 - v^2,\quad z = t^2 - u^2 - v^2$

(k) $x = \sqrt{u^2 + v^2},\quad y = \arctan (v/u),\quad z = z$

5. In Exercises 3 and 4 we were required to check the equality of certain variables. Give a general explanation why all these equalities must hold.

6. Let x, y, and z be coordinates on a three-dimensional manifold M. Let $w = \phi(x, y, z)$ on M. Let $z = f(x, y)$ determine a two-dimensional submanifold K of M, and let $y = g(x)$ determine a one-dimensional submanifold L of K. Describe the largest domain on which each of the following variables is defined.

(a) $\left(\dfrac{\partial w}{\partial x}\right)_{yz}$

(b) $\left(\dfrac{\partial w}{\partial x}\right)_{y}$

(c) $\left(\dfrac{\partial z}{\partial y}\right)_{x}$

(d) dw

(e) $D_z w$

(f) dx

(g) $\left(\dfrac{\partial w}{\partial x}\right)_{y} dx + \left(\dfrac{\partial w}{\partial y}\right)_{z} dy$

(h) $D_x w\, dx$

7. Given the situation in Exercise 6, each of the following equations is a consequence of the fundamental theorem on differentials. Explain why in each case, and in each case describe the domain on which the equation holds.

(a) $\left(\dfrac{\partial w}{\partial x}\right)_{yz} dx + \left(\dfrac{\partial w}{\partial y}\right)_{xz} dy + \left(\dfrac{\partial w}{\partial z}\right)_{xy} dz = \left(\dfrac{\partial w}{\partial x}\right)_{y} dx + \left(\dfrac{\partial w}{\partial y}\right)_{x} dy$

(b) $\left(\dfrac{\partial w}{\partial x}\right)_{yz} dx + \left(\dfrac{\partial w}{\partial y}\right)_{xz} dy + \left(\dfrac{\partial w}{\partial z}\right)_{xy} dz = D_x w\, dx$

(c) $\left(\dfrac{\partial z}{\partial x}\right)_{y} dx + \left(\dfrac{\partial z}{\partial y}\right)_{x} dy = D_x z\, dx$

8. Let
$$z = (x^3 + y^3)^{1/3}.$$

(a) Show that at $(0, 0)$,
$$\left(\dfrac{\partial z}{\partial x}\right)_{y} = \left(\dfrac{\partial z}{\partial y}\right)_{x} = 1$$
so that formally
$$dz = dx + dy.$$

(b) Substitute $x = t$, $y = t$, and show that on this submanifold
$$dz = \sqrt[3]{2}\, dt.$$

(c) Contrast the results in parts (a) and (b), and explain the discrepancy. [*Hint:* In item (ii), it is assumed that f has continuous third partial derivatives.]

(d) Sketch the surface involved here, and explain the difficulty geometrically. [*Hint:* Look for a tangent plane at the origin.]

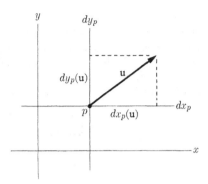

FIGURE 5–3

5–5 Geometric representations. Algebraically, each tangent space to an n-dimensional manifold is an n-dimensional vector space. If $u_1, u_2, \ldots, u_n$ is a coordinate set on the manifold, then

$$\frac{\partial}{\partial u_1}, \frac{\partial}{\partial u_2}, \ldots, \frac{\partial}{\partial u_n} \tag{37}$$

is a basis in the tangent space. Since, by definition,

$$du_i \left(\frac{\partial}{\partial u_j} \right) = \frac{\partial u_i}{\partial u_j} = \delta_{ij},$$

it appears that the differentials

$$du_1, du_2, \ldots, du_n$$

are the component variables with respect to the basis (37).

Let us agree that if the manifold is Euclidean n-space with $x_1, x_2, \ldots, x_n$ as rectangular coordinates, then the abstract tangent spaces of derivative operators will always be represented concretely by making

$$\frac{\partial}{\partial x_1}, \frac{\partial}{\partial x_2}, \ldots, \frac{\partial}{\partial x_n}$$

an orthonormal basis with $\partial/\partial x_i$ parallel to the x_i-axis in the manifold. Then, dx_i will always measure components in the x_i-direction; this is illustrated, for $n = 2$, in Fig. 5–3.

In Fig. 5–3, x and y are rectangular coordinates on the plane. The tangent plane T_p coincides with the original plane but has its origin at p, so dx_p and dy_p measure directed lengths of the projections of vectors onto the axes shown.

Let

$$X = \begin{bmatrix} x_1 \\ \vdots \\ x_n \end{bmatrix}, \qquad U = \begin{bmatrix} u_1 \\ \vdots \\ u_n \end{bmatrix},$$

with the understanding that the X-coordinates are rectangular and U is another coordinate set. The geometric significance of dX has been determined. To find that of dU, proceed as follows. First, note that, by the fundamental theorem on differentials,

$$dx_i = \sum_{j=1}^{n} \frac{\partial x_i}{\partial u_j} \, du_j.$$

This is the computation that verifies the matrix equation

$$\begin{bmatrix} dx_1 \\ \vdots \\ dx_n \end{bmatrix} = \begin{bmatrix} \dfrac{\partial x_1}{\partial u_1} & \cdots & \dfrac{\partial x_1}{\partial u_n} \\ \vdots & & \vdots \\ \dfrac{\partial x_n}{\partial u_1} & \cdots & \dfrac{\partial x_n}{\partial u_n} \end{bmatrix} \begin{bmatrix} du_1 \\ \vdots \\ du_n \end{bmatrix}. \tag{38}$$

The square matrix of partial derivatives that appears here will be denoted by

$$D_U X.$$

It is called the *Jacobian matrix* of X with respect to U. Incidentally, to distinguish a Jacobian matrix from its transpose, note that

$$D_U X = \left[\frac{\partial x_i}{\partial u_j} \right] = \left[dx_i \left(\frac{\partial}{\partial u_j} \right) \right].$$

In line with the mechanical rule described in Section 4–6, the vectors $\partial/\partial u_j$ determine the columns, and the component variables dx_i generate the rows.

In this notation, (38) becomes

$$dX = D_U X \, dU. \tag{39}$$

The importance of (39) cannot be overemphasized. It is the basic relation in the multidimensional calculus of variables. Differential calculus replaces nonlinear transformation by linear transformation. Given only a differentiable coordinate transformation from U to X, (39) yields in each tangent space a linear change of basis.

Recall (see equations (11) and (12), Section 4–2) that for the change of basis given by (39), the dX-components of the basis vectors for dU appear in the columns of $D_U X$. In general, this basis will not be orthonormal, but in many important cases the vectors will be orthogonal so that appropriate multipliers will yield an orthonormal basis. To insert such multipliers by matrix algebra, one proceeds as follows. Let a_j be the length of the vector given by column j of $D_U X$; that is,

$$a_j = \sqrt{\sum_{i=1}^{n} (\partial x_i/\partial u_j)^2}.$$

Form the diagonal matrix,

$$A = \begin{bmatrix} a_1 & & 0 \\ & \ddots & \\ 0 & & a_n \end{bmatrix}.$$

Then (check by direct computation)

$$A^{-1} = \begin{bmatrix} \dfrac{1}{a_1} & & 0 \\ & \ddots & \\ 0 & & \dfrac{1}{a_n} \end{bmatrix},$$

and the jth column of

$$D_U X A^{-1}$$

is that of $D_U X$ divided by a_j. So insert $A^{-1}A$ into (39) to get

$$dX = (D_U X A^{-1})(A \, dU).$$

Here $D_U X A^{-1}$ displays the new orthonormal basis in its columns; $A \, dU$ is a new orthonormal component matrix; and the rotation from the dX-axes to the $A \, dU$-axes is given by the orthogonal matrix $D_U X A^{-1}$.

Apply this procedure to the case of polar coordinates in the plane. Let

$$X = \begin{bmatrix} x \\ y \end{bmatrix}, \qquad U = \begin{bmatrix} r \\ \theta \end{bmatrix}.$$

Since

$$x = r \cos \theta, \qquad y = r \sin \theta,$$

$$D_U X = \begin{bmatrix} \dfrac{\partial x}{\partial r} & \dfrac{\partial x}{\partial \theta} \\ \dfrac{\partial y}{\partial r} & \dfrac{\partial y}{\partial \theta} \end{bmatrix} = \begin{bmatrix} \cos \theta & -r \sin \theta \\ \sin \theta & r \cos \theta \end{bmatrix}.$$

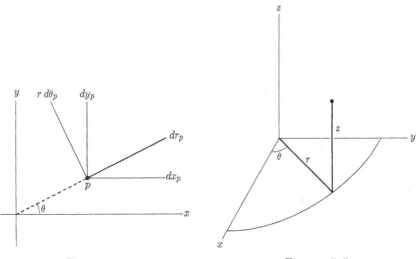

FIGURE 5–4   FIGURE 5–5

Column 1 has length 1, and column 2 has length r, and the two columns represent orthogonal vectors; so set

$$A = \begin{bmatrix} 1 & 0 \\ 0 & r \end{bmatrix}.$$

Then,

$$\begin{bmatrix} dx \\ dy \end{bmatrix} = dX = D_U X \, dU = D_U X A^{-1} A \, dU$$

$$- \begin{bmatrix} \cos\theta & -r\sin\theta \\ \sin\theta & r\cos\theta \end{bmatrix} \begin{bmatrix} 1 & 0 \\ 0 & 1/r \end{bmatrix} \begin{bmatrix} 1 & 0 \\ 0 & r \end{bmatrix} \begin{bmatrix} dr \\ d\theta \end{bmatrix}$$

$$= \begin{bmatrix} \cos\theta & -\sin\theta \\ \sin\theta & \cos\theta \end{bmatrix} \begin{bmatrix} dr \\ r\,d\theta \end{bmatrix}.$$

Thus, dr and $r\,d\theta$ are orthonormal component variables, and the rotation from the (dx, dy)-axes to the $(dr, r\,d\theta)$-axes is through an angle θ. A sketch is shown in Fig. 5–4.

In addition to the rectangular coordinate system, there are two important coordinate systems in three-space.

(i) *Cylindrical coordinates*, r, θ, z are polar coordinates in the xy-plane, together with the rectangular coordinate z (see Fig. 5–5). Clearly, the transformation equations are

$$x = r \cos\theta, \qquad y = r \sin\theta, \qquad z = z.$$

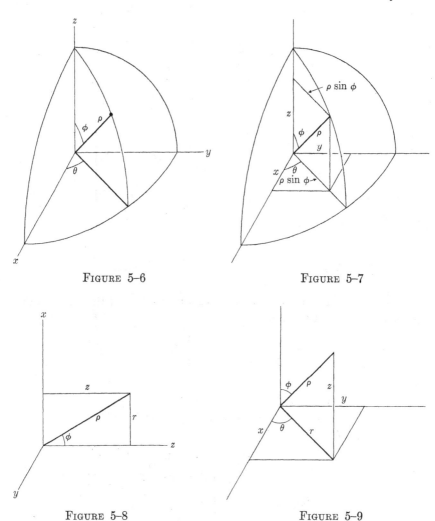

FIGURE 5–6   FIGURE 5–7

FIGURE 5–8   FIGURE 5–9

(ii) *Spherical coordinates*, ρ, ϕ, θ, are variables measuring, respectively, radial distance from the origin, angle from the positive z-axis (colatitude), and angle from the positive x-side of the xz-plane (longitude). (See Fig. 5–6.) It appears from Fig. 5–7 that

$$x = \rho \sin \phi \cos \theta,$$

$$y = \rho \sin \phi \sin \theta,$$

$$z = \rho \cos \phi.$$

By way of comparing cylindrical and spherical coordinates, note (Fig. 5–8) that in the plane on which $\theta = 0$—looked at from the positive y-side—(z, r) are rectangular coordinates and (ρ, ϕ) are polar coordinates. Thus, the following transformation equations are indicated:

$$r = \rho \sin \phi,$$

$$\theta = \theta,$$

$$z = \rho \cos \phi.$$

In the notation introduced here, there are seven symbols: x, y, z, r, θ, ρ, ϕ. Each of these has a fixed geometric definition—all seven are shown in Fig. 5–9. Although z and θ each appear in two of the standard coordinate systems, each has the same meaning in both systems. Furthermore, the orderings (x, y, z), (r, θ, z), (ρ, ϕ, θ) will generate right-handed differential systems. These points are mentioned because the student will find many sources of confusion in the literature. The most common ones are: (a) replacing ρ here by r and thus using r in a double role, meaning one thing when it is a cylindrical coordinate and another thing when it is a spherical coordinate; (b) interchanging the roles of ϕ and θ in the spherical system thus giving θ a double meaning, one in the cylindrical system and another in the spherical; (c) listing spherical coordinates in the order, radius, longitude, colatitude, thus generating a left-handed system, or at least, one with opposite orientation from x, y, z. The student is advised to check definitions carefully in each new work he reads.

A good way to get acquainted with a new coordinate system is to look at the level surfaces for the coordinates. It is left to the student to check that for the new coordinate variables introduced here the level surfaces may be characterized as follows:

r, cylinder,
θ, plane,
ρ, sphere,
ϕ, cone.

Finally, it should be noted that the only cases discussed at this time are those in which the manifold is the entire space. The picture of the tangent bundle is obscure here because the tangent spaces are just translations of the original manifold. However, for these flat manifolds, the geometric significance of differentials is relatively easy to study, and the present section is devoted *only* to these cases. In particular, an important example of a two-dimensional manifold set is a surface in three-space. Note that such examples are not covered by the present discussion; they will be studied later in Section 10–3.

EXAMPLES

1. Determine the geometric significance of the spherical-coordinate differentials. Let

$$X = \begin{bmatrix} x \\ y \\ z \end{bmatrix}, \qquad U = \begin{bmatrix} \rho \\ \phi \\ \theta \end{bmatrix}.$$

Here,

$$D_U X = \begin{vmatrix} \dfrac{\partial x}{\partial \rho} & \dfrac{\partial x}{\partial \phi} & \dfrac{\partial x}{\partial \theta} \\[2mm] \dfrac{\partial y}{\partial \rho} & \dfrac{\partial y}{\partial \phi} & \dfrac{\partial y}{\partial \theta} \\[2mm] \dfrac{\partial z}{\partial \rho} & \dfrac{\partial z}{\partial \phi} & \dfrac{\partial z}{\partial \theta} \end{vmatrix} = \begin{bmatrix} \sin \phi \cos \theta & \rho \cos \phi \cos \theta & -\rho \sin \phi \sin \theta \\ \sin \phi \sin \theta & \rho \cos \phi \sin \theta & \rho \sin \phi \cos \theta \\ \cos \phi & -\rho \sin \phi & 0 \end{bmatrix}.$$

The columns represent orthogonal vectors; for example, the scalar product of the first two is:

$$\rho \sin \phi \cos \phi \cos^2 \theta + \rho \sin \phi \cos \phi \sin^2 \theta - \rho \sin \phi \cos \phi$$
$$= \rho \sin \phi \cos \phi (\cos^2 \theta + \sin^2 \theta) - \rho \sin \phi \cos \phi = 0.$$

Checking the other two pairs is left to the student. The lengths of these vectors are as follows:

column 1: 1,
column 2: ρ,
column 3: $\rho \sin \phi$.

Hence, let

$$A = \begin{bmatrix} 1 & 0 & 0 \\ 0 & \rho & 0 \\ 0 & 0 & \rho \sin \phi \end{bmatrix},$$

then $D_U X A^{-1}$ is orthogonal, and

$$A \, dU = \begin{bmatrix} d\rho \\ \rho \, d\phi \\ \rho \sin \phi \, d\theta \end{bmatrix}$$

generates orthonormal component variables on the tangent spaces. The

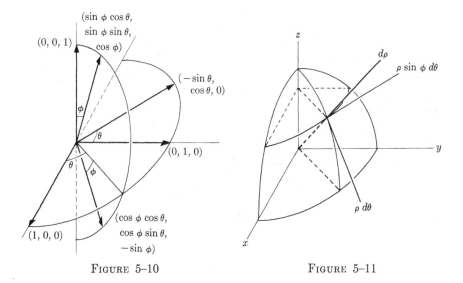

FIGURE 5–10 FIGURE 5–11

orthogonal transformation matrix

$$D_U X A^{-1} = \begin{bmatrix} \sin\phi\cos\theta & \cos\phi\cos\theta & -\sin\theta \\ \sin\phi\sin\theta & \cos\phi\sin\theta & \cos\theta \\ \cos\phi & -\sin\phi & 0 \end{bmatrix},$$ (40)

has determinant $+1$, thus it is a rotation matrix. The basis vectors for $d\rho$, $\rho\,d\phi$, $\rho\sin\phi\,d\theta$ are displayed in the columns of (40) and are sketched in Fig. 5–10. The axis system for these differentials is shown in Fig. 5–11.

2. To picture partial derivatives as quotients of differentials, note that a partial derivative involves holding one variable constant. Draw the one-dimensional manifold determined by holding this variable constant, and draw its tangent line. Then, draw a right triangle with a portion of this tangent line as hypotenuse, and label the sides of this right triangle with the appropriate distance-measuring differentials—dx, dy, dr, or $r\,d\theta$ depending on direction. The partial derivative of one coordinate with respect to another can be read off from such a triangle. For example,

$$\left(\frac{\partial x}{\partial r}\right)_\theta = \cos\theta \quad \text{(Fig. 5–12)}, \qquad \left(\frac{\partial x}{\partial \theta}\right)_r = -r\sin\theta \quad \text{(Fig. 5–13)},$$

$$\left(\frac{\partial r}{\partial x}\right)_y = \cos\theta \quad \text{(Fig. 5–14)}, \qquad \left(\frac{\partial r}{\partial \theta}\right)_x = x\sec\theta\tan\theta \quad \text{(Fig. 5–15)}.$$

FIGURE 5–12

FIGURE 5–13

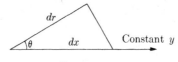

FIGURE 5–14

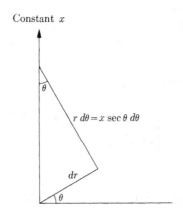

FIGURE 5–15

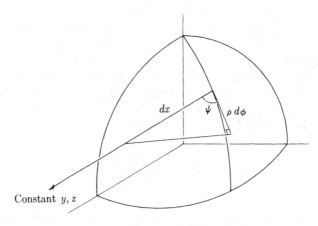

FIGURE 5–16

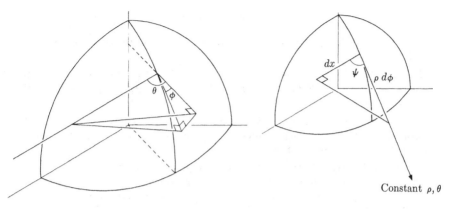

Constant ρ, θ

FIGURE 5–17	FIGURE 5–18

3. Picture geometrically as quotients of differentials

$$\left(\frac{\partial\phi}{\partial x}\right)_{yz} \qquad \text{and} \qquad \left(\frac{\partial\phi}{\partial x}\right)_{\rho\theta}.$$

Along a (y, z)-constant line, dx measures actual distances. On the other hand, $\rho\,d\phi$ measures projections of distances onto the tangent line to a vertical circle. Such projections are constructed by means of the right triangle shown in Fig. 5–16. It appears from this that if all differentials are evaluated at a point on the (y, z)-constant line, then

$$\frac{\rho\,d\phi}{dx} = \cos\psi.$$

To find $\cos\psi$ in terms of the coordinate variables, note (Fig. 5–17) that projection through the angle ψ is equivalent to projection through an angle θ followed by one through an angle ϕ (Fig. 5–17). Thus,

$$\cos\psi = \cos\phi\cos\theta, \qquad \text{and} \qquad \left(\frac{\partial\phi}{\partial x}\right)_{yz} = \frac{\cos\phi\cos\theta}{\rho}.$$

To find the other partial derivative called for, differentials must be evaluated on the tangent line to a (ρ, θ)-constant circle. On this line $\rho\,d\phi$ measures actual distances, and dx measures projections as indicated by the right triangle in Fig. 5–18. Thus,

$$\frac{dx}{\rho\,d\phi} = \cos\psi,$$

where ψ is the same angle as before; so,

$$\left(\frac{\partial\phi}{\partial x}\right)_{\rho\theta} = \frac{1}{\rho\cos\phi\cos\theta}.$$

EXERCISES

1. Determine the geometric significance of the cylindrical-coordinate differentials.

2. Let $x = u^2 - v^2$, $y = 2uv$ define a coordinate transformation on the plane. Find multiples of du and dv that will give rectangular components on the tangent planes.

3. Solve Exercise 1, Section 5–3, by purely geometric means. That is, picture appropriate distance-measuring differentials, and find each of the partial derivatives called for as a quotient of differentials.

4. Draw differential triangles to picture as quotients of differentials
   (a) each partial derivative of a spherical coordinate with respect to a rectangular coordinate; other rectangular coordinates constant,
   (b) each partial derivative of a spherical coordinate with respect to a rectangular coordinate; other spherical coordinates constant,
   (c) each partial derivative of a rectangular coordinate with respect to a spherical coordinate; other spherical coordinates constant,
   (d) each partial derivative of a rectangular coordinate with respect to a spherical coordinate; other rectangular coordinates constant.

5. Repeat the plan of Exercise 4 for cylindrical and spherical coordinates.

6. Let s measure arc length on a space curve C. By definition (Section 2–2), ds measures distances on the tangent lines to C.
   (a) From the results in this section show that $ds^2 = dx^2 + dy^2 + dz^2$.
   (b) By direct substitution in the result of part (a), find ds^2 in terms of cylindrical and spherical coordinates and their differentials.
   (c) Use the results of Exercise 1 and Example 1 to derive geometrically the results in part (b).

5–6 Foundations of the calculus of variables. The postulates for a manifold are so set up that if one says, "Let x and y be admissible coordinates on a one-dimensional manifold," then one essentially has *carte blanche* to do as he wishes with dx and dy. This is one of the strong points of the calculus of variables; algebra with differentials yields reliable results because any derivatives involved have already been assumed to exist.

There is, however, the obvious question, Given a specific model, how can one tell that it satisfies the postulates? For the most part, this question has been by-passed so far, although an answer (without proof or any indication thereof) was given in Section 2–3. The time has now come, not for a complete proof, but for a more detailed discussion of the matter.

Specifically, let M be the locus of $\phi(x, y) = 0$, and let L be the portion of M contained in some neighborhood. In terms of properties of ϕ, there are conditions guaranteeing that
   (i) x is continuous and one to one on L;
   (ii) the range of x (restricted to L) is an interval;
   (iii) $y = f(x)$ on L where f has three derivatives.

A check with the postulates (Section 2–3) will show that if these results are established, then L is a neighborhood on a manifold. The only question that remains is that of postulate (e). Is the whole locus M connected? The present discussion will deal with local properties only. Where connectedness is important (say, in connection with line integrals) it must be investigated for each specific example.

Informally, the result (iii) above seems to say that $\phi(x, y) = 0$ can be solved for y. Actually, a considerable portion of the whole problem can be phrased in terms of "solvability" of equations or systems of equations. First, consider the following situation. Let

$$u = \phi(x, y), \qquad v = \psi(x, y) \tag{41}$$

relate variables u and v to variables x and y over a portion of the plane. Assuming that the fundamental theorem on differentials applies, one has

$$du = \frac{\partial u}{\partial x}\,dx + \frac{\partial u}{\partial y}\,dy, \qquad dv = \frac{\partial v}{\partial x}\,dx + \frac{\partial v}{\partial y}\,dy. \tag{42}$$

This system of linear equations is solvable for dx and dy provided that

$$\begin{vmatrix} \dfrac{\partial u}{\partial x} & \dfrac{\partial u}{\partial y} \\[2mm] \dfrac{\partial v}{\partial x} & \dfrac{\partial v}{\partial y} \end{vmatrix} \neq 0. \tag{43}$$

That is, the system (42) is solvable on T_p provided that (43) holds at p. Now, it might be suspected that if (42) is solvable on T_p, then (41) is solvable in some neighborhood of p. This is, indeed, the case; and the following basic theorem (the proof is omitted) gives this and other pertinent information.

INVERSION THEOREM. Let x and y be coordinates at a point p_0 in a two-dimensional manifold. In some neighborhood N_0 for p_0, let

$$u = \phi(x, y), \qquad v = \psi(x, y),$$

where each of the functions ϕ and ψ has continuous first partial derivatives. Suppose that

$$\begin{vmatrix} \dfrac{\partial u}{\partial x} & \dfrac{\partial u}{\partial y} \\[2mm] \dfrac{\partial v}{\partial x} & \dfrac{\partial v}{\partial y} \end{vmatrix} \neq 0$$

at p_0. Given these conditions, the following results hold:

(a) There is a neighborhood N of p_0, and there are uniquely determined functions F and G such that over N,

$$x = F(u, v), \qquad y = G(u, v).$$

(b) If the functions ϕ and ψ have continuous partial derivatives of order k, then so do the functions F and G.

(c) There is an $\epsilon > 0$ such that the common domain of F and G contains all number pairs (a, b) for which

$$|u(p_0) - a| < \epsilon \qquad \text{and} \qquad |v(p_0) - b| < \epsilon.$$

Grant, now, that the plane with rectangular coordinates is a two-dimensional manifold. A postulational characterization of the plane would make this so. The inversion theorem may be used to establish (i), (ii), and (iii) above as follows. Set

$$u = \phi(x, y),$$

and let M be the locus of $u = 0$. Specialize ψ in the inversion theorem so that

$$v = \psi(x, y) = x.$$

Obviously, ψ has continuous partial derivatives; assume that ϕ does. To apply the inversion theorem, assume that at some point p_0 of M,

$$\begin{vmatrix} \dfrac{\partial u}{\partial x} & \dfrac{\partial u}{\partial y} \\[2mm] \dfrac{\partial v}{\partial x} & \dfrac{\partial v}{\partial y} \end{vmatrix} = \begin{vmatrix} \dfrac{\partial u}{\partial x} & \dfrac{\partial u}{\partial y} \\[2mm] 1 & 0 \end{vmatrix} = -\dfrac{\partial u}{\partial y} = -\phi_2(x, y) \neq 0.$$

By part (a) of the inversion theorem, there is a unique G such that

$$y = G(u, v) = G(u, x)$$

on some neighborhood of p_0. If L is the portion of M in this neighborhood, then $u = 0$ on L, and

$$y = G(0, x)$$

on L. Define f by $f(x) = G(0, x)$. By part (b) of the inversion theorem, f has three derivatives if ϕ has continuous third partial derivatives, and (iii) is established. Since x is a coordinate on the plane, it is already continuous. If a is a value assumed by x on L, then $y = f(a)$, and the point in question is uniquely determined; thus x is one to one on L, and (i) is established. Finally, the domain of G contains all number pairs $(0, a)$, where a is in the range of x on L. By part (c) of the inversion theorem, this domain includes all number pairs $(0, a)$, where $|x(p_0) - a| < \epsilon$; hence the range of x contains this interval, and if necessary, x may be further restricted so that its range is this interval. This establishes (ii) and completes the analytic characterization of a one-dimensional manifold.

Summary. If ϕ has continuous third partial derivatives and if at some point p_0 on the locus of $\phi(x, y) = 0$, $\phi_2(x, y) \neq 0$, then the part of the locus in some neighborhood of p_0 is a one-dimensional manifold with x as an admissible coordinate and y as a differentiable variable.

This procedure can be generalized to any number of dimensions. First, note the complete set of postulates for a k-dimensional manifold. Let M be a space in which distance is defined, and let p be a point of M. A set $x_1, x_2, \ldots, x_k$ of variables is called a *coordinate set in M at p* provided that the following conditions are satisfied:

   (a) The variables x_i have a common domain which is a subset of M containing a neighborhood of p in M.

   (b) The range of each x_i is an interval.

   (c) Each x_i is continuous and the mapping

$$q \rightarrow [x_1(q), x_2(q), \ldots, x_k(q)]$$

of points into k-tuples of numbers is one to one.

A *k-dimensional manifold* is a structure consisting of a point set M and a set V of variables defined on subsets of M for which the following conditions are satisfied:

   (d) For each point p in M, some set of k variables from V is a coordinate set in M at p.

   (e) Any two points of M can be connected by a finite chain of overlapping neighborhoods each of which is contained in the common domain of one of these coordinate sets.

   (f) If $y, x_1, x_2, \ldots, x_k$ is a set of $k + 1$ variables from V with some neighborhood common to all their domains, then there is a function f with continuous third partial derivatives such that

$$y = f(x_1, x_2, \ldots, x_k),$$

identically, on this neighborhood.

Rephrase the analytic characterization of the one-dimensional case as follows. Let $u = \phi(x, y)$ where ϕ has continuous third partial derivatives, and consider the locus of $u = 0$. The portion near a point where $\partial u/\partial y \neq 0$ is a manifold with the *other* variable, x, as coordinate.

The multidimensional generalization is as follows. Let $x_1, x_2, \ldots, x_n$, $t_1, t_2, \ldots, t_k$ be rectangular coordinates in $(n + k)$-space; let

$$u_i = \phi^i(x_1, x_2, \ldots, x_n, t_1, t_2, \ldots, t_k) \qquad i = 1, 2, \ldots, n, \qquad (44)$$

and let M be the locus of

$$u_i = 0 \qquad i = 1, 2, \ldots, n.$$

Suppose that each ϕ^i has continuous third partial derivatives, and suppose that at some point p_0 of M the $n \times n$ determinant

$$\det \left[\frac{\partial u_i}{\partial x_j} \right] \neq 0.$$

Then, some neighborhood of p_0 in M is a k-dimensional manifold with the remaining variables, $t_1, t_2, \ldots, t_k$, forming a coordinate set.

This is proved from the $(n + k)$-dimensional case of the inversion theorem by a procedure analogous to that used above for the case $n = k = 1$. We will not repeat the details of this procedure.

The following abbreviated notation is useful for summarizing this result. Take three matrices of variables with dimensions as noted:

$$U(n \times 1), \qquad X(n \times 1), \qquad T(k \times 1).$$

Assume that X and T together give a coordinate set on $(n + k)$-space and that the variables in U have continuous third partial derivatives with respect to these coordinates. Near a point where

$$U = 0 \qquad \text{and} \qquad \det (D_X U) \neq 0,$$

the locus of $U = 0$ is a k-dimensional manifold with T as an admissible coordinate set and X as a set of differentiable variables.

The determinant of partial derivatives,

$$\det (D_X U),$$

which has appeared here in several places, is the determinant of the Jacobian matrix $D_X U$ (see Section 5–5). It is usually called simply the *Jacobian* of $u_1, u_2, \ldots, u_n$ with respect to $x_1, x_2, \ldots, x_n$. The usual notation for this determinant is

$$\frac{\partial(u_1, u_2, \ldots, u_n)}{\partial(x_1, x_2, \ldots, x_n)}. \tag{45}$$

The notations $\det (D_X U)$ and $\det [\partial u_i / \partial x_j]$ will be the preferred ones here, because, if one uses (45) to denote the determinant, there is no obvious associated notation for the matrix $D_X U$ itself. In most treatments of calculus, this presents no problem because the importance of the Jacobian matrix is not generally recognized. It has already been pointed out (Section 5–5) that this matrix generates a linear transformation on the tangent space, and this fact will assume greater importance in later developments. However, the reader should be apprised of the fact that, in most of the literature, (45) is the notation for det $(D_X U)$ and there is no notation for $D_X U$.

EXERCISES

1. *Parametric equations.* Suppose that f and g have continuous third derivatives. Show that the locus of

$$x = f(t), \qquad y = g(t)$$

(in xyt-space) is a one-dimensional manifold with t as a coordinate and x and y as differentiable variables. *Hint:* Set

$$u = x - f(t), \qquad v = y - g(t),$$

$$U = \begin{bmatrix} u \\ v \end{bmatrix}, \qquad X = \begin{bmatrix} x \\ y \end{bmatrix}, \qquad T = [t].$$

2. *Composite functions.* Let

$$z = f(x, y),$$

and introduce the coordinate transformation

$$x = \phi(u, v), \qquad y = \psi(u, v);$$

then

$$z = g(u, v).$$

Show that if f, ϕ, and ψ have continuous third partial derivatives, then g also has continuous third partial derivatives. *Hint:* Set

$$w_1 = x - \phi(u, v), \qquad w_2 = y - \psi(u, v), \qquad w_3 = z - f(x, y),$$

$$U = \begin{bmatrix} w_1 \\ w_2 \\ w_3 \end{bmatrix}, \qquad X = \begin{bmatrix} x \\ y \\ z \end{bmatrix}, \qquad T = \begin{bmatrix} u \\ v \end{bmatrix}.$$

3. *Admissible coordinates.* Let

$$X = \begin{bmatrix} x \\ y \end{bmatrix}$$

be a coordinate set on a two-dimensional manifold. Define

$$U = \begin{bmatrix} u \\ v \end{bmatrix}$$

to be an admissible coordinate set provided that

$$x = \phi(u, v), \qquad y = \psi(u, v),$$

where ϕ and ψ have continuous third partial derivatives and provided that

$$\det D_U X \neq 0.$$

(a) Use the inversion theorem to show that this makes U a coordinate set (guarantees that it determines points uniquely).

(b) A variable z is called differentiable if

$$z = f(x, y),$$

where f has continuous third partial derivatives. Use Exercise 2 to show that the class of differentiable variables is the same for all admissible coordinate sets.

4. Use Exercise 3 and its three-dimensional analogue to determine at what points polar, cylindrical, and spherical coordinates are admissible. More particularly, list all points at which they are *not* admissible.

CHAPTER 6

APPLICATIONS

6–1 Geometry in polar coordinates. Suppose that $r = f(\theta)$ on a one-dimensional manifold M in the plane. In terms of derivatives of r with respect to θ, how does one find slope, arc length, and curvature of M?

The most effective tool for studying the geometry of a plane curve is a differential triangle. The one shown in Fig. 6–1 is a duplicate of the one given in Fig. 2–3. Now, Fig. 2–3 was arrived at by postulating that it represented ds correctly and then by deducing from the theory of one-dimensional manifolds that it also pictured dx and dy correctly. On the other hand, in Section 5–5 it was postulated that Fig. 6–1 pictured dx and dy correctly. Since the two pictures are the same, the two sets of assumptions (Sections 2–2 and 5–5) about geometric significance of differentials must be consistent.

Thus, one is justified in taking the geometric picture of dr and $d\theta$ derived in Section 5–5 and combining it with the picture of ds in Section 2–2. The result is Fig. 6–2, and this is the basis for the study of geometry of plane curves in terms of polar coordinates.

From Fig. 6–2 come the key formulas:

$$ds = \sqrt{1 + r^2 \left(\frac{d\theta}{dr}\right)^2}\, dr, \tag{1}$$

$$ds = \sqrt{r^2 + \left(\frac{dr}{d\theta}\right)^2}\, d\theta, \tag{2}$$

$$\cos\psi = \frac{dr}{ds}, \quad \sin\psi = r\frac{d\theta}{ds}, \quad \tan\psi = r\frac{d\theta}{dr}, \quad \cot\psi = \frac{1}{r}\frac{dr}{d\theta}.$$

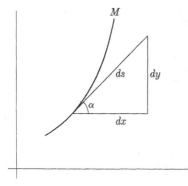

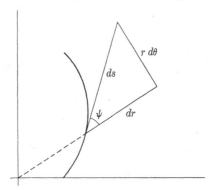

FIGURE 6–1   FIGURE 6–2

The distinction between (1) and (2) is one of orientation of the manifold. These expressions for ds may differ in sign. Note that here the derivatives are related to the angle ψ between the radial direction and the tangential direction to the curve. The old slope angle, α, is related to ψ by

$$\alpha = \psi + \theta.$$

As the examples below will show, one can now find, from a polar-coordinate representation of a curve, the same geometric quantities that have previously been found from rectangular representations. The analytic forms will, of course, be different.

EXAMPLES

1. Sketch the loci of

$$r = \sin \theta \quad \text{and} \quad r = \sin 2\theta,$$

and find their angles of intersection. A suggestion for sketching polar coordinate curves is as follows: (a) Find points at which r has maxima and minima. Frequently, these may be found by inspection. If necessary, set $dr/d\theta = 0$ to find them. (b) List all values of θ for which $r = 0$. These values indicate angles at which the curve passes through the origin. (c) Draw the indicated points and tangent lines through the origin and connect them in the order of increasing θ.

For $r = \sin \theta$, maximum r is at $(1, \pi/2)$, minimum r is at $(-1, 3\pi/2)$, and $r = 0$ for $\theta = 0$, π, and 2π. The points $(1, \pi/2)$ and $(-1, 3\pi/2)$ are the same point, $(0, 1)$ in rectangular coordinates; so, as θ progresses from 0 to 2π, the curve goes through this point twice, and it has a horizontal tangent at the origin. The result is the circle in Fig. 6–3.

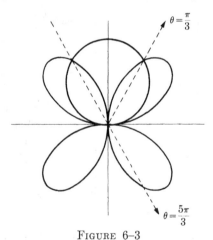

FIGURE 6–3

For $r = \sin 2\theta$, maximum r is at $(1, \pi/4)$ and $(1, 5\pi/4)$, minimum r is at $(-1, 3\pi/4)$ and $(-1, 7\pi/4)$, and $r = 0$ for $\theta = 0, \pi/2, \pi, 3\pi/2,$ and 2π. The maxima and minima for r give four extreme points on the curve. Between each pair of points, the curve loops through the origin with either a horizontal or vertical tangent. The result is the "four-leaved rose" curve in Fig. 6–3.

To find the points at which these curves intersect, solve the equations simultaneously:

$$\sin \theta = \sin 2\theta = 2 \sin \theta \cos \theta,$$

$$\sin \theta (1 - 2 \cos \theta) = 0,$$

$$\theta = 0, \frac{\pi}{3}, \pi, \frac{5\pi}{3}.$$

For the first curve,

$$dr = \cos \theta \, d\theta, \qquad \frac{d\theta}{dr} = \sec \theta;$$

$$\tan \psi_1 = \frac{r \, d\theta}{dr} = \sin \theta \sec \theta = \tan \theta.$$

For the second curve,

$$dr = 2 \cos 2\theta \, d\theta, \qquad \frac{d\theta}{dr} = \left(\frac{1}{2}\right) \sec 2\theta,$$

$$\tan \psi_2 = \frac{r \, d\theta}{dr} = \left(\frac{1}{2}\right) \sec 2\theta \sin 2\theta = \left(\frac{1}{2}\right) \tan 2\theta.$$

The angle from the tangent line to the first curve to the tangent line to the second curve is $\psi_2 - \psi_1$, and

$$\tan (\psi_2 - \psi_1) = \frac{\tan \psi_2 - \tan \psi_1}{1 + \tan \psi_1 \tan \psi_2}.$$

Evaluating this expression at the points of intersection of the curves, one finds that the angles of intersection are as follows:

$$\text{at the origin,} \qquad 0;$$

$$\text{at } \theta = \frac{\pi}{3}, \qquad \arctan 3\sqrt{3};$$

$$\text{at } \theta = \frac{5\pi}{3}, \qquad \arctan (-3\sqrt{3}).$$

The choice of positive tangential direction on the curves is purely arbitrary here. However, it is characterized by $dr > 0$ or by $d\theta > 0$, and it must be characterized the same way on both curves. It is left to the student to verify that, for either choice of positive differential, the above results give the angles from the first positive tangential direction to the second. In checking this, recall that if $r < 0$, the $dr > 0$ direction is inward.

2. The equation

$$r = e^{\alpha\theta}$$

describes the so-called equiangular spiral. Note that here

$$dr = ae^{\alpha\theta}\, d\theta = ar\, d\theta,$$

so

$$\tan\psi = \frac{r\, d\theta}{dr} = \frac{1}{a}.$$

That is, every radial line cuts the curve at the same angle.

3. Find the points on the cardioid on which

$$r = 1 - \cos\theta,$$

at which there are horizontal or vertical tangents. For a horizontal tangent,

$$0 = dy = d\,(r\sin\theta) = d\,[\sin\theta(1 - \cos\theta)]$$

$$= d\,[\sin\theta - (\tfrac{1}{2})\sin 2\theta] = (\cos\theta - \cos 2\theta)\, d\theta,$$

$$2\cos^2\theta - \cos\theta - 1 = 0; \qquad \cos\theta = 1, -\tfrac{1}{2}; \qquad \theta = 0, \frac{2\pi}{3}, \frac{4\pi}{3}, 2\pi.$$

Similarly, for a vertical tangent,

$$0 = dx = d\,[\cos\theta(1 - \cos\theta)] = (\sin 2\theta - \sin\theta)\, d\theta,$$

$$\theta = 0, \frac{\pi}{3}, \pi, \frac{5\pi}{3}, 2\pi.$$

There are three vertical and two horizontal tangent points, as shown in Fig. 6–4. The origin is a singular point ($dx = dy = 0$). Behavior of the curve at the origin can be determined by noting that

$$dr = \sin\theta\, d\theta,$$

$$\tan\psi = \frac{r\, d\theta}{dr} = \frac{1 - \cos\theta}{\sin\theta} = \tan\frac{\theta}{2}.$$

Thus, as $\theta \to 0$,

$$\alpha = \psi + \theta \to 0.$$

There is a horizontal tangent cusp point at the origin.

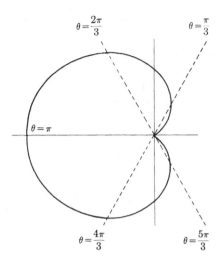

$\theta = \dfrac{2\pi}{3}$

$\theta = \dfrac{\pi}{3}$

$\theta = \pi$

$\theta = \dfrac{4\pi}{3}$

$\theta = \dfrac{5\pi}{3}$

FIGURE 6-4

4. Find the length of the cardioid in Example 3. Here $dr = \sin\theta\, d\theta$; so

$$ds^2 = dr^2 + r^2\, d\theta^2 = [\sin^2\theta + (1 - \cos\theta)^2]\, d\theta^2$$

$$= 2(1 - \cos\theta)\, d\theta = 4\sin^2\frac{\theta}{2}\, d\theta^2.$$

Thus, the length of the curve is

$$\int_{\theta=0}^{\theta=2\pi} ds = \int_0^{2\pi} 2\left|\sin\frac{\theta}{2}\right| d\theta = \int_0^{2\pi} 2\sin\frac{\theta}{2}\, d\theta = 8.$$

EXERCISES

1. From the relations $dy/dx = \tan\alpha$ and $\psi = \alpha - \theta$, make direct substitutions for dx and dy to derive analytically the formula $\tan\psi = r\, d\theta/dr$.

2. Sketch each of the following pairs of curves and find the angles of intersection.

   (a) $r\sin\theta = 2$, $r = 5\cos\theta$ (b) $r = \cos\theta$, $r = \cos 2\theta$

   (c) $r\cos\theta = 2$, $r = \csc^2(\theta/2)$ (d) $r = \sin\theta$, $r = 1 - \sin\theta$

   (e) $r = 3\cos\theta$, $r = 1 + \cos\theta$ (f) $r = \cos\theta$, $r = \sin 2\theta$

   (g) $r^2\sin 2\theta = 1$, $r^2 = 4\sin 2\theta$ (h) $r = 1 - \cos\theta$, $r = 1 + \cos\theta$

3. Suppose that the two curves $r = f_1(\theta)$ and $r = f_2(\theta)$ intersect at (r_0, θ_0). Show that they are perpendicular at that point provided that

$$r_0^2 + f_1'(\theta_0)f_2'(\theta_0) = 0.$$

4. Show that the loci of each of the following pairs of equations intersect at right angles.

(a) $r = \cos\theta$, $r = \sin\theta$

(b) $r = a\theta$, $r\theta = a$

(c) $r = 1 - \cos\theta$, $r = 1 + \cos\theta$

(d) $r^2 \sin 2\theta = a^2$, $r^2 \cos 2\theta = b^2$

(e) $r = a \sec^2(\theta/2)$, $r = b \csc^2(\theta/2)$

(f) $r = \dfrac{1}{1 + \cos\theta}$, $r = \dfrac{1}{1 - \cos\theta}$

5. Let f and g be functions such that

$$f(t_0) = g(t_0) \neq 0.$$

(a) Show that the loci of

$$y = f(x), \qquad y = g(x)$$

intersect at $x = t_0$ at an angle whose tangent is

$$\frac{f'(t_0) - g'(t_0)}{1 + f'(t_0)g'(t_0)}.$$

(b) Show that the loci of

$$r = f(\theta), \qquad r = g(\theta)$$

intersect at $\theta = t_0$ at an angle whose tangent is

$$\frac{f(t_0)g'(t_0) - g(t_0)f'(t_0)}{f'(t_0)g'(t_0) + f(t_0)g(t_0)}.$$

(c) Let $f = \sin$, $g = \cos$. Sketch the rectangular sine and cosine curves and the polar sine and cosine curves, and compare their angles of intersection.

6. Show that, on the tangent bundle to any curve,

$$x \, ds = \cos\theta \sqrt{r^2 \, dr^2 + r^4 \, d\theta^2}, \qquad y \, ds = \sin\theta \sqrt{r^2 \, dr^2 + r^4 \, d\theta^2}.$$

7. For the lemniscate on which

$$r^2 = \cos 2\theta,$$

show that

$$r^2 \, dr^2 + r^4 \, d\theta^2 = 4 \, d\theta^2.$$

Then use Exercise 6 to find

(a) the surface area obtained by revolving the lemniscate about the line on which $\theta = 0$,

(b) the surface area obtained by revolving it about the line on which $\theta = \pi/2$.

8. Sketch both the lemniscates

$$r^2 = \cos 2\theta \qquad \text{and} \qquad r^2 = \sin 2\theta.$$

For the second lemniscate, find the same quantities asked for in Exercise 7 for the first.

9. For each of the following equations, find the length of the locus.

    (a) $r = a \sin \theta$ (b) $r = a(1 + \cos \theta)$

    (c) $r = a \sin^3 (\theta/3),\ 0 \leq \theta \leq \pi$

    (d) $r = a (\sin \theta - \cos \theta)$ (e) $r = 3 \cos \theta + 4 \sin \theta$

10. Show that, for each constant $\alpha > 0$, the locus of

$$r = \theta^{-\alpha}, \qquad \pi \leq \theta < \infty,$$

is a spiral that goes around the origin an infinite number of times. Show that for $\alpha > 1$, this spiral has finite length, and for $\alpha \leq 1$, it has infinite length.

11. Show that the curvature of the locus of $r = f(\theta)$ is given by

$$\frac{r^2 + 2 (D_\theta r)^2 - r\, D_\theta^2 r}{[r^2 + (D_\theta r)^2]^{3/2}}.$$

[*Hint:* Curvature is $d\alpha/ds$, and $\alpha = \psi + \theta$.]

12. Find the radius of curvature at (r, θ) for the locus of each of the following equations.

    (a) $r = a \sin \theta$ (b) $r = e^{a\theta}$

    (c) $r = a\theta$ (d) $r = a(1 - \cos \theta)$

13. A ray is revolving about the origin at an angular speed ω. It strikes the locus of $r = f(\theta)$ at the point p. Show that p is moving in the plane with a speed $r\omega \csc \psi$.

14. Let v be the speed of the point p in Exercise 13. Show that the distance of p from the origin is changing at the rate $v \cos \psi$.

15. A planet follows an elliptic path

$$r = \frac{m}{(1 - \epsilon \cos \theta)}$$

in such a way that

$$\frac{d\theta}{dt} = \frac{h}{m^2} (1 - \epsilon \cos \theta)^2.$$

Find its speed in terms of the constants h, m, and the variable θ.

16. A cam in the shape of a limaçon, the locus of

$$r = 2 + \cos \theta,$$

is revolving with constant angular speed ω about the origin. The follower moves in a vertical slot (Fig. 6–5). In terms of ϕ and ω, find the speed of the follower when the cam has turned through an angle ϕ. [*Hint:* Take a moving frame of reference that revolves with the cam.]

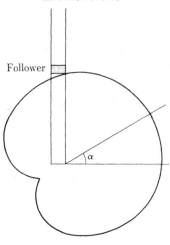

Figure 6–5

6–2 Implicit relations, one-dimensional loci. It was pointed out in Section 5–6 that if f has continuous third partial derivatives and if

$$f_2(x, y) \neq 0,$$

at some point p_0 on the locus of

$$f(x, y) = 0, \tag{3}$$

then the portion of this locus in some neighborhood of p_0 is a one-dimensional manifold with x as a coordinate and y as a differentiable variable.

This being the case, on the tangent bundle to this manifold, one has

$$df(x, y) = f_1(x, y)\, dx + f_2(x, y)\, dy = 0,$$

so, on the manifold itself,

$$D_x y = \frac{dy}{dx} = \frac{-f_1(x, y)}{f_2(x, y)}. \tag{4}$$

If $f_1(x, y) \neq 0$, then y is an admissible coordinate, and

$$D_y x = \frac{dx}{dy} = \frac{-f_2(x, y)}{f_1(x, y)}. \tag{5}$$

Zeros for $D_x y$ and $D_y x$ are, respectively, horizontal and vertical tangent points on the locus; so these results may be used to sketch the locus of (3).

Note that, in general, the equation $dy/dx = 0$ is an equation in x and y. This may be solved with (3) to locate the desired horizontal tangent points on the curve.

The basic theorem requires that f_1 and f_2 be continuous and that at least one of them be different from zero. Points on the locus of (3) where these conditions fail are called *singular points* on the curve. Specifically, singular points are points on the locus at which both partial derivatives of f vanish or at which at least one of them is discontinuous. In the neighborhood of such a point the locus is not necessarily a manifold, and there is no general characterization of the behavior of the curve at such a point. It may have a unique tangent line; it may have a cusp; it may have several branches; there may be no curve at all (the singular point may be an isolated point of the locus).

Curve sketching, then, may be reduced to the following routine. Locate the zeros of f_1 and f_2 on the locus of (3). Zeros for f_1 only are horizontal tangent points; zeros for f_2 only are vertical tangent points. Common zeros and points of discontinuity for either partial derivative are singular points. Locate and characterize all such points, and see whether the nature of the curve is indicated by this information.

There is one further consideration that is sometimes useful. A polynomial equation of even (odd) degree has an even (odd) number of real roots. Thus, for polynomial equations in x and y, there is some clue to the number of branches. For each vertical line ($x = $ constant), the number of intersections with the curve is restricted by the degree in y of the given equation, and for each horizontal line, the number of intersections is restricted by the degree in x. Sometimes this information helps to determine the behavior of the curve at singular points.

Before turning to applications of (4) and (5) above to the computation of derivatives and sketching of curves, it is well to note a few points of importance in the theory.

(i) Classically, the name *implicit function theorem* is given to the assertion that if f has continuous partial derivatives and if at some point p_0 on the locus of $f(x, y) = 0$, $f_2(x, y) \neq 0$, then there is a unique function g such that $y = g(x)$ over the portion of the locus in some neighborhood of p_0. In many treatments of calculus this theorem is regarded as justification for the derivation of (4) given above. The information that x is a full-fledged coordinate at p_0 (has an interval for a range) is missing here. Unless this is established, use of the calculus of variables is not really justified.

(ii) It must be emphasized that the implicit function theorem, like the complete theorem characterizing a manifold, is a strictly *local* theorem. It does not say that the entire locus of $f(x, y) = 0$ can be described by a relation of the form $y = g(x)$, nor does it say that over the entire locus explicit relations are uniquely determined. For example, take the locus of

$$x^2 + y^2 = 1.$$

This locus is a circle and cannot be defined by a single equation $y = f(x)$. Furthermore, in the large, there is no unique solution for y in terms of x. There are infinitely many solutions for y in terms of x, such as

$$y = \sqrt{1 - x^2},$$

$$y = -\sqrt{1 - x^2},$$

$$y = \begin{cases} \sqrt{1 - x^2} & \text{for } x \text{ rational,} \\ -\sqrt{1 - x^2} & \text{for } x \text{ irrational,} \end{cases}$$

to name three. What the implicit function theorem does say is that given any point on the circle except the vertical tangent point, there is a neighborhood of this point in which the equation of the circle determines y uniquely in terms of x.

(iii) The phrase "neighborhood of p_0" appearing in these theorems refers to a two-dimensional neighborhood. For example, in the case of the circle, the restriction $|x| < \epsilon$ does not yield a unique solution of the implicit equation. It still leaves all those suggested above. For appropriate ϵ and δ, the simultaneous restrictions $|x| < \epsilon$ and $|y - 1| < \delta$ do yield the unique solution $y = \sqrt{1 - x^2}$. Similarly, if $x^2 + y^2 = 1$, $|x| < \epsilon$, and $|y + 1| < \delta$, then $y = -\sqrt{1 - x^2}$.

EXAMPLES

1. Consider the equation

$$x^2 - y^2 = 0.$$

Using either formula (4) above or the so-called implicit differentiation technique described in Section 2–2, one finds that

$$\frac{dy}{dx} = \frac{x}{y}.$$

This result is meaningless at the origin; but if $u = x^2 - y^2$, then both first partials of u vanish at the origin; hence the implicit function theorem does not apply there. The locus is two straight lines intersecting at the origin and thus it is not a manifold. For any point on the locus other than the origin, there is a neighborhood in which the locus is a manifold.

2. Consider the equation

$$f(x, y) = (x^3 - x^2y - xy^2 + y^3)^{1/3} = 0.$$

Here $f(x, 0) = x$; so $f_1(0, 0) = 1$. Also, $f(0, y) = y$; hence $f_2(0, 0) = 1$. However,

$$x^3 - x^2y - xy^2 + y^3 = (x - y)^2(x + y);$$

thus the locus is the same as in Example 1 and is not a manifold in any neighborhood of the origin. Neither partial derivative of $f(x, y)$ vanishes at the origin, but the partials of f are not continuous there.

3. Sketch the locus of

$$x^5 + y^5 + 5xy = 0. \tag{6}$$

Let $f(x, y) = x^5 + y^5 + 5xy$; then

$$f_1(x, y) = 5x^4 + 5y, \qquad f_2(x, y) = 5y^4 + 5x.$$

These are continuous everywhere; so there are no singular points introduced by discontinuities. Set $f_1(x, y) = 0$ to get

$$y = -x^4. \tag{7}$$

Substitute (7) in (6), and solve for x:

$$x^5 - x^{20} - 5x^5 = 0;$$

$$x = 0, -4^{1/15}.$$

By (7) the corresponding values of y are

$$y = 0, -4^{4/15}.$$

Similarly, the conditions $f(x, y) = f_2(x, y) = 0$ yield

$$y = 0, -4^{1/15}, \qquad x = 0, -4^{4/15}.$$

Thus, a complete list of critical points is

$$(0, 0), \qquad\qquad \text{singular point,}$$
$$(-4^{1/15}, -4^{4/15}), \qquad \text{horizontal tangent,}$$
$$(-4^{4/15}, -4^{1/15}), \qquad \text{vertical tangent.}$$

At each level, both horizontal and vertical (except through the critical points), the curve has an odd number of branches. To satisfy all these conditions, a curve of the type pictured in Fig. 6–6 is indicated.

4. Sketch the locus of

$$(x^2 + y^2)^2 - 2(x^2 - y^2) = 0. \tag{8}$$

Let f be the indicated function; then

$$f_1(x, y) = 4x(x^2 + y^2) - 4x, \qquad f_2(x, y) = 4y(x^2 + y^2) + 4y.$$

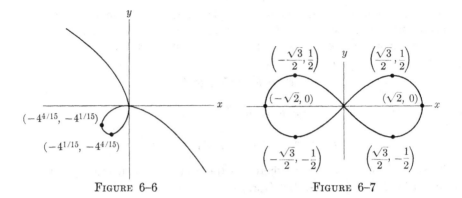

FIGURE 6-6   FIGURE 6-7

If $f_1(x, y) = 0$, then

$$x = 0 \quad \text{or} \quad x^2 + y^2 = 1.$$

Substituting the conditions above into (8), one gets

$$x = 0: \qquad y^4 + 2y^2 = 0, \quad y = 0,$$

$$x^2 + y^2 = 1: \quad 1 - 2(x^2 - 1 + x^2) = 0,$$

$$x = \pm \frac{\sqrt{3}}{2}, \qquad y = \pm \frac{1}{2}.$$

Note that all four combinations of signs are admissible in this last solution. The only root of the equation

$$f_2(x, y) = 0 \quad \text{is} \quad y = 0.$$

Substitution of $y = 0$ into (6) yields

$$x^4 - 2x^2 = 0;$$

$$x = 0, \pm\sqrt{2}.$$

Again the partial derivatives of f are continuous everywhere; thus a complete list of critical points is

$$(\pm\sqrt{3}/2, \pm\tfrac{1}{2}), \qquad \text{four horizontal tangent points,}$$

$$(\pm\sqrt{2}, 0), \qquad \text{two vertical tangent points,}$$

$$(0, 0), \qquad \text{singular point.}$$

The sketch, then, must be something like Fig. 6-7.

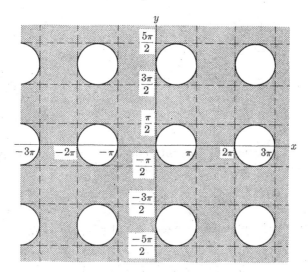

FIGURE 6–8

5. Sketch the locus of

$$\sin x + \cos y - 1 = 0. \tag{9}$$

If f is the indicated function, then

$$f_1(x, y) = \cos x, \qquad f_2(x, y) = -\sin y.$$

The condition $f_1(x, y) = 0$ yields

$$\cos x = 0, \qquad x = \frac{(2n + 1)\pi}{2}.$$

However, to satisfy (9), one must have $\sin x \geq 0$ and $\cos y \geq 0$, for if either of these were negative, the other would have to be greater than 1. Thus, the only solutions of $f_1(x, y) = 0$ that satisfy (9) are

$$x = \frac{(4n + 1)\pi}{2}, \qquad y = \frac{(2n + 1)\pi}{2}.$$

Similarly, the solutions of $f_2(x, y) = 0$ that satisfy (9) are

$$y = 2n\pi, \qquad x = n\pi.$$

None of these points coincide, and f_1 and f_2 are continuous everywhere, and hence there are no singular points. It helps to shade in the excluded regions, those for which $\sin x < 0$ or $\cos y < 0$. Once this is done and the horizontal and vertical tangent points are located, the sketch is fairly obvious (Fig. 6–8).

6. Sketch the locus of

$$\sqrt{x} - \sqrt{y} = 1. \qquad (10)$$

Let $f(x, y) = \sqrt{x} - \sqrt{y} - 1$; then

$$f_1(x, y) = \frac{1}{2\sqrt{x}}, \qquad f_2(x, y) = \frac{-1}{2\sqrt{y}}.$$

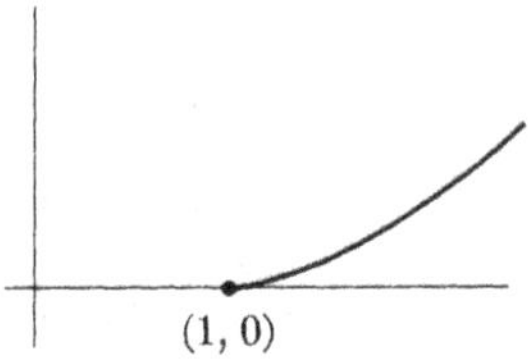

(1, 0)

FIGURE 6-9

These have no zeros but are discontinuous at $x = 0$ and $y = 0$, respectively. However, to satisfy (10), one must have

$$y > 0 \qquad \text{and} \qquad x > 1;$$

thus the only critical point is a singular point at $(1, 0)$. Now,

$$\frac{dy}{dx} = -\frac{f_1(x, y)}{f_2(x, y)} = \frac{\sqrt{y}}{\sqrt{x}} = 0$$

at $(1, 0)$; so there seems to be a horizontal tangent at this point. Note, however, that this is an end point of the curve. A sketch is shown in Fig. 6-9.

7. Find dy/dx on the locus of

$$x^2 + xy + y^2 = 3,$$

and determine at which points, if any, the result may not be valid. Purely formally,

$$(2x + y)\, dx + (x + 2y)\, dy = 0,$$

and hence

$$\frac{dy}{dx} = -\frac{2x + y}{x + 2y}.$$

This result is valid, provided that $x + 2y \neq 0$. The points on the locus at which $x + 2y = 0$ are found by substituting $x = -2y$ in the equation of the locus:

$$4y^2 - 2y^2 + y^2 = 3,$$

$$y^2 = 1;$$

$$y = 1 \quad \text{or} \quad -1,$$

$$x = -2 \quad \text{or} \quad 2.$$

Thus the result fails at $(-2, 1)$ and $(2, -1)$.

Exercises

1. From each of the following equations find dy/dx, and determine at what points (if any) the result may not be valid.

(a) $xy = 0$

(b) $(x^2/a^2) + (y^2/b^2) = 1$

(c) $(x^2/a^2) - (y^2/b^2) = 1$

(d) $xy = 1$

(e) $\sqrt{x} + \sqrt{y} = \sqrt{a}$

(f) $x^{2/3} + y^{2/3} = a^{2/3}$

(g) $x^3 + y^3 = 3axy$

(h) $x^3 + 3x^2y + y^3 = 1$

(i) $x^3 - 3xy + y^3 = 1$

(j) $x^4 + 4x^3y + y^4 = 6$

(k) $x + 2\sqrt{xy} + y = 4$

(l) $x^2 + 4\sqrt{xy} + y^2 = 6$

(m) $x^3 - xy + y^2 = 0$

(n) $x^2 - 2x\sqrt{xy} - y^2 = 0$

(o) $y^2(1 + x) = x^2(1 - x)$

(p) $y^3 = x^2 + y^2$

(q) $(x^2 + y^2)^2 = 2(x^2 - y^2)$

(r) $x^3 - y^3 = 3xy$

(s) $(y^2 - 2x^2)^2 = x^5$

(t) $y(1 + e^{1/x}) = x$

(u) $y^2(2 - x) = x^3$

(v) $(y - 2x)^2 = x^5$

(w) $\sin x + \cos y = 1$

(x) $xy + \ln(xy) = 1$.

2. Show that each of the following equations has a solution $y = f(x)$ near the point indicated, and in each case find $D_x y$ at the point indicated.

(a) $x^3 + y^3 - x^2y = 5$ (2, 1)

(b) $(x + y)^3 - x^2y = 25$ (1, 2)

(c) $\sqrt{3x} + \sqrt{2y} = 7$ (3, 8)

(d) $x \cos(xy) = 0$ (1, $\pi/2$)

(e) $xy + \ln(xy) = 1$ (1, 1)

(f) $x^5 + y^5 + xy = 3$ (1, 1)

(g) $\sin x + \cos y = 1$ ($\pi/6$, $\pi/3$)

(h) $y = \cos(x - y)$ (1, 1)

(i) $\cos x = \ln(x + y)$ (0, e)

(j) $\arctan(x/y) + \ln\sqrt{x^2 + y^2} = 0$ (0, 1)

3. Sketch the locus of each of the following equations.

(a) $x^3 + y^3 = 3xy$

(b) $x^3 + y^3 + 3xy - 0$

(c) $y^2 = x^2 - x^3$

(d) $\sqrt{x} + \sqrt{y} = 2$

(e) $x^{1/3} + y^{1/3} = 1$

(f) $x^{2/3} + y^{2/3} = 1$

(g) $x^{4/3} + y^{4/3} = 1$

(h) $x^2 - y^3 + y^2 + 2y = 0$

(i) $x^2 - y^4(5 + y) = 0$

(j) $x + y + xy = 0$

(k) $x + y - xy = 0$

(l) $\sin x - \cos y = 1$

(m) $\cosh x + \cosh y = 8/3$

(n) $e^{x+y} + y - x = 0$

(o) $(y - x^2)^2 - x^5 = 0$

4. Show that for each of the following functions f, the locus of $f(x, y) = 0$ has a singular point at the origin. Show that no two of these curves behave in the same manner in the neighborhood of the singular point.

(a) $f(x, y) = x^2 - y^2$

(b) $f(x, y) = x^3 - y^3$

(c) $f(x, y) = y^2 - x^3$

(d) $f(x, y) = y^3 - x^2$

(e) $f(x, y) = (x + y)(x - y)(x - 2y)$

6–3 Constrained maxima and minima, one dimension. The general pattern of procedure for applied maximum and minimum problems may be summarized as follows. A variable z is to be maximized or minimized; thus it is found in terms of other variables:

$$z = f(x, y). \tag{11}$$

The problem then imposes an auxiliary condition,

$$\phi(x, y) = 0. \tag{12}$$

The procedure is purely mechanical. Solve (12) for y in terms of x; substitute into (11) to get z in terms of x alone; compute dz/dx and set it equal to zero, etc.

The present section will deal with generally the same type problem, but a few new features will be added to the discussion.

First, a more careful investigation of the underlying theory is called for. Let x and y be coordinates in the plane, and let z be defined over the plane by (11). Under suitable conditions, (12) defines a one-dimensional manifold in the plane. Let x, y, and z be restricted to this manifold. It is then required to locate maxima and/or minima for this restricted variable z. The process of restricting the variable z to the manifold (12) is referred to as "imposing a constraint," hence the phrase constrained maximum-minimum problem.

Frequently the constraint involves not only equation (12) but an additional restriction, such as $a \leq x \leq b$. For example, the side of a rectangle inscribed in a circle must lie between zero and the diameter of the circle. Such additional restrictions usually come from the physical significance of the variables, and they must be watched for diligently. The usual effect of such a restriction is to replace the entire locus of (12), which is frequently a manifold without boundary points, by a smaller manifold which has boundary points.

The theory may now be summarized as follows. A variable z is defined on a manifold M (one-dimensional in this section, multidimensional in Section 6–4), and one is required to find the maximum and/or minimum values of z on M. The present discussion is limited to the cases in which z is differentiable at every interior point of M. Now, the idea is not to derive a procedure for finding a point and proving that it is a maximum point for z. Rather, it is to develop a method for finding all points that *might* be maxima or minima. If *all* such possibilities are located and z is evaluated at each, the maximum and minimum obviously appear. The possible max-min points are of two kinds:

(i) *Boundary points of M.* In the one-dimensional case, there are at most two boundary points of M, and they should be automatically included in the list. Analysis of boundary points in higher dimension is discussed in Section 6–4.

(ii) *Stationary points for z on M.* By definition, p is a stationary point for z on M if $dz_p = 0$ on the tangent to M at p.

Justification of the theory means proving that the list of possibilities is complete. Note that this is equivalent to saying that if p is not a boundary point of M and dz_p is not identically zero on the tangent to M at p, then p is neither a maximum nor a minimum point for z. This last assertion is easily proved. If dz_p is not identically zero, then for some coordinate u at p,

$$D_u z(p) \neq 0. \tag{13}$$

Since p is not a boundary point, there are points q_1 and q_2 on opposite sides of p, and it follows from (13) that for some such q_1 and q_2,

$$z(p) - z(q_1) > 0 \qquad \text{and} \qquad z(q_2) - z(p) > 0.$$

Thus, z has values both greater than and less than $z(p)$.

The first paragraph of this section outlines a direct method of attack on the problem of finding stationary points. The following method is frequently easier to apply, particularly in the multidimensional cases. From (11) and the definition of stationary point, it follows that

$$dz = f_1(x, y)\, dx + f_2(x, y)\, dy = 0 \tag{14}$$

over the tangent to the locus of (12) at a stationary point. However, over the entire tangent bundle for the locus of (12),

$$\phi_1(x, y)\, dx + \phi_2(x, y)\, dy = 0. \tag{15}$$

Hence, (14) and (15) both hold on the tangent at a stationary point. If (15) is multiplied by any constant λ and added to (14), the result

$$[f_1(x, y) + \lambda\phi_1(x, y)]\, dx + [f_2(x, y) + \lambda\phi_2(x, y)]\, dy = 0 \tag{16}$$

still holds on the tangent at every stationary point. Since (16) holds for any λ, let λ be chosen so that

$$f_2(x, y) + \lambda\phi_2(x, y) = 0. \tag{17}$$

This is possible provided that

$$\phi_2(x, y) \neq 0. \tag{18}$$

Equation (17) reduces (16) to

$$[f_1(x, y) + \lambda\phi_1(x, y)]\, dx = 0,$$

but given (18), x is a coordinate on the locus of (12); thus dx is not identically zero and it follows that

$$f_1(x, y) + \lambda\phi_1(x, y) = 0. \tag{19}$$

Summary. If (18) holds at a stationary point, then (17) may be made to hold, and (19) follows.

Suppose (18) fails at some stationary point. If

$$\phi_1(x, y) \neq 0 \tag{20}$$

there, then (19) will determine λ, y is a coordinate, and (17) follows.

Equations (17) and (19) are called the Lagrange equations, and λ is called a Lagrange multiplier. The method of the Lagrange multiplier is as follows. Take the three equations

$$f_1(x, y) + \lambda\phi_1(x, y) = 0,$$
$$f_2(x, y) + \lambda\phi_2(x, y) = 0, \tag{21}$$
$$\phi(x, y) = 0.$$

These three equations in x, y, and λ must hold at every stationary point where either (18) or (20) holds. Therefore, if all solutions of the system (21) are found, the only other possible stationary points are those at which

$$\phi_1(x, y) = \phi_2(x, y) = 0.$$

These points are (see Section 6–2) the singular points on the locus of (12).

Thus, the following outline lists all possible max-min points for z:

    (i) the boundary points of M,
    (ii) stationary points for z on M,
        (a) solutions of the Lagrange equations (21),
        (b) singular points on the locus of (12).

Note, finally, that the complete set (21) has as many equations as there are unknowns, although, in general, these are nonlinear and may have more than one solution. Often, however, eliminating the multiplier between the first two equations gives a relation between the other variables that serves to solve the problem.

EXAMPLES

1. A wire of length a is cut into two pieces. One is bent into a square, the other is bent into a circle. Find the proportions so that the sum of the areas of the square and circle will be a minimum, a maximum. Let the square have side s and the circle have radius r. The variable whose extreme values are sought is

$$A = s^2 + \pi r^2.$$

The constraint is

$$4s + 2\pi r = a.$$

Thus, the Lagrange equations are

$$2s + 4\lambda = 0, \qquad 2\pi r + 2\pi\lambda = 0.$$

Solve these for λ to get

$$-\frac{s}{2} = \lambda = -r.$$

Thus,

$$s = 2r$$

is the condition for a stationary point. Since

$$\frac{\partial}{\partial s}(4s + 2\pi r) = 4,$$

there are no singular points. Now, in this problem, there is an additional constraint that does not show in the given equation. This is that s must lie between 0 and $a/4$, or equivalently, that r must lie between 0 and $a/2\pi$. Thus, the manifold in question has two boundary points. By appropriate substitutions, A can be evaluated at these boundary points and at the stationary point with the following results.

Point	s	r	A
Stationary	$\dfrac{a}{\pi + 4}$	$\dfrac{a}{2(\pi + 4)}$	$\dfrac{a^2}{4(\pi + 4)}$
Boundary	0	$\dfrac{a}{2\pi}$	$\dfrac{a^2}{4\pi}$
Boundary	$\dfrac{a}{4}$	0	$\dfrac{a^2}{16}$

Comparison of these values of A shows that the stationary point is the minimum and the $s = 0$ boundary point is the maximum. That is, to get minimum total area make the side of the square equal to the diameter of the circle, and to get maximum total area put all the wire into the circle.

Note that here, as in many constrained maximum-minimum problems, it is unnecessary to use any second-derivative criterion to determine the nature of the stationary point.

2. The strength of a beam with rectangular cross section is proportional to its breadth and to the square of its depth. Find the shape of the strongest beam that can be cut from a log with circular cross section. Let x and y be

coordinates measuring from the center of the log. The rectangle will have dimensions $2x$ by $2y$, so one is required to maximize

$$S = K(2x)(2y)^2 = 8Kxy^2,$$

subject to the constraint

$$x^2 + y^2 = a^2.$$

The Lagrange equations are

$$8Ky^2 + 2\lambda x = 0,$$

$$16Kxy + 2\lambda y = 0.$$

Solving the equations above for λ yields

$$-\frac{4Ky^2}{x} = \lambda = -8Kx,$$

or

$$y^2 = 2x^2,$$

provided that $x \neq 0$ and $y \neq 0$. However, the additional constraints are

$$0 \leq x \leq a \quad \text{and} \quad 0 \leq y \leq a,$$

so the exceptional values occur at boundary points. Finally, there are no singular points because $2x$ and $2y$ cannot both be zero with $x^2 + y^2 = a^2$. Now, $S = 0$ at each boundary point and $S > 0$ on the rest of the locus; so the stationary point at which $y^2 = 2x^2$ must be the maximum. The strongest beam has depth $\sqrt{2}$ times its breadth.

3. Show that the shortest line segment from the origin to the locus of $xy^2 = 1$ intersects the curve at right angles. The distance from the origin to (x, y) is $\sqrt{x^2 + y^2}$. This is a minimum if and only if its square is, so minimize

$$z = x^2 + y^2$$

subject to the constraint

$$xy^2 = 1.$$

Since z is positive and tends to infinity at each extremity of the curve, the stationary points must be minima. The Lagrange equations are

$$2x + \lambda y^2 = 0, \quad 2y + 2\lambda xy = 0.$$

At a singular point one would have $y^2 = 2xy = 0$, and this is impossible with $xy^2 = 1$. Now, the slope of the line from $(0, 0)$ to (x, y) is y/x, so

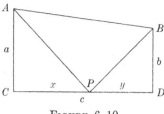

FIGURE 6–10

solve the Lagrange equations for y/x by transposing the second term in each equation and by dividing the second by the first:

$$\frac{y}{x} = \frac{-2xy}{-y^2} = \frac{2x}{y}.$$

This must hold at the stationary point. However, given $xy^2 = 1$,

$$y^2\, dx + 2xy\, dy = 0,$$

$$\frac{dy}{dx} = \frac{-y}{2x}.$$

The tangent line and the line from the origin have negative reciprocal slopes at the stationary point.

EXERCISES

1. A wire of length a is cut into two pieces, one of which is bent into a square, the other into an equilateral triangle. Find the proportions that will make the sum of the areas of the square and triangle (a) a minimum, and (b) a maximum.

2. Solve Exercise 1 for the case in which the triangle is to be an isosceles right triangle.

3. Solve Exercise 1 for the case in which both pieces are bent into triangles, one right isosceles, the other equilateral.

4. In Fig. 6–10 find the proportion x/y to minimize, and then maximize, each of the following quantities.

    (a) $AP + PB$ (b) $(AP)^2 + (PB)^2$

    (c) $(AP)^2 + 2(PB)^2$ (d) $(AP)^2 - (PB)^2$

    (e) area ACP + area PDB (f) area APB

5. Find the point on the locus of $y = \sqrt{x}$ which is nearest the point $(c, 0)$,

    (a) if $c \geq \frac{1}{2}$,   (b) if $c < \frac{1}{2}$.

6. The intensity of illumination at any point is proportional to the strength of the light source and varies inversely as the square of the distance from the source. Given two sources of relative strengths a and b and a point p on the line joining them, find the relative distances of p from the two sources so that the total intensity will be a minimum.

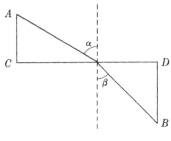

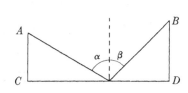

FIGURE 6–11   FIGURE 6–12

7. A box is to have a square base and vertical sides, no top. For a fixed volume, find the proportions for minimum area.

8. In each of the following statements one figure is to be inscribed in another. Regard the circumscribing figure as fixed, and find the proportions of the inscribed figure to give the indicated maximum.

    (a) Rectangle of maximum area inscribed in a circle

    (b) Right circular cylinder of maximum volume inscribed in a sphere

    (c) Rectangle of maximum area inscribed in a semicircle

    (d) Right circular cylinder of maximum volume inscribed in a hemisphere

9. For a cone with a given slant height, find the vertex angle for maximum volume.

10. In Fig. 6–11, light travels with velocity c_1 above the line CD and with velocity c_2 below it. Assuming that it follows a straight-line path within each medium, show that the condition

$$\frac{\sin \alpha}{c_1} = \frac{\sin \beta}{c_2}$$

yields a stationary point for the total time of travel from A to B. *Note.* The theory that light will follow a path which yields a stationary value for time elapsed is known as *Fermat's principle.* The above formula is known as *Snell's law of refraction.*

11. Use Fermat's principle to prove the law of reflection. If the light is reflected by the line CD (Fig. 6–12)—so that the velocity remains constant—then the stationary-time path is given by the condition

$$\sin \alpha = \sin \beta.$$

12. Use Fermat's principle to prove the focal property of the parabola. If light travels from the focus $(c, 0)$ to (a, b) via the parabola, (Fig. 6–13) then the stationary-time path is the one for which the segment from the parabola to (a, b) is horizontal.

13. Generalize Example 3. The shortest line segment from (a, b) to the locus of

$$\phi(x, y) = 0$$

intersects this curve at right angles.

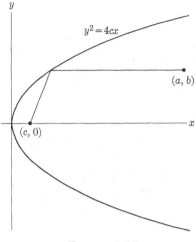

FIGURE 6–13

6–4 Constrained maxima and minima, several dimensions. Suppose that

$$w = f(x_1, x_2, \ldots, x_n),$$

and suppose maxima and minima for w are sought subject to constraints

$$\phi^i(x_1, x_2, \ldots, x_n) = 0, \qquad i = 1, 2, \ldots, k.$$

Formally, the Lagrange multiplier system is set up in the following manner. Introduce $\lambda_1, \lambda_2, \ldots, \lambda_k$ (one multiplier for each constraint). Then, set up the n equations

$$f_j(x_1, x_2, \ldots, x_n) + \sum_{i=1}^{k} \lambda_i \phi_j^i(x_1, x_2, \ldots, x_n), \qquad j = 1, 2, \ldots, n.$$

There are n of these and k constraint equations in the $n + k$ unknowns, $x_1, x_2, \ldots, x_n, \lambda_1, \lambda_2, \ldots, \lambda_k$. As before, these equations must hold at a stationary point for w provided that the stationary point is not a singular point on the locus of the constraint equations. Therefore, finding all solutions of the Lagrange equations and all singular points will serve to locate all stationary points.

To see how the singular points are characterized, recall the master theorem of Section 5–6 on the characterization of manifolds. If the Jacobian determinant of the k variables $\phi^i(x_1, x_2, \ldots, x_n)$ with respect to k of the x's is not zero, then the locus is (locally) a manifold of dimension $(n - k)$ with the remaining x's as coordinates. It does not matter which x's emerge as coordinates. As long as the locus is an $(n - k)$-dimensional manifold in the neighborhood of a point, the point is nonsingular.

To put this in another way, let

$$X = \begin{bmatrix} x_1 \\ \vdots \\ x_n \end{bmatrix}, \qquad U = \begin{bmatrix} \phi^1(x_1, x_2, \ldots, x_n) \\ \vdots \\ \phi^k(x_1, x_2, \ldots, x_n) \end{bmatrix};$$

then

$$D_X U$$

is a $k \times n$ matrix of partial derivatives. At a singular point every $k \times k$ matrix formed by taking k columns of $D_X U$ has a zero determinant.

The plan of proof that these conditions locate all stationary points is exactly the same as the one used in Section 6–3 for the special case $n = 2$, $k = 1$. Set

$$df(x_1, x_2, \ldots, x_n) + \sum_{i=1}^{k} \lambda_i \, d\phi^i(x_1, x_2, \ldots, x_n) = 0,$$

and look at the coefficients of $dx_1, dx_2, \ldots, dx_n$. Setting k of these coefficients equal to zero will determine the λ's provided that the determinant of the system is not zero. Happily, the nonvanishing of this determinant guarantees that the remaining x's are coordinates. Their differentials are not identically zero; hence the corresponding coefficients must be, and the complete set of Lagrange equations emerges.

If a multidimensional manifold has a boundary, it will be a manifold of dimension one lower, obtained by adding one more constraint equation. So, the analysis of boundary points is another Lagrange multiplier problem. The analysis never progresses more than one step because a boundary manifold never has a boundary itself. No proof of this is suggested here, but look at familiar examples. A solid sphere has a spherical surface for boundary, but the spherical surface has none. A disc has a circle for boundary, but the circle has none.

EXAMPLES

1. The conditions

$$x + y + z = 1, \qquad x^2 + y^2 \le 1$$

define an ellipse, together with its interior (see Fig. 6–14). Find the points on this figure that are closest to and farthest from the origin. Stationary points are required for

$$x^2 + y^2 + z^2.$$

With $\phi(x, y, z) = x + y + z - 1$, no first partial of ϕ vanishes, thus Lagrange multipliers may be used to locate all stationary points. At

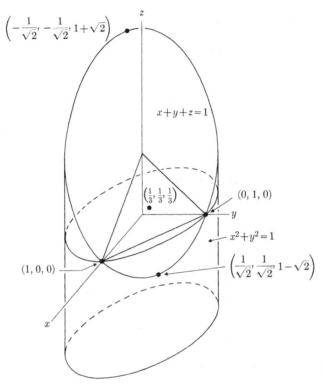

$\left(-\dfrac{1}{\sqrt{2}}, -\dfrac{1}{\sqrt{2}}, 1+\sqrt{2}\right)$

$x+y+z=1$

$\left(\dfrac{1}{3}, \dfrac{1}{3}, \dfrac{1}{3}\right)$

$(0, 1, 0)$

$x^2+y^2=1$

$(1, 0, 0)$

$\left(\dfrac{1}{\sqrt{2}}, \dfrac{1}{\sqrt{2}}, 1-\sqrt{2}\right)$

FIGURE 6–14

interior points of the figure, the stationary points are characterized by the equations

$$2x + \lambda = 0,$$
$$2y + \lambda = 0,$$
$$2z + \lambda = 0,$$
$$x + y + z = 1.$$

This system is easily solved:

$$x = y = z = \frac{-\lambda}{2}, \qquad x + y + z = \frac{-3\lambda}{2} = 1,$$

$$\lambda = -\tfrac{2}{3}, \qquad x = y = z = \tfrac{1}{3}.$$

Thus, in the interior there is a single stationary point at $(\tfrac{1}{3}, \tfrac{1}{3}, \tfrac{1}{3})$.

The boundary is characterized by two constraints:

$$x + y + z = 1, \qquad x^2 + y^2 = 1.$$

The Lagrange equations are

$$2x + \lambda + 2x\mu = 0,$$
$$2y + \lambda + 2y\mu = 0,$$
$$2z + \lambda = 0,$$
$$x + y + z = 1,$$
$$x^2 + y^2 = 1.$$

Subtract the first equation from the second:

$$2(y - x) + 2\mu(y - x) = 0,$$
$$2(y - x)(1 + \mu) = 0.$$

Thus,

$$y = x \quad \text{or} \quad \mu = -1.$$

The root $y = x$ substituted in the fifth equation gives

$$x = y = \frac{1}{\sqrt{2}} \quad \text{or} \quad x = y = -\frac{1}{\sqrt{2}};$$

the fourth equation pairs these results with

$$z = 1 - \sqrt{2} \quad \text{and} \quad z = 1 + \sqrt{2},$$

respectively. Thus two points are found:

$$\left(\frac{1}{\sqrt{2}}, \frac{1}{\sqrt{2}}, 1 - \sqrt{2}\right), \quad \left(-\frac{1}{\sqrt{2}}, -\frac{1}{\sqrt{2}}, 1 + \sqrt{2}\right).$$

To pursue the possibility $\mu = -1$, substitute in the first equation to get

$$\lambda = 0;$$

then by the third equation

$$z = 0,$$

and the last two read

$$x + y = 1, \quad x^2 + y^2 = 1.$$

This system has the solutions

$$x^2 + (1 - x)^2 = 1, \quad 2x^2 - 2x = 0,$$
$$x = 0 \quad \text{or} \quad 1, \quad y = 1 - x = 1 \quad \text{or} \quad 0.$$

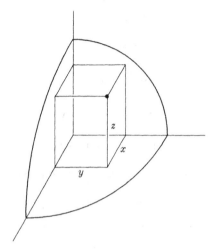

FIGURE 6-15

Two more points appear:

$$(0, 1, 0), \qquad (1, 0, 0).$$

The Jacobian determinant of the constraint system with respect to x and y is

$$\begin{vmatrix} 1 & 1 \\ 2x & 2y \end{vmatrix} = 2y - 2x.$$

This determinant vanishes if $y = x$, but this possibility has already been investigated. Actually, there are no singular points, but this need not be proved. If there are any, they satisfy $y = x$ and have already been listed. In all, then, there are five stationary points for $x^2 + y^2 + z^2$ on this locus, one in the interior and four on the boundary. If the five values of $x^2 + y^2 + z^2$ are computed and compared, it appears that the minimum occurs at $(\frac{1}{3}, \frac{1}{3}, \frac{1}{3})$ and the maximum at $(-1/\sqrt{2}, -1/\sqrt{2}, 1 + \sqrt{2})$.

2. Find the relative dimensions of the rectangular parallelepiped of maximum volume that can be inscribed in a sphere. Figure 6-15 shows the relevant variables. The parallelepiped will have volume $8xyz$, and the constraints are $x^2 + y^2 + z^2 = a^2, 0 \leq x \leq a, 0 \leq y \leq a, 0 \leq z \leq a$. The Lagrange equations are

$$8yz + 2\lambda x = 0,$$

$$8xz + 2\lambda y = 0,$$

$$8xy + 2\lambda z = 0,$$

$$x^2 + y^2 + z^2 = a^2.$$

From the first equation,

$$\lambda = -\frac{4yz}{x}.$$

Substituting this in the second, one has

$$8xz - \frac{8y^2z}{x} = 0,$$

$$8z(x^2 - y^2) = 0.$$

Similarly, from the first and third equations, one has

$$8y(x^2 - z^2) = 0,$$

and from the second and third,

$$8x(y^2 - z^2) = 0.$$

If x, y, or z is zero, one has a boundary point on the manifold, and the volume of the parallelepiped is zero. So, these solutions may be discarded in looking for a maximum. The only alternative is

$$x^2 - y^2 = x^2 - z^2 = y^2 - z^2 = 0,$$

and since each variable must be positive, this yields

$$x = y = z.$$

That is, the point at which the parallelepiped is a cube is a stationary point for the volume. Finally, the partial derivatives of $x^2 + y^2 + z^2$ are $2x$, $2y$, and $2z$; and these do not all vanish at any one point on the sphere. Thus, the Lagrange method yields all the stationary points, and the cube is the parallelepiped of maximum volume.

3. Find the maxima and minima for $x^3 + y^3 + z^3$ on the solid sphere $x^2 + y^2 + z^2 \leq 1$. The interior is three-dimensional; there are no constraints, hence no multipliers. The Lagrange equations are

$$3x^2 = 0, \qquad 3y^2 = 0, \qquad 3z^2 = 0.$$

These give $(0, 0, 0)$ as a stationary point. The boundary is the locus of $x^2 + y^2 + z^2 = 1$, and the Lagrange equations are

$$3x^2 + 2\lambda x = 0,$$

$$3y^2 + 2\lambda y = 0,$$

$$3z^2 + 2\lambda z = 0.$$

Each equation has two solutions:

$$x = 0, \qquad x = \frac{-2\lambda}{3},$$

$$y = 0, \qquad y = \frac{-2\lambda}{3},$$

$$z = 0, \qquad z = \frac{-2\lambda}{3}.$$

Choosing the first possibility each time yields the point previously found (which is not even on the boundary). So, try

$$x = 0, \qquad y = z = \frac{-2\lambda}{3};$$

since $x^2 + y^2 + z^2 = 1$, this gives

$$2\left(\frac{4\lambda^2}{9}\right) = 1, \qquad \lambda = \pm \frac{3}{2\sqrt{2}},$$

$$y = z = \frac{-2\lambda}{3} = \mp \frac{1}{\sqrt{2}}.$$

So, there are two points:

$$\left(0, \frac{1}{\sqrt{2}}, \frac{1}{\sqrt{2}}\right), \qquad \left(0, \frac{-1}{\sqrt{2}}, \frac{-1}{\sqrt{2}}\right).$$

Similarly, $y = 0$, $x = z = -2\lambda/3$ gives two points:

$$\left(\frac{1}{\sqrt{2}}, 0, \frac{1}{\sqrt{2}}\right), \qquad \left(\frac{-1}{\sqrt{2}}, 0, \frac{-1}{\sqrt{2}}\right),$$

and $z = 0$, $x = y = -2\lambda/3$ gives

$$\left(\frac{1}{\sqrt{2}}, \frac{1}{\sqrt{2}}, 0\right), \qquad \left(\frac{-1}{\sqrt{2}}, \frac{-1}{\sqrt{2}}, 0\right).$$

Now, try

$$x = y = 0, \qquad z = -2\lambda/3.$$

This gives $z^2 = 1$, and yields two points. Symmetric conditions yield four others. In all, these combinations produce the six points:

$$(0, 0, 1), \quad (0, 0, -1), \quad (0, 1, 0), \quad (0, -1, 0), \quad (1, 0, 0), \quad (-1, 0, 0).$$

The final possibility is

$$x = y = z = \frac{-2\lambda}{3},$$

$$3\left(\frac{4\lambda^2}{9}\right) = 1,$$

$$\lambda = \pm \frac{\sqrt{3}}{2},$$

$$x = y = z = \mp \frac{1}{\sqrt{3}}.$$

Two points:

$$\left(\frac{1}{\sqrt{3}}, \frac{1}{\sqrt{3}}, \frac{1}{\sqrt{3}}\right), \quad \left(\frac{-1}{\sqrt{3}}, \frac{-1}{\sqrt{3}}, \frac{-1}{\sqrt{3}}\right).$$

Again there are no singular points, but it is sufficient to note that

$$\frac{\partial}{\partial x}(x^2 + y^2 + z^2) = 2x$$

and that $x = 0$ has already been used. The following chart now locates the maxima and minima:

Points	Value of $x^3 + y^3 + z^3$
(0, 0, 0)	0
$(1/\sqrt{3}, 1/\sqrt{3}, 1/\sqrt{3})$	$1/\sqrt{3}$
$(-1/\sqrt{3}, -1/\sqrt{3}, -1/\sqrt{3})$	$-1/\sqrt{3}$
$(1/\sqrt{2}, 1/\sqrt{2}, 0)$, etc.	$1/\sqrt{2}$
$(-1/\sqrt{2}, -1/\sqrt{2}, 0)$, etc.	$-1/\sqrt{2}$
(1, 0, 0), etc.	1
(−1, 0, 0), etc.	−1

Thus, there are three maximum points:

$$(1, 0, 0), \quad (0, 1, 0), \quad (0, 0, 1)$$

and three minimum points:

$$(-1, 0, 0), \quad (0, -1, 0), \quad (0, 0, -1).$$

4. The locus of

$$(y - x^2)(y - 2x^2) = 0$$

is the two indicated parabolas. Clearly, y assumes its minimum value on this locus at the origin. However, the Lagrange multiplier equations for stationary y subject to this constraint would be

$$0 + \lambda(-6xy + 8x^3) = 0, \qquad 1 + \lambda(2y - 3x^2) = 0.$$

These are not satisfied for $x = y = 0$. The explanation is that the origin is a singular point on this locus, and at singular points, the Lagrange multiplier conditions are not necessary for stationarity.

EXERCISES

1. Find the maximum and minimum points for each of the following variables on the indicated figure.

    (a) $x + y$ on the sphere $x^2 + y^2 + z^2 = 1$

    (b) xyz on the curve $x^2 + y^2 = 1, y = z$

    (c) xy on the plane circle $x^2 + y^2 \leq 1$

    (d) $x + y + z$ on the solid sphere $x^2 + y^2 + z^2 \leq 1$

    (e) xyz on the solid sphere $x^2 + y^2 + z^2 \leq 1$

    (f) xyz on the plane triangle $x + y + z = 4, x \geq 1, y \geq 1, z \geq 1$

2. A box in the shape of a rectangular parallelepiped is to be made from a given amount of material. Find the proportions for maximum volume

    (a) if it has a top,   (b) if it has no top.

3. Find the points nearest the origin on the curve on which

$$x^2 - xy + y^2 - z^2 = 1 \qquad \text{and} \qquad x^2 + y^2 = 1.$$

4. Prove that the product of the sines of the angles of a triangle is a maximum when the triangle is equilateral.

5. A figure consists of a right circular cylinder surmounted by a right circular cone. If the surface area (including base) is fixed, how should the height and radius of the cylinder and the vertex angle of the cone be related for maximum volume?

6. Consider a cross section of the figure in Exercise 4 (rectangle surmounted by a triangle). For a given perimeter, find proportions for maximum area.

7. A manufacturer markets competing items and can sell u of the first and v of the second at prices x and y, respectively, where

$$u = 250(y - x), \qquad v = 32{,}000 + 250(x - 2y).$$

Manufacturing costs for the first item are $50 each and for the second, $60 each. Find the selling prices for maximum profit.

8. Let w be defined on the surface $\phi(x, y, z) = 0$ and measure distance from the point (a, b, c). Show that the line joining (a, b, c) to a stationary point for w on the surface is always normal to the surface.

9. The Lagrange multiplier equations give necessary conditions for stationarity provided one of the indicated Jacobians is different from zero. Show that they give sufficient conditions even if all these Jacobians are zero.

6–5 Chain rules. Take three matrices of variables,

$$X = \begin{bmatrix} x_1 \\ x_2 \\ \vdots \\ x_n \end{bmatrix}, \qquad Y = \begin{bmatrix} y_1 \\ y_2 \\ \vdots \\ y_k \end{bmatrix}, \qquad Z = \begin{bmatrix} z_1 \\ z_2 \\ \vdots \\ z_m \end{bmatrix},$$

and let dX, dY, dZ be the corresponding matrices of differentials. Let

$$D_X Y = \left[\frac{\partial y_i}{\partial x_j} \right],$$

with $D_X Z$, $D_Y Z$, etc., defined in a similar manner. Here $D_X Y$ is a $k \times n$ matrix; $D_X Z$ is $m \times n$; $D_Y Z$ is $m \times k$; etc.

Now, suppose the entries in X form a coordinate set on an n-dimensional manifold M_n. If each entry in Y is a differentiable variable on this manifold, then by the fundamental theorem on differentials,

$$dy_i = \sum_{j=1}^{n} \frac{\partial y_i}{\partial x_j} dx_j, \qquad i = 1, 2, \ldots, k,$$

and the matrix form for this set of k linear equations is simply

$$dY = D_X Y \, dX. \tag{22}$$

If, now, the entries in Y form a coordinate set on a k-dimensional manifold M_k and if each entry in Z is differentiable there, then, similarly,

$$dZ = D_Y Z \, dY. \tag{23}$$

If the z's are also differentiable on M_n, then

$$dZ = D_X Z \, dX. \tag{24}$$

Suppose that $M_n = M_k$ or that one of these is a submanifold of the other. On the tangent bundle for the smaller manifold, (22), (23), and (24) all hold, thus, there,

$$D_X Z \, dX = dZ = D_Y Z \, dY = D_Y Z \, D_X Y \, dX. \tag{25}$$

Since (25) is an identity on the smaller tangent bundle, the coefficients of

dX must be equal in the extreme left and right members of (25); thus on the smaller manifold

$$D_X Z = D_Y Z \, D_X Y. \tag{26}$$

This is a chain rule for partial derivatives, or perhaps it is better to say that (26) is a whole collection of chain rules. For example, pick out the entry $\partial z_i/\partial x_j$ in $D_X Z$. Recalling the rules for matrix multiplication, one has from (26)

$$\frac{\partial z_i}{\partial x_j} = \sum_{r=1}^{k} \frac{\partial z_i}{\partial y_r} \frac{\partial y_r}{\partial x_j}. \tag{27}$$

The matrix equation (26) represents nm such formulas. If $n = k = m = 1$, then (27) reduces to

$$\frac{dz_1}{dx_1} = \frac{dz_1}{dy_1} \frac{dy_1}{dx_1},$$

the familiar one-dimensional chain rule.

In a way, (26) is a master formula for chain rules, but there is a slight risk involved in committing it to memory. Matrix multiplication is not commutative, so the order of the factors on the right-hand side of (26) must be preserved. A safer procedure is to return to (25). The plan in (25) is as follows. Two expressions for dZ are obtained from the fundamental theorem; into one of these a substitution from the fundamental theorem is made to get two expressions for dZ in terms of dX. Equating the two coefficients of dX then yields the chain rules.

EXAMPLES

1. Let Z consist of the single variable z, and let

$$X = \begin{bmatrix} x \\ y \end{bmatrix}, \qquad U = \begin{bmatrix} r \\ \theta \end{bmatrix}.$$

From the standard relations, $x = r \cos \theta$, $y = r \sin \theta$, one has

$$D_U X = \begin{bmatrix} \cos \theta & -r \sin \theta \\ \sin \theta & r \cos \theta \end{bmatrix}.$$

Now

$$D_U Z \, dU = dZ = D_X Z \, dX = D_X Z \, D_U X \, dU,$$

thus

$$D_U Z = D_X Z \, D_U X,$$

or

$$\begin{bmatrix} \dfrac{\partial z}{\partial r} & \dfrac{\partial z}{\partial \theta} \end{bmatrix} = \begin{bmatrix} \dfrac{\partial z}{\partial x} & \dfrac{\partial z}{\partial y} \end{bmatrix} \begin{bmatrix} \cos \theta & -r \sin \theta \\ \sin \theta & r \cos \theta \end{bmatrix}.$$

The above expression is matrix form for two equations:

$$\frac{\partial z}{\partial r} = \frac{\partial z}{\partial x} \cos \theta + \frac{\partial z}{\partial y} \sin \theta,$$

$$\frac{\partial z}{\partial \theta} = - \frac{\partial z}{\partial x} r \sin \theta + \frac{\partial z}{\partial y} r \cos \theta.$$

2. In thermodynamics, it is assumed that four variables, p (pressure), v (volume), u (internal energy), and t (temperature), are defined on a two-dimensional manifold in such a way that any pair forms an admissible coordinate set. Consider a change of coordinates from (v, t) to (v, u). Physically, assume first that volume and temperature are controlled by the experimenter; then convert to a system in which volume and energy are controlled. Relate the partial derivatives of pressure in these two systems. Set up the following notation:

$$X = \begin{bmatrix} v \\ u \end{bmatrix}, \qquad Y = \begin{bmatrix} v \\ t \end{bmatrix}, \qquad Z = [p].$$

In writing out partial derivatives here, it is essential to use subscripts to indicate which variable is constant. However, with this done, the formula

$$D_X Z = D_Y Z \, D_X Y$$

becomes

$$\left[\left(\frac{\partial p}{\partial v} \right)_u \quad \left(\frac{\partial p}{\partial u} \right)_v \right] = \left[\left(\frac{\partial p}{\partial v} \right)_t \quad \left(\frac{\partial p}{\partial t} \right)_v \right] \begin{bmatrix} \left(\frac{\partial v}{\partial v} \right)_u & \left(\frac{\partial v}{\partial u} \right)_v \\ \left(\frac{\partial t}{\partial v} \right)_u & \left(\frac{\partial t}{\partial u} \right)_v \end{bmatrix}$$

$$= \left[\left(\frac{\partial p}{\partial v} \right)_t \quad \left(\frac{\partial p}{\partial t} \right)_v \right] \begin{bmatrix} 1 & 0 \\ \left(\frac{\partial t}{\partial v} \right)_u & \left(\frac{\partial t}{\partial u} \right)_v \end{bmatrix}.$$

This yields the two chain rules

$$\left(\frac{\partial p}{\partial v} \right)_u = \left(\frac{\partial p}{\partial v} \right)_t + \left(\frac{\partial p}{\partial t} \right)_v \left(\frac{\partial t}{\partial v} \right)_u, \tag{28}$$

$$\left(\frac{\partial p}{\partial u} \right)_v = \left(\frac{\partial p}{\partial t} \right)_v \left(\frac{\partial t}{\partial u} \right)_v. \tag{29}$$

Formula (29) is practically obvious. For v constant, the coordinate change is one-dimensional; hence, so is the chain rule. Formula (28) illustrates the occasional necessity for "full-dress" partial derivative notation. Written without subscripts, it would involve unequal variables, both denoted by $\partial p / \partial v$.

EXERCISES

1. Let $Z = [z]$, $X = [x]$, and

$$Y = \begin{bmatrix} u \\ v \end{bmatrix}.$$

(a) Write out the chain rule (26) explicitly in terms of derivatives and partial derivatives for this special case.

(b) Write dz in terms of du and dv by means of the fundamental theorem and divide by dx to obtain again the result in part (a).

(c) What chain rules are being used in Exercise 1, Section 5–4?

2. In each of the following, find $(\partial w/\partial r)_s$ and $(\partial w/\partial s)_r$ by chain rules and also by first getting w in terms of r and s.

(a) $w = \sqrt{x^2 + y^2 + z^2}$; $x = e^r \cos s$, $y = e^r \sin s$, $z = e^s$

(b) $w = \ln (x^2 + y^2 + z^2)$; $x = r + s$, $y = r - s$, $z = rs$

(c) $w = xyz$; $x = r + s$, $y = r - s$, $z = r/s$

3. Let w depend on the three variables x, y, and z. Each of the following entries lists one or more variables. In each case, suppose that x, y, and z depend on the variable(s) listed. Derive a chain rule giving the derivative(s) of w with respect to the new variable(s) in terms of the derivatives of w with respect to x, y, and z and the derivatives of x, y, and z with respect to the new variable(s).

(a) t (b) x (c) t, x (d) u, v

(e) x, y (f) x, y, t (g) x, u, v (h) t, u, v

4. Let $x_1, x_2, \ldots, x_6$ be variables giving the market prices of six different basic commodities. Assume market conditions under which any three of these prices can be set arbitrarily to determine the other three prices. The partial derivative

$$\left(\frac{\partial x_1}{\partial x_2} \right)_{x_3 x_4}$$

is called the *marginal price* of commodity 1 with respect to the price of commodity 2 for constant prices of commodities 3 and 4. Find a chain rule relating each of the following pairs of marginal prices.

(a) $\left(\dfrac{\partial x_1}{\partial x_2} \right)_{x_3 x_4}$ and $\left(\dfrac{\partial x_1}{\partial x_2} \right)_{x_3 x_5}$

(b) $\left(\dfrac{\partial x_1}{\partial x_2} \right)_{x_3 x_4}$ and $\left(\dfrac{\partial x_1}{\partial x_2} \right)_{x_5 x_6}$

5. Let x, y, z, and t be the position and time coordinates. In a hydrodynamics problem, let $w = f(x, y, z, t)$; that is, w is a variable associated with the moving fluid which depends on position and time. Suppose that the flow is described by $x = \phi_1(t)$, $y = \phi_2(t)$, $z = \phi_3(t)$ which give the position of a certain particle of the fluid at time t.

(a) There are now two "time derivatives" of w:

$$\frac{dw}{dt} \quad \text{and} \quad \left(\frac{\partial w}{\partial t}\right)_{xyz}$$

How are these determined by the given analytic setup?

(b) Describe the physical significance of each of these two derivatives. In general, they are unequal; this should be intuitively clear from their physical description.

(c) Find a chain rule connecting these two derivatives.

6–6 Inversion. Suppose

$$X = \begin{bmatrix} x_1 \\ x_2 \\ \vdots \\ x_n \end{bmatrix}, \quad Y = \begin{bmatrix} y_1 \\ y_2 \\ \vdots \\ y_n \end{bmatrix}$$

with each set of entries forming an admissible coordinate set on an n-dimensional manifold. By the fundamental theorem on differentials,

$$dY = D_X Y \, dX \tag{30}$$

and also

$$dX = D_Y X \, dY. \tag{31}$$

However, solving (31) for dY yields

$$dY = (D_Y X)^{-1} \, dX,$$

so comparison with (30) shows that

$$(D_Y X)^{-1} = D_X Y. \tag{32}$$

If $n = 1$, (32) is the simple inversion formula for derivatives, $D_x y = 1/D_y x$. However, in higher dimensions, it says that to get partial derivatives "going the other way" one must invert a matrix. It was noted in Section 4–4 that

$$A^{-1} = \frac{\text{adj } A}{\det A}. \tag{33}$$

It was also noted that, for a 2×2 matrix, the adjoint is very simply described:

$$\text{adj} \begin{bmatrix} a & b \\ c & d \end{bmatrix} = \begin{bmatrix} d & -b \\ -c & a \end{bmatrix}; \tag{34}$$

reverse the main diagonal entries and change the signs of the others. In higher dimensions, a formula is probably not worthwhile. If the adjoint is needed, compute it directly from the definition. However, for certain Jacobian matrices there are labor-saving tricks for finding the inverse (see Example 3, below).

EXAMPLES

1. Let z be a variable on the plane and let x, y, r, θ be the usual rectangular and polar coordinates. Find $\partial z/\partial x$ and $\partial z/\partial y$ in terms of r, θ, $\partial z/\partial r$, and $\partial z/\partial \theta$. Introduce the usual matrices

$$X = \begin{bmatrix} x \\ y \end{bmatrix}, \qquad U = \begin{bmatrix} r \\ \theta \end{bmatrix}, \qquad Z = [z].$$

One is required to find expressions for $D_X Z$; so get a chain rule for this:

$$D_X Z \, dX = dZ = D_U Z \, dU = D_U Z \, D_X U \, dX,$$

$$D_X Z = D_U Z \, D_X U. \tag{35}$$

Now, $D_U Z = [\partial z/\partial r \quad \partial z/\partial \theta]$, and these entries are admissible in the answer. However, from

$$x = r \cos \theta,$$

$$y = r \sin \theta,$$

one computes directly

$$D_U X = \begin{bmatrix} \cos \theta & -r \sin \theta \\ \sin \theta & r \cos \theta \end{bmatrix},$$

and [35] calls for $D_X U$. So, apply (32) and (33) to get

$$D_X U = (D_U X)^{-1} = \frac{\text{adj }(D_U X)}{\text{det }(D_U X)}.$$

Using (34) to get adj($D_U X$), one has

$$D_X U = \frac{\begin{bmatrix} r \cos \theta & r \sin \theta \\ -\sin \theta & \cos \theta \end{bmatrix}}{\begin{vmatrix} \cos \theta & -r \sin \theta \\ \sin \theta & r \cos \theta \end{vmatrix}} = \frac{\begin{bmatrix} r \cos \theta & r \sin \theta \\ -\sin \theta & \cos \theta \end{bmatrix}}{r},$$

$$= \begin{bmatrix} \cos \theta & \sin \theta \\ \dfrac{-\sin \theta}{r} & \dfrac{\cos \theta}{r} \end{bmatrix}.$$

Therefore, by (35),

$$\begin{bmatrix} \dfrac{\partial z}{\partial x} & \dfrac{\partial z}{\partial y} \end{bmatrix} = \begin{bmatrix} \dfrac{\partial z}{\partial r} & \dfrac{\partial z}{\partial \theta} \end{bmatrix} \begin{bmatrix} \cos\theta & \sin\theta \\ -\dfrac{\sin\theta}{r} & \dfrac{\cos\theta}{r} \end{bmatrix};$$

$$\frac{\partial z}{\partial x} = \frac{\partial z}{\partial r}\cos\theta - \frac{\partial z}{\partial \theta}\frac{\sin\theta}{r},$$

$$\frac{\partial z}{\partial y} = \frac{\partial z}{\partial r}\sin\theta + \frac{\partial z}{\partial \theta}\frac{\cos\theta}{r}.$$

2. In the thermodynamics setup of Example 2, Section 6–5, an important relation to be satisfied is

$$\left(\frac{\partial u}{\partial v}\right)_t - t\left(\frac{\partial p}{\partial t}\right)_v + p = 0. \tag{36}$$

Rewrite this in the system with u and v as coordinates. Here, partial derivatives of u and p must be transformed; hence let

$$Z = \begin{bmatrix} p \\ u \end{bmatrix},$$

and, as in Example 2, Section 6–5, let

$$X = \begin{bmatrix} v \\ u \end{bmatrix}, \qquad Y = \begin{bmatrix} v \\ t \end{bmatrix}.$$

(These are the two coordinate sets.) Now, (36) contains entries from $D_Y Z$ for which substitutions must be made; so get chain rules for them:

$$D_Y Z \, dY = dZ = D_X Z \, dX = D_X Z \, D_Y X \, dY,$$

$$D_Y Z = D_X Z \, D_Y X. \tag{37}$$

Here is a chain rule, but it will not accomplish the desired purpose because partial derivatives in $D_Y X$ are with respect to v and t, and only those with respect to v and u are wanted. So, rewrite (37) as

$$D_Y Z = D_X Z \, (D_X Y)^{-1}. \tag{38}$$

Specifically, now,

$$D_X Y = \begin{bmatrix} \left(\dfrac{\partial v}{\partial v}\right)_u & \left(\dfrac{\partial v}{\partial u}\right)_v \\ \left(\dfrac{\partial t}{\partial v}\right)_u & \left(\dfrac{\partial t}{\partial u}\right)_v \end{bmatrix} = \begin{bmatrix} 1 & 0 \\ \left(\dfrac{\partial t}{\partial v}\right)_u & \left(\dfrac{\partial t}{\partial u}\right)_v \end{bmatrix};$$

hence

$$(D_X Y)^{-1} = \frac{\begin{bmatrix} \left(\dfrac{\partial t}{\partial u}\right)_v & 0 \\[2mm] -\left(\dfrac{\partial t}{\partial v}\right)_u & 1 \end{bmatrix}}{\begin{vmatrix} 1 & 0 \\[2mm] \left(\dfrac{\partial t}{\partial v}\right)_u & \left(\dfrac{\partial t}{\partial u}\right)_v \end{vmatrix}} = \frac{\begin{bmatrix} \left(\dfrac{\partial t}{\partial u}\right)_v & 0 \\[2mm] -\left(\dfrac{\partial t}{\partial v}\right)_u & 1 \end{bmatrix}}{\left(\dfrac{\partial t}{\partial u}\right)_v},$$

and (38) becomes

$$\begin{bmatrix} \left(\dfrac{\partial p}{\partial v}\right)_t & \left(\dfrac{\partial p}{\partial t}\right)_v \\[2mm] \left(\dfrac{\partial u}{\partial v}\right)_t & \left(\dfrac{\partial u}{\partial t}\right)_v \end{bmatrix} = \frac{1}{\left(\dfrac{\partial t}{\partial u}\right)_v} \begin{bmatrix} \left(\dfrac{\partial p}{\partial v}\right)_u & \left(\dfrac{\partial p}{\partial u}\right)_v \\[2mm] 0 & 1 \end{bmatrix} \begin{bmatrix} \left(\dfrac{\partial t}{\partial u}\right)_v & 0 \\[2mm] -\left(\dfrac{\partial t}{\partial v}\right)_u & 1 \end{bmatrix}.$$

The complete matrix product need not be computed because the only chain rules needed are for $(\partial u/\partial v)_t$ and $(\partial p/\partial t)_v$. Computing appropriate entries in the product, one has

$$\left(\frac{\partial u}{\partial v}\right)_t = \frac{-(\partial t/\partial v)_u}{(\partial t/\partial u)_v}, \qquad \left(\frac{\partial p}{\partial t}\right)_v = \frac{(\partial p/\partial u)_v}{(\partial t/\partial u)_v}.$$

Substituting these into (36) yields

$$\left(\frac{\partial t}{\partial v}\right)_u + t\left(\frac{\partial p}{\partial u}\right)_v - p\left(\frac{\partial t}{\partial u}\right)_v = 0.$$

3. Let

$$x = \rho \sin \phi \cos \theta, \qquad y = \rho \sin \phi \sin \theta, \qquad z = \rho \cos \phi;$$

define,

$$X = \begin{bmatrix} x \\ y \\ z \end{bmatrix}, \qquad U = \begin{bmatrix} \rho \\ \phi \\ \theta \end{bmatrix};$$

compute $D_X U$. By direct computation,

$$D_U X = \begin{bmatrix} \sin \phi \cos \theta & \rho \cos \phi \cos \theta & -\rho \sin \phi \sin \theta \\[2mm] \sin \phi \sin \theta & \rho \cos \phi \sin \theta & \rho \sin \phi \cos \theta \\[2mm] \cos \phi & -\rho \sin \phi & 0 \end{bmatrix},$$

and by (32)

$$D_X U = (D_U X)^{-1}.$$

Now, Section 5–5 explains how to invert a Jacobian matrix. In Section 5–5, to find rectangular components related to dU, one factors

$$D_U X = RA,$$

where R is orthogonal and A is a diagonal matrix. Once this factorization is seen, inversion is simple because the inverse of an orthogonal matrix is its transpose, and the inverse of a diagonal matrix is the diagonal matrix of reciprocals. For the present example,

$$D_U X = \begin{bmatrix} \sin\phi\cos\theta & \cos\phi\cos\theta & -\sin\theta \\ \sin\phi\sin\theta & \cos\phi\sin\theta & \cos\theta \\ \cos\phi & -\sin\phi & 0 \end{bmatrix} \begin{bmatrix} 1 & 0 & 0 \\ 0 & \rho & 0 \\ 0 & 0 & \rho\sin\phi \end{bmatrix};$$

so

$$D_X U = (D_U X)^{-1} = \begin{bmatrix} 1 & 0 & 0 \\ 0 & \dfrac{1}{\rho} & 0 \\ 0 & 0 & \dfrac{1}{\rho\sin\phi} \end{bmatrix} \begin{bmatrix} \sin\phi\cos\theta & \sin\phi\sin\theta & \cos\phi \\ \cos\phi\cos\theta & \cos\phi\sin\theta & -\sin\phi \\ -\sin\theta & \cos\theta & 0 \end{bmatrix}$$

$$= \begin{bmatrix} \sin\phi\cos\theta & \sin\phi\sin\theta & \cos\phi \\ \dfrac{\cos\phi\cos\theta}{\rho} & \dfrac{\cos\phi\sin\theta}{\rho} & \dfrac{-\sin\phi}{\rho} \\ \dfrac{-\sin\theta}{\rho\sin\phi} & \dfrac{\cos\theta}{\rho\sin\phi} & 0 \end{bmatrix}.$$

EXERCISES

1. Let w be a variable on the plane, and let x and y be the rectangular coordinates. A variable of some importance (see Section 7–2) is

$$\left(\frac{\partial w}{\partial x}\right)^2 + \left(\frac{\partial w}{\partial y}\right)^2. \tag{39}$$

For each of the following changes of coordinate variable, find the variable (39) in terms of u, v, $\partial w/\partial u$, and $\partial w/\partial v$.

(a) $x = u - v,\ y = u + v$ (b) $x = u^2 - v^2,\ y = 2uv$

(c) $x = e^u \cos v,\ y = e^u \sin v$ (d) $x = 2u + 3v,\ y = 4u - 5v$

(e) $x = u \cosh v,\ y = u \sinh v$

2. Let w be given in three-space in terms of x, y, and z. For each of the following changes of coordinate variable find the variable

$$\left(\frac{\partial w}{\partial x}\right)_{yz}^2 + \left(\frac{\partial w}{\partial y}\right)_{xz}^2 + \left(\frac{\partial w}{\partial z}\right)_{xy}^2$$

in terms of the new variables and partials of w with respect to them.

  (a) $x = r \cos \theta$, $y = r \sin \theta$, $z = z$
  (b) $x = \rho \sin \phi \cos \theta$, $y = \rho \sin \phi \sin \theta$, $z = \rho \sin \phi$
  (c) $x = t + u + v$, $y = t + u - v$, $z = t - u - v$

3. In Example 2, the relation

$$\left(\frac{\partial u}{\partial v}\right)_t - t\left(\frac{\partial p}{\partial t}\right)_v + p = 0$$

was converted into a system with u and v as coordinates. Rewrite this basic equation of thermodynamics in each of the following coordinate systems.

  (a) p, v (b) p, t (c) p, u (d) u, t

6–7 Implicit relations, multidimensional loci. The basic theorem on characterization of manifolds was stated in Section 5–6. It may be summarized in the following manner. Let

$$X = \begin{bmatrix} x_1 \\ x_2 \\ \vdots \\ x_n \end{bmatrix}, \qquad T = \begin{bmatrix} t_1 \\ t_2 \\ \vdots \\ t_k \end{bmatrix}, \qquad U = \begin{bmatrix} u_1 \\ u_2 \\ \vdots \\ u_n \end{bmatrix}.$$

Assume that X and T together form a coordinate set on $(n + k)$-space, and assume that each entry in U is a differentiable variable on this $(n + k)$ dimensional manifold. Let M be the locus of $U = 0$, and assume that at some point p_0 of M,

$$\det D_X U \neq 0.$$

Then, some neighborhood of p_0 in M is a k-dimensional manifold with the entries in T forming a coordinate set, and with the entries in X as differentiable variables.

The present section is concerned with the following problem. Since U consists of differentiable variables on $(n + k)$-space, the matrices of partial derivatives

$$D_X U \quad \text{and} \quad D_T U \tag{40}$$

are defined. One concludes from the theorem that

$$D_T X \tag{41}$$

is defined, and that the question at issue here is, How does one find (41) from (40)?

First, consider an additional item in matrix algebra. Let A, B, C, and D be matrices. Assume that A and B have the same number of rows, and denote by

$$[A, B]$$

the matrix obtained by placing the two arrays side by side. The comma in this notation is inserted only to eliminate any possible confusion with the matrix product AB. Let C and D have the same number of columns and denote by

$$\begin{bmatrix} C \\ D \end{bmatrix}$$

the matrix obtained by placing the two arrays one over the other. Assuming that the dimensions are correct, it can be shown that

$$[A, B]\begin{bmatrix} C \\ D \end{bmatrix} = AC + BD.$$

That is, matrix multiplication may be performed by "blocks." The student would do well to check this by direct computation for the case in which each of A, B, C, and D is 2×2. The general proof is quite straightforward but will not be given here.

To return to the main problem, note first that

$$D_{\begin{bmatrix} X \\ T \end{bmatrix}}U = [D_X U, D_T U] \qquad \text{and} \qquad d\begin{bmatrix} X \\ T \end{bmatrix} = \begin{bmatrix} dX \\ dT \end{bmatrix}.$$

Since M is the locus of $U = 0$, clearly

$$D_{\begin{bmatrix} X \\ T \end{bmatrix}}U = 0$$

on M. Thus, on the tangent bundle for M,

$$0 = D_{\begin{bmatrix} X \\ T \end{bmatrix}}U \, d\begin{bmatrix} X \\ T \end{bmatrix} = [D_X U, D_T U]\begin{bmatrix} dX \\ dT \end{bmatrix}$$

$$= [D_X U, D_T U]\begin{bmatrix} D_T X \, dT \\ dT \end{bmatrix}$$

$$= D_X U \, D_T X \, dT + D_T U \, dT.$$

Thus, on M,
$$D_X U \, D_T X = -D_T U,$$
so
$$D_T X = -(D_X U)^{-1} \, D_T U$$

$$= \frac{-(\text{adj } D_X U) \, D_T U}{\det D_X U}. \tag{42}$$

Happily, this denominator was assumed not zero, so (42) solves the problem. It gives the matrix (41) in terms of those in (40).

Caution. Do not confuse the results obtained in this section with the chain-rule formulas of Section 6–5. A standard chain rule reads

$$D_Y Z \, D_X Y = D_X Z,$$

whereas a basic formula in the discussion of implicit relations is

$$D_X U \, D_T X = -D_T U.$$

Actually the difference between the two is not nearly so surprising as the unfortunate similarity, for the variables are related in entirely different ways in the two problems.

Once the details of a general formula have been checked, a good way to remember it is to reduce it to the one-dimensional case. For a simple chain rule:

$$z = f(y), \qquad y = g(x), \qquad D_y z \, D_x y = D_x z.$$

For an implicit relation:

$$y = \phi(x, t) = 0, \qquad du = \frac{\partial u}{\partial x} \, dx + \frac{\partial u}{\partial t} \, dt = 0.$$

Divide by dt:

$$\frac{\partial u}{\partial x} \frac{dx}{dt} + \frac{\partial u}{\partial t} = 0, \qquad \frac{\partial u}{\partial x} \frac{dx}{dt} = -\frac{\partial u}{\partial t}.$$

EXAMPLES

1. Let M be the locus of

$$x^2 + y^2 + z^2 - t^2 = 0, \qquad 2xy - z^2 = 0.$$

Find the partial derivatives of x and y with respect to z and t on M. Set

$$u = x^2 + y^2 + z^2 - t^2, \qquad v = 2xy - z^2,$$

$$X = \begin{bmatrix} x \\ y \end{bmatrix}, \qquad T = \begin{bmatrix} z \\ t \end{bmatrix}, \qquad U = \begin{bmatrix} u \\ v \end{bmatrix}.$$

Now, (42) applies:

$$D_T U = \begin{bmatrix} 2z & -2t \\ -2z & 0 \end{bmatrix}, \qquad D_X U = \begin{bmatrix} 2x & 2y \\ 2y & 2x \end{bmatrix},$$

$$\text{adj } D_X U = \begin{bmatrix} 2x & -2y \\ -2y & 2x \end{bmatrix}, \qquad \det D_X U = 4(x^2 - y^2);$$

thus

$$\begin{bmatrix} \dfrac{\partial x}{\partial z} & \dfrac{\partial x}{\partial t} \\[2mm] \dfrac{\partial y}{\partial z} & \dfrac{\partial y}{\partial t} \end{bmatrix} = \frac{-1}{4(x^2 - y^2)} \begin{bmatrix} 2x & -2y \\ -2y & 2x \end{bmatrix} \begin{bmatrix} 2z & -2t \\ -2z & 0 \end{bmatrix}$$

$$= \frac{-1}{4(x^2 - y^2)} \begin{bmatrix} 4z(x + y) & -4xt \\ -4z(x + y) & 4yt \end{bmatrix}.$$

That is,

$$\frac{\partial x}{\partial z} = \frac{-z}{x - y}, \qquad \frac{\partial x}{\partial t} = \frac{xt}{x^2 - y^2},$$

$$\frac{\partial y}{\partial z} = \frac{z}{x - y}, \qquad \frac{\partial y}{\partial t} = \frac{-yt}{x^2 - y^2}.$$

These are not defined where $x^2 - y^2 = 0$, but (z, t) is not an admissible coordinate set at these points.

2. Let M be the locus of

$$u = f(x, y), \qquad x = g(z, t, u). \tag{43}$$

Find the dimension of M, choose an appropriate number of variables as coordinates, and find chain rules giving the partial derivatives of the remaining variables with respect to these coordinates in terms of partial derivatives that may be computed directly from (43). First, the dimension of M is three because it is the locus of two equations in five variables. Let x, y, and z be the coordinates on M. To recast the problem in the notation of the master equation (42), let

$$v = f(x, y) - u, \qquad w = g(z, t, u) - x,$$

$$X = \begin{bmatrix} t \\ u \end{bmatrix}, \qquad T = \begin{bmatrix} x \\ y \\ z \end{bmatrix}, \qquad U = \begin{bmatrix} v \\ w \end{bmatrix}.$$

Now,

$$D_T U = \begin{bmatrix} f_1(x, y) & f_2(x, y) & 0 \\ -1 & 0 & g_1(z, t, u) \end{bmatrix},$$

$$D_X U = \begin{bmatrix} 0 & -1 \\ g_2(z, t, u) & g_3(z, t, u) \end{bmatrix}, \qquad \det D_X U = g_2(z, t, u),$$

$$\text{adj } D_X U = \begin{bmatrix} g_3(z, t, u) & 1 \\ -g_2(z, t, u) & 0 \end{bmatrix}.$$

Thus, by (42),

$$\begin{bmatrix} \dfrac{\partial t}{\partial x} & \dfrac{\partial t}{\partial y} & \dfrac{\partial t}{\partial z} \\[2mm] \dfrac{\partial u}{\partial x} & \dfrac{\partial u}{\partial y} & \dfrac{\partial u}{\partial z} \end{bmatrix} = \dfrac{-1}{g_2(z, t, u)} \begin{bmatrix} g_3(z, t, u) & 1 \\ -g_2(z, t, u) & 0 \end{bmatrix} \begin{bmatrix} f_1(x, y) & f_2(x, y) & 0 \\ -1 & 0 & g_1(z, t, u) \end{bmatrix}$$

$$= \begin{bmatrix} \dfrac{1 - f_1(x, y)g_3(z, t, u)}{g_2(z, t, u)} & \dfrac{-f_2(x, y)g_3(z, t, u)}{g_2(z, t, u)} & \dfrac{-g_1(z, t, u)}{g_2(z, t, u)} \\[3mm] f_1(x, y) & f_2(x, y) & 0 \end{bmatrix}.$$

EXERCISES

1. Show that if $X = [x]$, $T = [t]$, $U = [f(x, t)]$, the formula

$$D_T X = -(D_X U)^{-1} D_T U$$

reduces to formula (5) of Section 6–2.

2. Each of the following pairs of equations determines (locally) a manifold on which x and y are given in terms of u and v. In each case find the partial derivatives of x and y with respect to u and v.

    (a) $x + 2y - 3u - 2v = 0$, $\quad 2x + y + u + v = 0$

    (b) $x^2 + y^2 + u^2 - 2v^2 = 0$, $\quad x^2 - y^2 + u^2 + v^2 = 0$

    (c) $xe^y - uv + y = 0$, $\quad ye^v - xv + x = 0$

    (d) $x^2 + y + u = 0$, $\quad y^2 - x - v = 0$

3. Let $w = xy$. Find the partial derivatives of w with respect to u and v on each of the manifolds in Exercise 2.

4. Under suitable conditions each of the following sets of equations defines (locally) a manifold. In each case determine the dimensionality of the manifold. Let the last letters of the alphabet (an appropriate number of them) designate the coordinates on this manifold, and find chain rules giving partial derivatives

of the other variables with respect to these coordinates in terms of partials coming from the given equations.

(a) $f(x, y, z) = 0$ (b) $f(x, y, z) = 0$, $g(x, y, z) = 0$

(c) $z = f(x, y)$, $x = g(y, z)$ (d) $f(w, x, y, z) = 0$, $z = g(x, y)$

(e) $v = f(y, z)$, $u = g(x, y)$, $z = \phi(u, v)$

(f) $f(u, v, x, y, z) = 0$, $g(u, v, x, y, z) = 0$, $\phi(u, v, x, y, z) = 0$

(g) $z = f(u, v, w)$, $y = g(u, v, w)$, $x = \phi(u, v, w)$

(h) $x = f(u, v)$, $y = g(u, v)$, $z = \phi(u, v)$

(i) $f(u, v, x, y, z) = 0$, $g(u, v, x, y, z) = 0$

(j) $x = f(u)$, $y = g(u)$, $z = \phi(u)$

CHAPTER 7

VECTOR DIFFERENTIAL CALCULUS

7-1 Introduction. The vector position variable $\mathbf{r}$ is given by

$$\mathbf{r} = x\mathbf{i} + y\mathbf{j} + z\mathbf{k}.$$

Let $\mathbf{w}$ be another vector variable on three-space. The present chapter will be concerned with the differential calculus of the mapping

$$\mathbf{r} \to \mathbf{w}. \tag{1}$$

Let X be the matrix of orthonormal component variables,

$$X = \begin{bmatrix} x \\ y \\ z \end{bmatrix}.$$

The mapping (1) may be described in terms of these components:

$$X(\mathbf{r}) \to X(\mathbf{w}). \tag{2}$$

It must be borne in mind, however, that (1) is the basic mapping here, and (2) is only one description of it. In terms of another set of components, there might be quite a different matrix-to-matrix mapping.

Now, dX is the component matrix for the tangent spaces; and vector differentials have been defined by the rule

$$dX(d\mathbf{w}) = d(X(\mathbf{w})). \tag{3}$$

In words, the $\mathbf{i}$, $\mathbf{j}$, $\mathbf{k}$ components of $d\mathbf{w}$ are the differentials of the $\mathbf{i}$, $\mathbf{j}$, $\mathbf{k}$ components of $\mathbf{w}$. As noted in Example 8, Section 5-2, this definition of $d\mathbf{w}$ is independent of the choice of basis. That is, if $Y = BX$ where B is a matrix of constants, then $dY = B\,dX$; so

$$dY(d\mathbf{w}) = B\,dX(d\mathbf{w}) = B\,d(X(\mathbf{w})) = d(BX(\mathbf{w})) = d(Y(\mathbf{w})). \tag{4}$$

It is essential to note that (4) depends on the fact that B is constant. More generally, if $Y = BX$, then

$$dY = B\,dX + (dB)X,$$

and the computation of vector differentials in components generated by curvilinear coordinates is a little more complicated (see Section 7-5).

For the present, however, consider only the rectangular case. Let (2) be a description of (1); by the fundamental theorem on differentials,

$$d(X(\mathbf{w})) = D_{X(\mathbf{r})}X(\mathbf{w})\, d(X(\mathbf{r})),$$

or in the light of (3),

$$dX(d\mathbf{w}) = D_{X(\mathbf{r})}X(\mathbf{w})\, dX(d\mathbf{r}). \tag{5}$$

Now, (5) is a matrix description of a linear transformation T such that $d\mathbf{w} = T\, d\mathbf{r}$. As seen in (5), the matrix of T in X-components is $D_{X(\mathbf{r})}X(\mathbf{w})$, so T itself will be denoted by

$$D_{\mathbf{r}}\mathbf{w}.$$

Then,

$$d\mathbf{w} = D_{\mathbf{r}}\mathbf{w}\, d\mathbf{r}. \tag{6}$$

There will also appear vector-to-scalar mappings,

$$\mathbf{r} \to u. \tag{7}$$

In component form,

$$X(\mathbf{r}) \to u, \tag{8}$$

or (in more familiar guise) $u = f(x, y, z)$. Again one has, from the fundamental theorem,

$$du = D_{X(\mathbf{r})}u\, dX(d\mathbf{r}), \tag{9}$$

where

$$D_{X(\mathbf{r})}u = \left[\frac{\partial u}{\partial x} \quad \frac{\partial u}{\partial y} \quad \frac{\partial u}{\partial z}\right].$$

Now, (9) describes a linear transformation of vectors into scalars. Denote this transformation by

$$D_{\mathbf{r}}u;$$

then (9) is the matrix form of

$$du = D_{\mathbf{r}}u\, d\mathbf{r}. \tag{10}$$

Summary. A vector-to-vector mapping (1) is given; it is differentiable but probably not linear. However, on the tangent spaces, it generates the linear vector-to-vector transformation seen in (6). The "derivative of a vector with respect to a vector" appearing in (6) is a symbol for this linear transformation. The mapping (1) may be written in terms of components as in (2). In that case, the linear transformation in (6) is described by the matrix equation (5).

If, instead of (1), one has the vector-to-scalar mapping (7), then there is generated on the tangent spaces a linear vector-to-scalar transformation. Notation for this transformation assumes the form of a "derivative of a scalar with respect to a vector." If (7) is written in terms of components as it is in (8), then the matrix equation (9) describes the linear transformation displayed directly in (10).

The present chapter will be concerned with the linear transformations

$$D_r u \quad \text{and} \quad D_r w$$

and with various operators associated with and generated by these transformations. The operators concerned are called *gradient*, *divergence*, and *curl*. The plan of the chapter is as follows. First, the transformations will be represented in the matrix forms (5) and (9), and the gradient, divergence, and curl operators will be described in terms of $(\mathbf{i}, \mathbf{j}, \mathbf{k})$-components. It is in this form that they are most commonly encountered in practice. However, there will be a return to the vector forms (6) and (10), and it will be shown how gradient, divergence, and curl can be given coordinate-free definitions in terms of the linear transformations $D_r \mathbf{w}$ and $D_r u$ themselves. From these more basic definitions one can then find expressions for the operators in other coordinate systems.

EXERCISE

1. In the above discussion each of the following is a matrix, a scalar, a vector, or a linear transformation on a vector space. Classify each. For each matrix give the dimensions. For each linear transformation give the domain and range spaces.

(a) r
(b) X
(c) $r(\mathbf{w})$
(d) $\mathbf{w}$
(e) $D_r u$
(f) $D_{X(r)} u$
(g) dX
(h) $dX(d\mathbf{r})$
(i) $dX(d\mathbf{w})$
(j) $d\mathbf{r}$
(k) $D_r \mathbf{w}$
(l) $\mathbf{r}$
(m) $D_{X(r)} X(\mathbf{w})$
(n) $X(\mathbf{w})$
(o) du

7–2 Gradients. If equation (9) is written out in detail, it reads

$$du = \begin{bmatrix} \dfrac{\partial u}{\partial x} & \dfrac{\partial u}{\partial y} & \dfrac{\partial u}{\partial z} \end{bmatrix} \begin{bmatrix} dx \\ dy \\ dz \end{bmatrix}$$

$$= \frac{\partial u}{\partial x} dx + \frac{\partial u}{\partial y} dy + \frac{\partial u}{\partial z} dz. \tag{11}$$

Thus, it assumes the familiar form of a scalar product. Specifically, if **grad** u is a vector defined by

$$\mathbf{grad}\ u = \frac{\partial u}{\partial x}\mathbf{i} + \frac{\partial u}{\partial y}\mathbf{j} + \frac{\partial u}{\partial z}\mathbf{k}, \tag{12}$$

then (11) may be written

$$du = (\mathbf{grad}\ u) \cdot d\mathbf{r}. \tag{13}$$

The vector **grad** u is called the *gradient of* u. For the present, (12) may be taken as the definition of **grad** u, though it will appear later (Section 7–4) that (13) serves better than (12) as a definition.

Let C be a curve in three-space; let s be the arc-length variable on C; let p be a point of C. On T_p^1, the tangent line to C at p, ds_p measures distances (see Section 2–2). However, in T_p^3, the tangent to three-space at p, $d\mathbf{r}_p$ is the position vector with respect to p as the origin. Thus, for any point q on T_p^1, $d\mathbf{r}_p(q)$ is the vector from p to q, and $ds_p(q)$ is the directed (scalar) distance from p to q. Hence, over the tangent bundle to C,

$$d\mathbf{r} = \mathbf{T}\ ds, \tag{14}$$

where at each point of C, $\mathbf{T}$ gives the unit tangent vector in the direction of positive ds.

In the light of (14), one may divide (13) by ds to get

$$\frac{du}{ds} = (\mathbf{grad}\ u) \cdot \mathbf{T} \tag{15}$$

on C. The differential quotient du/ds is called the *directional derivative of* u *in the direction of* C, and (15) shows that this directional derivative is equal to the component of **grad** u in the positive tangential direction to C. Figure 7–1 shows a geometric construction of du/ds from **grad** u.

For any constant c, the locus of

$$u = c$$

is called a *level surface for* u. On the tangent plane to such a level surface, clearly,

$$du = 0.$$

Thus, by (13), if $d\mathbf{r}$ is tangent to a level surface for u, then

$$(\mathbf{grad}\ u) \cdot d\mathbf{r} = 0;$$

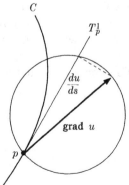

FIGURE 7–1

hence **grad** u is normal (perpendicular) to the level surface. It follows at once from (15) that in this normal direction

$$\frac{du}{ds} = |\mathbf{grad}\ u| = \sqrt{\left(\frac{\partial u}{\partial x}\right)^2 + \left(\frac{\partial u}{\partial y}\right)^2 + \left(\frac{\partial u}{\partial z}\right)^2}.$$

Furthermore, the directional derivative is a maximum in this direction (Fig. 7–1).

Summary. At any point, the maximal rate of change of u is in the direction of the gradient, and this direction is always perpendicular to the level surface for u.

There are many applications of this bit of theory. The following statements are a few samples.

(i) *Analytic geometry.* Given the equation of a surface in the form

$$f(x, y, z) = 0,$$

one sets $u = f(x, y, z)$; then **grad** u gives the normal direction at any point. Given such a normal vector, the equation of the tangent plane may be written down at once (see Section 4–9). If p is a point on the surface and $(\mathbf{grad}\ u)_p$ is the value of **grad** u at p, then in terms of

$$\mathbf{r} = x\mathbf{i} + y\mathbf{j} + z\mathbf{k},$$

the equation of the tangent plane at p is

$$(\mathbf{grad}\ u)_p \cdot [\mathbf{r} - \mathbf{r}(p)] = 0.$$

(ii) *Electrostatics.* Let $u(p)$ be the electrostatic potential at p. The effect of the electrostatic field on a charged particle will be to accelerate it in the direction of maximum rate of decrease of potential. This is the direction of $-\mathbf{grad}\ u$, and, indeed, in properly chosen units, the force on the particle is exactly $-\mathbf{grad}\ u$. Level surfaces for u are called *equipotential surfaces*, and the present theory shows that the force is always normal to the equipotential surface.

(iii) *Lagrange multipliers.* At a stationary point for z on the locus of $u = v = 0$, the locus is parallel to a level surface for z; hence **grad** z is perpendicular to the locus. However, the locus is the intersection of level surfaces for u and v; so **grad** u and **grad** v are always perpendicular to the locus. Thus, at a stationary point, these three gradient vectors are coplanar; that is, they are linearly dependent:

$$\mathbf{grad}\ z + \lambda\ \mathbf{grad}\ u + \mu\ \mathbf{grad}\ v = 0.$$

These are the Lagrange equations.

EXAMPLES

1. Let

$$u = xyz;$$

find the directional derivative of u at the point with coordinates $(1, -2, 4)$ in the direction toward the point $(2, 4, -1)$. Here

$$\mathbf{grad}\ u = yz\mathbf{i} + xz\mathbf{j} + xy\mathbf{k},$$

and hence at $(1, -2, 4)$,

$$\mathbf{grad}\ u = -8\mathbf{i} + 4\mathbf{j} - 2\mathbf{k}.$$

The vector from the first of these points to the second is

$$(2 - 1)\mathbf{i} + [4 - (-2)]\mathbf{j} + (-1 - 4)\mathbf{k} = \mathbf{i} + 6\mathbf{j} - 5\mathbf{k},$$

and the unit vector in this direction is

$$\mathbf{T} = \frac{\mathbf{i} + 6\mathbf{j} - 5\mathbf{k}}{\sqrt{1^2 + 6^2 + 5^2}} = \frac{\mathbf{i} + 6\mathbf{j} - 5\mathbf{k}}{\sqrt{62}}.$$

Thus,

$$\frac{du}{ds} = (\mathbf{grad}\ u) \cdot \mathbf{T} = \frac{-8 + 24 + 10}{\sqrt{62}} = \frac{26}{\sqrt{62}}.$$

2. Let

$$u = x^2 + y^2 - z^2,$$

and let C be the curve defined by the parametric equations

$$x = 2t, \qquad y = t^2, \qquad z = 2t^3.$$

Find an expression for du/ds in the direction tangent to C at an arbitrary point of C.

First,

$$\mathbf{grad}\ u = 2x\mathbf{i} + 2y\mathbf{j} - 2z\mathbf{k}.$$

On the tangent bundle to C,

$$d\mathbf{r} = dx\mathbf{i} + dy\mathbf{j} + dz\mathbf{k} = (2\mathbf{i} + 2t\mathbf{j} + 6t^2\mathbf{k})\ dt,$$

and

$$ds = \sqrt{dx^2 + dy^2 + dz^2} = 2\sqrt{1 + t^2 + 9t^4}\ dt,$$

and therefore the unit tangent is

$$\mathbf{T} = \frac{d\mathbf{r}}{ds} = \frac{\mathbf{i} + t\mathbf{j} + 3t^2\mathbf{k}}{\sqrt{1 + t^2 + 9t^4}}.$$

Thus,

$$\frac{du}{ds} = (\mathbf{grad}\ u) \cdot \mathbf{T} = \frac{2x + 2yt - 6zt^2}{\sqrt{1 + t^2 + 9t^4}} = \frac{4t + 2t^3 - 12t^5}{\sqrt{1 + t^2 + 9t^4}}.$$

3. Find the equation of the tangent plane to the locus of

$$z = x^2 + y^2$$

at the point with coordinates $(1, -1, 2)$. Rewrite the equation of the surface in the form

$$x^2 + y^2 - z = 0;$$

then let $u = x^2 + y^2 - z$, and

$$\mathbf{grad}\ u = 2x\mathbf{i} + 2y\mathbf{j} - \mathbf{k} = 2\mathbf{i} - 2\mathbf{j} - \mathbf{k}$$

at the point in question. Thus, the required tangent plane has the equation

$$2(x - 1) - 2(y + 1) - (z - 2) = 0;$$

$$2x - 2y - z = 2.$$

Incidentally, the normal line to the given surface at the given point has equations

$$\frac{x - 1}{2} = \frac{y + 1}{-2} = \frac{z - 2}{-1}.$$

EXERCISES

1. From each of the following equations, find, in terms of x, y, and z, the gradient of w, the directional derivative of w in the direction of the gradient, and the directional derivative of w in the direction away from the origin.

(a) $w = xyz$ (b) $w = \sqrt{x^2 + y^2 + z^2}$

(c) $w = \arctan{(y/x)}$ (d) $w = xy/z$

(e) $w = \ln{(x^2 + y^2 + z^2)}$ (f) $w = 1/\sqrt{x^2 + y^2 + z^2}$

(g) $w = \arccos{(z/\sqrt{x^2 + y^2 + z^2})}$ (h) $w = xy + xz + yz$

2. Each of parts (a) through (h) gives parametric equations of a curve in three-space. In each case, take the variable w from the corresponding part of

Exercise 1, and find the directional derivative of w in the positive tangential direction to the given curve.

    (a) $x = t, \ y = t^2, \ z = t^3$

    (b) $x = 5 \sin t, \ y = 4 \cos t, \ z = 3 \cos t$

    (c) $x = \cos t, \ y = \sin t, \ z = t$

    (d) $x = t, \ y = t^2, \ z = 1/t$

    (e) $x = t, \ y = 2t, \ z = 3t$

    (f) $x = 1 - t, \ y = 2 - t, \ z = 3 - t$

    (g) $x = t \cos t, \ y = t \sin t, \ z = 2t$

    (h) $x = y = z$

3. In each of the following exercises coordinates of two points are given. In each case take the variable w from the corresponding part of Exercise 1, and find the directional derivative of w at the first point in the direction toward the second.

    (a) $(1, 1, 1), \ (2, 3, 4)$ (b) $(1, 2, 2), \ (0, 0, 2)$

    (c) $(1, 1, 2), \ (2, 2, -1)$ (d) $(1, 1, 2), \ (2, 1, 1)$

    (e) $(2, -1, 3), \ (-1, 2, 3)$ (f) $(2, 1, -2), \ (0, 0, 0)$

    (g) $(1, 1, 1), \ (2, 2, -1)$ (h) $(1, -1, 2), \ (1, 2, -1)$

4. Each of the following parts gives the equation of a surface and coordinates of a point on it. In each case find the equation of the tangent plane to the given surface at the given point.

    (a) $xyz = 1, \ (1, 1, 1)$ (b) $x^2 + y^2 + z^2 = 6, \ (1, 2, 1)$

    (c) $xy + xz + yz = 3, \ (1, 1, 1)$ (d) $x^3 + y^3 - z^3 = 1, \ (1, 1, 1)$

    (e) $z = x^2 + 2y^2, \ (1, 1, 3)$ (f) $z = \sin (x + y), \ (\pi/4, \pi/4, 1)$

    (g) $z = e^{xy}, \ (1, 1, e)$ (h) $z = \arctan (y/x), \ (1, 1, \pi/4)$

5. For each part of Exercise 4, find equations of the line normal to the given surface at the given point.

6. Show that if at a given point three directional derivatives of a variable in three noncoplanar directions are specified, all directional derivatives of this variable at the given point are determined.

7. Show that the gradient of $1/\sqrt{x^2 + y^2 + z^2}$ is directed toward the origin and has magnitude $1/(x^2 + y^2 + z^2)$. Discuss this result in terms of forces and potentials.

7-3 Divergence and curl. The symbol ∇, pronounced *del*, is commonly used in vector analysis to denote the vector derivative operator

$$\nabla = \mathbf{i} \frac{\partial}{\partial x} + \mathbf{j} \frac{\partial}{\partial y} + \mathbf{k} \frac{\partial}{\partial z}.$$

In this notation

$$\mathbf{grad} \ u = \nabla u,$$

and this is probably the most commonly used notation for gradient.

Now, let

$$\mathbf{w} = w_1\mathbf{i} + w_2\mathbf{j} + w_3\mathbf{k}$$

be a vector variable on three-space. Purely formally, there seem to be two ways to operate on $\mathbf{w}$ with $\boldsymbol{\nabla}$:

$$\boldsymbol{\nabla} \cdot \mathbf{w} = \frac{\partial w_1}{\partial x} + \frac{\partial w_2}{\partial y} + \frac{\partial w_3}{\partial z},$$

$$\boldsymbol{\nabla} \times \mathbf{w} = \left(\frac{\partial w_3}{\partial y} - \frac{\partial w_2}{\partial z}\right)\mathbf{i} + \left(\frac{\partial w_1}{\partial z} - \frac{\partial w_3}{\partial x}\right)\mathbf{j} + \left(\frac{\partial w_2}{\partial x} - \frac{\partial w_1}{\partial y}\right)\mathbf{k}.$$

The operators $\boldsymbol{\nabla} \cdot$ and $\boldsymbol{\nabla} \times$ are called, respectively, *divergence* and *curl*. They are also denoted by div and **curl**; that is

$$\text{div } \mathbf{w} = \boldsymbol{\nabla} \cdot \mathbf{w}, \qquad \textbf{curl } \mathbf{w} = \boldsymbol{\nabla} \times \mathbf{w}.$$

Note that div is written in lightface to indicate a scalar operator, while **curl**, like **grad**, is set in boldface type to indicate a vector operator.

The use of the formal operator $\boldsymbol{\nabla}$ makes div and **curl** seem like the "natural" derivative operators on a vector variable. It will appear in Section 7–4 that from a purely mathematical standpoint these operators are not as natural as they seem in the present context. Nevertheless, there are quite cogent physical reasons for considering div and **curl**, and for this reason they deserve some attention. The physical significance of divergence and curl (along with a reason for the names) will appear in the study of vector integral calculus (Section 10–5). The purpose of the present section is essentially only to introduce the symbols.

A gradient is a vector variable, and thus one may form its divergence and curl. Formally,

$$\textbf{curl grad } u = \boldsymbol{\nabla} \times \boldsymbol{\nabla}u,$$

and one begins to suspect that the result is zero because $\boldsymbol{\nabla} \times \boldsymbol{\nabla}$ looks like a vector product of "parallel vectors." Of course, it is no such thing, because $\boldsymbol{\nabla}$ is an operator, not a vector. However, the analogy does hold; direct computation shows that each component of $\boldsymbol{\nabla} \times \boldsymbol{\nabla}u$ assumes the form

$$\frac{\partial^2 u}{\partial x_i\,\partial x_j} - \frac{\partial^2 u}{\partial x_j\,\partial x_i},$$

and these are zero if u has continuous second partial derivatives. The result,

$$\textbf{curl grad } u = 0,$$

is quite useful because it says that for $\mathbf{w}$ to be a gradient, it must be true that

$$\boldsymbol{\nabla} \times \mathbf{w} = 0.$$

That is, not every vector variable is a gradient; only those with zero curl will qualify. It will be shown in Section 10–6, the converse is also true. If $\nabla \times \mathbf{w} = 0$, then there is a scalar u such that $\mathbf{w} = \nabla u$.

The divergence of a gradient is also quite important;

$$\text{div grad } u = \nabla \cdot \nabla u = \nabla^2 u$$

is called the *Laplacian* of u. The symbol ∇^2 (abbreviation of $\nabla \cdot \nabla$) is commonly used for the Laplacian operator. In terms of partial derivatives,

$$\nabla^2 u = \frac{\partial^2 u}{\partial x^2} + \frac{\partial^2 u}{\partial y^2} + \frac{\partial^2 u}{\partial z^2}.$$

A *harmonic* variable is one whose Laplacian vanishes, that is, u is harmonic provided that

$$\nabla^2 u = 0.$$

The equation $\nabla^2 u = 0$ is called the *Laplace differential equation*.

Now, $\mathbf{w}$ is a gradient if and only if $\nabla \times \mathbf{w} = 0$; and assuming $\mathbf{w} = \nabla u$, u is harmonic if and only if $\nabla \cdot \mathbf{w} = 0$. So, the gradient of a harmonic variable is characterized by the fact that its divergence and curl both vanish.

EXAMPLES

1. The vector

$$\mathbf{w} = \frac{-x\mathbf{i} - y\mathbf{j} - z\mathbf{k}}{(x^2 + y^2 + z^2)^{3/2}}$$

is directed toward the origin and has magnitude

$$\sqrt{\frac{x^2 + y^2 + z^2}{(x^2 + y^2 + z^2)^3}} = \frac{1}{x^2 + y^2 + z^2}.$$

Thus, it describes an inverse-square, attractive force field, say the gravitational field due to a mass point at the origin.

Now

$$\frac{\partial w_3}{\partial y} = \frac{3yz}{(x^2 + y^2 + z^2)^{5/2}} = \frac{\partial w_2}{\partial z},$$

$$\frac{\partial w_1}{\partial z} = \frac{3xz}{(x^2 + y^2 + z^2)^{5/2}} = \frac{\partial w_3}{\partial x},$$

$$\frac{\partial w_2}{\partial x} = \frac{3xy}{(x^2 + y^2 + z^2)^{5/2}} = \frac{\partial w_1}{\partial y},$$

so

$$\nabla \times \mathbf{w} = 0.$$

Therefore, $\mathbf{w} = \nabla u$ for an appropriate u; and indeed, it is not hard to find u. Assuming $\mathbf{w} = \nabla u$, one has

$$\frac{\partial u}{\partial x} = w_1 = \frac{-x}{(x^2 + y^2 + z^2)^{3/2}},$$

thus, informally,

$$u = \int \frac{-x\,dx}{(x^2 + y^2 + z^2)^{3/2}},$$

in which y and z are to be treated as constants. Computing this primitive, one has

$$u = \int \frac{-x\,dx}{(x^2 + y^2 + z^2)^{3/2}} = \frac{1}{\sqrt{x^2 + y^2 + z^2}}.$$

This is a tentative answer, subject to a check that

$$\frac{\partial u}{\partial y} = \frac{-y}{(x^2 + y^2 + z^2)^{3/2}}, \qquad \frac{\partial u}{\partial z} = \frac{-z}{(x^2 + y^2 + z^2)^{3/2}}.$$

Happily, these results check out; so a u has been found for which $\mathbf{w} = \nabla u$. This u is the gravitational potential for the given force field. Further discussion of the problem of finding a variable from its gradient will appear in Section 10–6.

To return to $\mathbf{w}$, however, note that

$$\frac{\partial w_1}{\partial x} = \frac{-(x^2 + y^2 + z^2)^{3/2} + 3x^2(x^2 + y^2 + z^2)^{1/2}}{(x^2 + y^2 + z^2)^3}$$

$$= \frac{2x^2 - y^2 - z^2}{(x^2 + y^2 + z^2)^{5/2}};$$

similarly,

$$\frac{\partial w_2}{\partial y} = \frac{2y^2 - x^2 - z^2}{(x^2 + y^2 + z^2)^{5/2}}, \qquad \frac{\partial w_3}{\partial z} = \frac{2z^2 - x^2 - y^2}{(x^2 + y^2 + z^2)^{5/2}},$$

and hence

$$\nabla \cdot \mathbf{w} = \frac{\partial w_1}{\partial x} + \frac{\partial w_2}{\partial y} + \frac{\partial w_3}{\partial z} = 0.$$

Now $\nabla \cdot \mathbf{w} = \nabla^2 u$; so the potential

$$u = \frac{1}{\sqrt{x^2 + y^2 + z^2}}$$

is harmonic.

2. The electrostatic field due to an infinitely long, charged wire results in an attractive force toward, and perpendicular to, the wire, with magnitude inversely proportional to the square of the distance from the wire. If the wire is the z-axis, the force vector is

$$\mathbf{w} = \frac{-x\mathbf{i} - y\mathbf{j}}{(x^2 + y^2)^{3/2}}.$$

In this case,

$$\frac{\partial w_3}{\partial y} = 0 = \frac{\partial w_2}{\partial z},$$

$$\frac{\partial w_1}{\partial z} = 0 = \frac{\partial w_3}{\partial x},$$

$$\frac{\partial w_2}{\partial x} = \frac{-3xy}{(x^2 + y^2)^{5/2}} = \frac{\partial w_1}{\partial y};$$

so again,

$$\nabla \times \mathbf{w} = 0,$$

and there is a potential u such that $\mathbf{w} = \nabla u$. Here,

$$u = \frac{1}{\sqrt{x^2 + y^2}}.$$

Now, however,

$$\frac{\partial w_1}{\partial x} = \frac{-(x^2 + y^2)^{3/2} + 3x^2(x^2 + y^2)^{1/2}}{(x^2 + y^2)^3} = \frac{2x^2 - y^2}{(x^2 + y^2)^{5/2}},$$

$$\frac{\partial w_2}{\partial y} = \frac{2y^2 - x^2}{(x^2 + y^2)^{5/2}},$$

$$\frac{\partial w_3}{\partial z} = 0,$$

and thus

$$\nabla \cdot \mathbf{w} = \frac{\partial w_1}{\partial x} + \frac{\partial w_2}{\partial y} + \frac{\partial w_3}{\partial z} = \frac{x^2 + y^2}{(x^2 + y^2)^{5/2}} = \frac{1}{(x^2 + y^2)^{3/2}}.$$

The potential is not harmonic in this problem.

EXERCISES

1. From each of the following equations, compute $\nabla \cdot \mathbf{w}$ and $\nabla \times \mathbf{w}$.
   (a) $\mathbf{w} = yz\mathbf{i} + xz\mathbf{j} + xy\mathbf{k}$
   (b) $\mathbf{w} = (\mathbf{i} + \mathbf{j} + \mathbf{k})/(x^2 + y^2 + z^2)^{3/2}$
   (c) $\mathbf{w} = e^x \cos y\, \mathbf{i} + 2e^x \sin y\, \mathbf{j} + e^z \mathbf{k}$

2. Compute $\nabla \cdot \mathbf{w}$ and $\nabla \times \mathbf{w}$ from each of the following equations and compare with the corresponding part of Exercise 1.

(a) $\mathbf{w} = (yz + x)\mathbf{i} + (xz - 2y)\mathbf{j} + (xy + z)\mathbf{k}$

(b) $\mathbf{w} = \dfrac{(x + 1)\mathbf{i} + (y + 1)\mathbf{j} + (z + 1)\mathbf{k}}{(x^2 + y^2 + z^2)^{3/2}}$

(c) $\mathbf{w} = 2e^z \cos y\ \mathbf{i} + e^z \sin y\ \mathbf{j} + e^z\ \mathbf{k}$

3. In each of the following exercises show that u is harmonic.

(a) $u = x^2 - 2y^2 + z^2$

(b) $u = (x^2 + y^2 + z^2)^{-1/2}$

(c) $u = e^x \cos y$

4. In the light of Exercise 3, explain the pairing off of answers in Exercises 1 and 2.

5. The formula for curl is often given in determinant form:

$$\nabla \times \mathbf{w} = \begin{vmatrix} \mathbf{i} & \dfrac{\partial}{\partial x} & w_1 \\[2mm] \mathbf{j} & \dfrac{\partial}{\partial y} & w_2 \\[2mm] \mathbf{k} & \dfrac{\partial}{\partial z} & w_3 \end{vmatrix} .$$

Explain how this formula is to be interpreted, and apply it to Exercise 1(a).

7–4 Coordinate-free definitions. Return now to the ideas introduced in Section 7–1. Let $\mathbf{r}$ be the position vector, and let u and $\mathbf{w}$ be, respectively, scalar and vector variables on three-space. The derivatives $D_r u$ and $D_r \mathbf{w}$ are linear transformations, mapping vectors into scalars and vectors, respectively. Specifically, they operate on the tangent spaces:

$$du = D_r u\ d\mathbf{r}, \quad d\mathbf{w} = D_r \mathbf{w}\ d\mathbf{r}.$$

Now, any linear transformation of vectors into scalars assumes the form of scalar multiplication by some fixed vector. That is, if T is such a transformation, then there is some $\mathbf{v}_0$ such that

$$T\mathbf{v} = \mathbf{v}_0 \cdot \mathbf{v}$$

for every $\mathbf{v}$. To see this, let

$$T\mathbf{i} = a, \qquad T\mathbf{j} = b, \qquad T\mathbf{k} = c,$$

and set

$$\mathbf{v}_0 = a\mathbf{i} + b\mathbf{j} + c\mathbf{k}.$$

Now, suppose $\mathbf{v} = v_1\mathbf{i} + v_2\mathbf{j} + v_3\mathbf{k}$; then, since T is linear,

$$T\mathbf{v} = v_1 T\mathbf{i} + v_2 T\mathbf{j} + v_3 T\mathbf{k} = v_1 a + v_2 b + v_3 c = \mathbf{v}_0 \cdot \mathbf{v}.$$

Thus, any transformation of the type of $D_r u$ has a fixed vector that represents it as a scalar product. This yields a coordinate-free definition of gradient; define **grad** u as the vector such that

$$D_r u \, d\mathbf{r} = (\mathbf{grad} \, u) \cdot d\mathbf{r} \qquad (16)$$

over each tangent space. Comparison of (10), Section 7–1, and (13), Section 7–2, shows that this agrees with the definition previously given in terms of $(\mathbf{i}, \mathbf{j}, \mathbf{k})$-components.

The $(\mathbf{i}, \mathbf{j}, \mathbf{k})$-component definition of div $\mathbf{w}$ may be rephrased in the following way. Introduce the matrix

$$D_{X(r)} X(\mathbf{w})$$

that represents the linear transformation

$$D_r \mathbf{w}$$

with respect to the basis $\mathbf{i}, \mathbf{j}, \mathbf{k}$. Then, div $\mathbf{w}$ is the sum of the main diagonal entries in the matrix $D_{X(r)} X(\mathbf{w})$. For any square matrix A, the sum of the main diagonal entries is called the *trace* of A and written

$$\mathrm{tr}\ A.$$

In these terms, the definition of Section 7–3 reads

$$\mathrm{div}\ \mathbf{w} = \mathrm{tr}\ D_{X(r)} X(\mathbf{w}).$$

Now, a new orthonormal basis gives a new matrix representation for $D_r \mathbf{w}$, but it turns out that all such matrices have the same trace. So, this trace common to all the matrices for $D_r \mathbf{w}$ may be called the *trace of the linear transformation* itself, and there emerges the coordinate-free definition

$$\mathrm{div}\ \mathbf{w} = \mathrm{tr}\ D_r \mathbf{w}. \qquad (17)$$

In brief outline, the proof that the trace is invariant under change of basis is as follows. Let A be a square matrix; the determinant

$$\det \{A - \lambda[\delta_{ij}]\}$$

is a polynomial in λ (called the *characteristic polynomial of A*). In display form, this determinant may be written

$$
\begin{vmatrix}
a_{11} - \lambda & a_{12} & \cdots & a_{1n} \\
a_{21} & a_{22} - \lambda & \cdots & a_{2n} \\
\vdots & \vdots & & \vdots \\
a_{n1} & a_{n2} & & a_{nn} - \lambda
\end{vmatrix}.
$$

Now, in expanding this, the only way to get λ^{n-1} is to take the term

$$(a_{11} - \lambda)(a_{22} - \lambda) \cdots (a_{nn} - \lambda)$$

and consider from its expansion a term with one factor a_{ii} and the remaining factors $-\lambda$. Thus, the coefficient of λ^{n-1} in the characteristic polynomial of A is

$$(-1)^{n-1} \operatorname{tr} A.$$

If A represents a linear transformation with respect to one basis, and if R is the orthogonal matrix that transforms components to those for a new basis, then with respect to the new basis, RAR^{-1} represents the given transformation. Note now that

$$
\begin{aligned}
\det \{RAR^{-1} - \lambda[\delta_{ij}]\} &= \det \{RAR^{-1} - \lambda R[\delta_{ij}]R^{-1}\} \\
&= \det \{R(A - \lambda[\delta_{ij}])R^{-1}\} \\
&= \det R \det \{A - \lambda[\delta_{ij}]\} \det (R^{-1}) \\
&= \det \{A - \lambda[\delta_{ij}]\}.
\end{aligned}
$$

That is, A and RAR^{-1} have the same characteristic polynomial; and since the trace is a particular coefficient in this polynomial, they have the same trace.

To define curl without coordinates requires a few more items from matrix theory. Let A be a square matrix, and form matrices A^+ and A^- as follows:

$$A^+ = \tfrac{1}{2}(A + A'), \qquad A^- = \tfrac{1}{2}(A - A').$$

Note that

$$(A^+)' = \tfrac{1}{2}(A + A')' = \tfrac{1}{2}(A' + A) = A^+,$$

while

$$(A^-)' = \tfrac{1}{2}(A - A')' = \tfrac{1}{2}(A' - A) = -A^-.$$

That is, entries in A^+ symmetric with respect to the main diagonal are equal, but such entries in A^- are negatives, one of the other. For this reason A^+ is called a *symmetric matrix* and A^- is called a *skew-symmetric matrix.* From the definition of A^+ and A^-, it follows at once that

$$A = A^+ + A^-,$$

and this is called the decomposition of A into its symmetric and skew-symmetric parts.

Let A and $RAR^{-1} = RAR'$ (because R is orthogonal) be the matrices for a linear transformation with respect to two different orthonormal bases. Then,

$$2(RAR')^+ = RAR' + (RAR')' = RAR' + RA'R'$$
$$= R(A + A')R' = 2RA^+R;$$

similarly,

$$2(RAR')^- = 2RA^-R'.$$

That is, change of basis preserves symmetric and skew-symmetric parts of the matrix. So, one can decompose the transformation itself and write

$$T = T^+ + T^-,$$

where T^+ comes from A^+ and T^- comes from A^-. The decomposition thus determined will be the same for all orthonormal bases.

Turn, now, to three-space. The general 3×3 skew-symmetric matrix may be written

$$\begin{bmatrix} 0 & -a & b \\ a & 0 & -c \\ -b & c & 0 \end{bmatrix},$$

and so, in general,

$$A^-X = \begin{bmatrix} 0 & -a & b \\ a & 0 & -c \\ -b & c & 0 \end{bmatrix} \begin{bmatrix} x \\ y \\ z \end{bmatrix} = \begin{bmatrix} bz - ay \\ ax - cz \\ cy - bx \end{bmatrix}. \tag{18}$$

This last result is the matrix of components of

$$(c\mathbf{i} + b\mathbf{j} + a\mathbf{k}) \times (x\mathbf{i} + y\mathbf{j} + z\mathbf{k});$$

thus if A is the matrix of a linear transformation T, then there is a $\mathbf{v}_0$ such that

$$T^-\mathbf{v} = \mathbf{v}_0 \times \mathbf{v}$$

for all $\mathbf{v}$. The vector $\mathbf{v}_0$ is exhibited above in $(\mathbf{i}, \mathbf{j}, \mathbf{k})$-components; but since T^- is the same for all bases, so is $\mathbf{v}_0$.

Finally, then, curl may be defined as gradient was, by the relation

$$2(D_r\mathbf{w})^- \, d\mathbf{r} = (\textbf{curl } \mathbf{w}) \times d\mathbf{r}. \tag{19}$$

To see that this agrees with the previous definition of curl, form

$$2[D_{X(r)}X(\mathbf{w})]^- = D_{X(r)}X(\mathbf{w}) - [D_{X(r)}X(\mathbf{w})]'$$

$$= \begin{bmatrix} 0 & \dfrac{\partial w_1}{\partial y} - \dfrac{\partial w_2}{\partial x} & \dfrac{\partial w_1}{\partial z} - \dfrac{\partial w_3}{\partial x} \\[3mm] \dfrac{\partial w_2}{\partial x} - \dfrac{\partial w_1}{\partial y} & 0 & \dfrac{\partial w_2}{\partial z} - \dfrac{\partial w_3}{\partial y} \\[3mm] \dfrac{\partial w_3}{\partial x} - \dfrac{\partial w_1}{\partial z} & \dfrac{\partial w_3}{\partial y} - \dfrac{\partial w_2}{\partial z} & 0 \end{bmatrix}.$$

Now, note (18) above. Multiplication by this matrix yields a vector product in which the components of the first factor are read off from the matrix by means of the following scheme:

$$\begin{bmatrix} . & . & 2 \\ 3 & . & . \\ . & 1 & . \end{bmatrix}. \tag{20}$$

Taking the entries from $2[D_{X(r)}X(\mathbf{w})]^-$ in this order, one has, by (19),

$$\textbf{curl } \mathbf{w} = \left(\frac{\partial w_3}{\partial y} - \frac{\partial w_2}{\partial z}\right)\mathbf{i} + \left(\frac{\partial w_1}{\partial z} - \frac{\partial w_3}{\partial x}\right)\mathbf{j} + \left(\frac{\partial w_2}{\partial x} - \frac{\partial w_1}{\partial y}\right)\mathbf{k};$$

this is the familiar formula.

In terms of the formal operator ∇, gradient, divergence, and curl seem to be the "natural" derivative operators in vector analysis. Actually, the linear transformations $D_r\mathbf{u}$ and $D_r\mathbf{w}$ are the basic derivatives, and the definitions (16), (17), and (19) of this section show the specialized way in which gradient, divergence, and curl are related to these fundamental operators.

The special nature of divergence and curl is further emphasized by the following considerations. Let x and u be scalar variables on a one-dimensional manifold; then u may be recovered from

$$D_x u$$

in the sense that if $D_x u$ is given in terms of x, then u is determined uniquely in terms of x except for an additive constant. Similarly, u may be recovered from

$$D_r u,$$

and **w** may be recovered from

$$D_r\mathbf{w}$$

(see Exercises 1 and 2 below). Furthermore, since **grad** u characterizes $D_r u$ completely, u may be recovered from **grad** u. However, **w** cannot be recovered from div **w** or **curl w** or both (see Exercise 3 below). In the light of (17) and (19) above, this is not too surprising. Curl characterizes only the skew-symmetric part of the derivative, and divergence gives only the trace. With only these two operators singled out, most of the symmetric part of the derivative is lost. That is, specifying div **w** and **curl w** does not tell enough about $D_r\mathbf{w}$ to determine **w**.

EXERCISES

1. Suppose $du = 0$ over the entire tangent bundle to three-space.
    (a) Show that $du/ds = 0$ identically on any curve in three-space.
    (b) Use part (a) and the Law of the Mean to prove that $u = $ constant.
    (c) Show that if **grad** $u = $ **grad** v, then $u = v + C$. [*Hint:* What is $d(u - v)$?]

2. Show that if $D_r\mathbf{w} = D_r\mathbf{v}$, then $\mathbf{w} = \mathbf{v} + C$. [*Hint:* $d\mathbf{w}$ may be written $dw_1\mathbf{i} + dw_2\mathbf{j} + dw_3\mathbf{k}$; apply the argument of Exercise 1 to each component.]

3. Let

$$\mathbf{w} = \frac{-x\mathbf{i} - y\mathbf{j} - z\mathbf{k}}{(x^2 + y^2 + z^2)^{3/2}}.$$

It was shown in Example 1, Section 7–3, that div **w** $= 0$ and **curl w** $= 0$. Compute the complete matrix $D_{X(r)}X(\mathbf{w})$, and show that it is not identically zero.

4. Give conditions for u to be harmonic in terms of the matrix for D_r**grad** u with respect to an arbitrary basis.

5. Each of the following expressions describes a linear vector-to-scalar transformation T. In each case find a vector $\mathbf{v}_0$ such that $T\mathbf{v} = \mathbf{v}_0 \cdot \mathbf{v}$ for all **v**.

    (a) $T\mathbf{i} = 1$, $T\mathbf{j} = 2$, $T\mathbf{k} = 3$.   (b) $T\mathbf{v} = [2\mathbf{i}\ \mathbf{j}\ -3\mathbf{k}]X(\mathbf{v})$.

6. Though the present theory is three-dimensional, the following two-dimensional examples serve to make more concrete the idea of decomposing a linear transformation into its symmetric and skew-symmetric parts.
    (a) Find the symmetric and skew-symmetric parts of an α-rotation. Sketch the image of **i** under each of these, and show geometrically that they add to the α-rotation of **i**.
    (b) Let T be given in $(\mathbf{i}, \mathbf{j})$-components by

$$\begin{bmatrix} 2 & 3 \\ -5 & 1 \end{bmatrix}.$$

Find matrices for T^+ and T^-. Compute $T(\mathbf{i} + 2\mathbf{j})$, $T^+(\mathbf{i} + 2\mathbf{j})$, $T^-(\mathbf{i} + 2\mathbf{j})$ and sketch them.

7. Let

$$A = \begin{bmatrix} 2 & 2 \\ 3 & 1 \end{bmatrix}.$$

(a) Find the characteristic equation for A and note how tr A appears as a coefficient.
(b) Let R be the matrix for a rotation by $\pi/6$. Compute RAR^{-1}; find the characteristic equation of this matrix, and compare with part (a).

7–5 Curvilinear coordinates. Let

$$Y = \begin{bmatrix} r \\ s \\ t \end{bmatrix}$$

be a matrix of curvilinear coordinates (cylindrical or spherical, for example). Then, dY is a matrix of component variables in the tangent spaces, and (Section 5–5) if, as is frequently the case, the basis for dY is orthogonal, then there is a diagonal matrix A such that $A\,dY$ is a matrix of rectangular component variables. For the cylindrical and spherical systems the $A\,dY$-matrices should be familiar:

$$\begin{bmatrix} dr \\ r\,d\theta \\ dz \end{bmatrix}, \quad \begin{bmatrix} d\rho \\ \rho\,d\phi \\ \rho\sin\phi\,d\theta \end{bmatrix}.$$

In any case,

$$A\,dY(d\mathbf{r}) = A\,dY,$$

because $d\mathbf{r}$ is the position vector on the tangent space, and a position vector on a vector space is merely an identity mapping. Thus, if one finds a matrix B such that

$$du = BA\,dY,$$

then B is a matrix for the linear transformation $D_\mathbf{r}u$ and thus displays **grad** u. Similarly, to find curvilinear coordinate expressions for div **w** and **curl w**, find a matrix C such that

$$A\,dY(d\mathbf{w}) = CA\,dY, \tag{21}$$

and use the definitions of Section 7–4.

The gradient problem is easy:

$$du = D_Y u\,dY = D_Y u A^{-1}A\,dY;$$

so $D_Y u A^{-1}$ displays the components of **grad** u.

For a vector variable $\mathbf{w}$, the problem is a little more complicated. First, it is necessary to pose the problem correctly. The curvilinear coordinates Y are not component variables; on the other hand, their differentials dY are not curvilinear, but affine. Hence, the differentials of curvilinear coordinates Y are component variables. The problem, then, is as follows. There is an orthonormal basis

$$\mathbf{b}_r, \quad \mathbf{b}_s, \quad \mathbf{b}_t \tag{22}$$

for $A\, dY$. Given $\mathbf{w}$ in terms of this basis, find $d\mathbf{w}$ in terms of the same basis, and write the result in the form (21).

Now (see Section 5–5), the basis (22) comes from $\mathbf{i}, \mathbf{j}, \mathbf{k}$ by a rotation whose matrix is $D_Y X A^{-1}$; that is,

$$dX = (D_Y X A^{-1})(A\, dY)$$

by the fundamental theorem, and this describes the change of basis. Thus set

$$R = D_Y X A^{-1},$$

and define Z by

$$Z = R^{-1}X; \tag{23}$$

then Z is a matrix of component variables with respect to the basis (22). The problem then reads, "Given $Z(\mathbf{w})$, find $A\, dY(d\mathbf{w})$." Recall that

$$dX(d\mathbf{w}) = d\big(X(\mathbf{w})\big);$$

this is the definition of $d\mathbf{w}$. The same cannot be done with Z because the rotation matrix R in (23) is not constant. Instead, one has

$$Z(\mathbf{w}) = R^{-1}X(\mathbf{w}),$$

and computing differentials, entry at a time, in these matrices yields

$$d\big(Z(\mathbf{w})\big) = R^{-1}\, d\big(X(\mathbf{w})\big) + d(R^{-1})X(\mathbf{w})$$
$$= R^{-1}\, dX(d\mathbf{w}) + d(R^{-1})RZ(\mathbf{w})$$
$$= A\, dY(d\mathbf{w}) + d(R^{-1})RZ(\mathbf{w});$$

$$A\, dY(d\mathbf{w}) = d\big(Z(\mathbf{w})\big) - d(R^{-1})RZ(\mathbf{w}). \tag{24}$$

This is the hard part of the problem. Strangely enough, when the right side of (24) is computed for a specific problem, it is easily reduced by inspection to the form called for on the right side of (21). This procedure gives the matrix C from which div $\mathbf{w}$ and **curl** $\mathbf{w}$ are found.

EXAMPLES

1. Let

$$Y = \begin{bmatrix} \rho \\ \phi \\ \theta \end{bmatrix};$$

recall (Section 5–5) that

$$A\,dY = \begin{bmatrix} d\rho \\ \rho\,d\phi \\ \rho\sin\phi\,d\theta \end{bmatrix}.$$

The rotation matrix R was found in Example 1, Section 5–5; but it is not difficult to compute:

$$R = D_Y X A^{-1} = \begin{bmatrix} \dfrac{\partial x}{\partial \rho} & \dfrac{\partial x}{\partial \phi} & \dfrac{\partial x}{\partial \theta} \\[2mm] \dfrac{\partial y}{\partial \rho} & \dfrac{\partial y}{\partial \phi} & \dfrac{\partial y}{\partial \theta} \\[2mm] \dfrac{\partial z}{\partial \rho} & \dfrac{\partial z}{\partial \phi} & \dfrac{\partial z}{\partial \theta} \end{bmatrix} \begin{bmatrix} 1 & 0 & 0 \\[2mm] 0 & \dfrac{1}{\rho} & 0 \\[2mm] 0 & 0 & \dfrac{1}{\rho\sin\phi} \end{bmatrix}$$

$$= \begin{bmatrix} \sin\phi\cos\theta & \cos\phi\cos\theta & -\sin\theta \\ \sin\phi\sin\theta & \cos\phi\sin\theta & \cos\theta \\ \cos\phi & -\sin\phi & 0 \end{bmatrix}.$$

A rotation matrix is easy to invert; merely take the transpose:

$$R^{-1} = R' = \begin{bmatrix} \sin\phi\cos\theta & \sin\phi\sin\theta & \cos\phi \\ \cos\phi\cos\theta & \cos\phi\sin\theta & -\sin\phi \\ -\sin\theta & \cos\theta & 0 \end{bmatrix}.$$

Thus,

$$d(R^{-1}) = \begin{bmatrix} \cos\phi\cos\theta\,d\phi & \cos\phi\sin\theta\,d\phi & -\sin\phi\,d\phi \\ -\sin\phi\sin\theta\,d\theta & +\sin\phi\cos\theta\,d\theta & \\ -\sin\phi\cos\theta\,d\phi & -\sin\phi\sin\theta\,d\phi & -\cos\phi\,d\phi \\ -\cos\phi\sin\theta\,d\theta & +\cos\phi\cos\theta\,d\theta & \\ -\cos\theta\,d\theta & -\sin\theta\,d\theta & 0 \end{bmatrix},$$

and

$$d(R^{-1})\,R = \begin{bmatrix} 0 & d\phi & \sin\phi\,d\theta \\ -d\phi & 0 & \cos\phi\,d\theta \\ -\sin\phi\,d\theta & -\cos\phi\,d\theta & 0 \end{bmatrix}.$$

Now, let

$$Z(\mathbf{w}) = \begin{bmatrix} w_1 \\ w_2 \\ w_3 \end{bmatrix},$$

then

$$d(R^{-1})\,RZ(\mathbf{w}) = \begin{bmatrix} 0 + w_2\,d\phi + w_3\sin\phi\,d\theta \\ 0 - w_1\,d\phi + w_3\cos\phi\,d\theta \\ 0 + 0 - (w_1\sin\phi + w_2\cos\phi)\,d\theta \end{bmatrix}$$

$$= \begin{bmatrix} 0 & \dfrac{w_2}{\rho} & \dfrac{w_3}{\rho} \\ 0 & -\dfrac{w_1}{\rho} & \dfrac{w_3\cot\phi}{\rho} \\ 0 & 0 & -\dfrac{w_1}{\rho} - \dfrac{w_2\cot\phi}{\rho} \end{bmatrix} \begin{bmatrix} d\rho \\ \rho\,d\phi \\ \rho\sin\phi\,d\theta \end{bmatrix}. \qquad (25)$$

The formula (24) also calls for

$$d(Z(\mathbf{w})) = \begin{bmatrix} \dfrac{\partial w_1}{\partial\rho}\,d\rho + \dfrac{\partial w_1}{\partial\phi}\,d\phi + \dfrac{\partial w_1}{\partial\theta}\,d\theta \\ \dfrac{\partial w_2}{\partial\rho}\,d\rho + \dfrac{\partial w_2}{\partial\phi}\,d\phi + \dfrac{\partial w_2}{\partial\theta}\,d\theta \\ \dfrac{\partial w_3}{\partial\rho}\,d\rho + \dfrac{\partial w_3}{\partial\phi}\,d\phi + \dfrac{\partial w_3}{\partial\theta}\,d\theta \end{bmatrix}$$

$$= \begin{bmatrix} \dfrac{\partial w_1}{\partial\rho} & \dfrac{1}{\rho}\dfrac{\partial w_1}{\partial\phi} & \dfrac{1}{\rho\sin\phi}\dfrac{\partial w_1}{\partial\theta} \\ \dfrac{\partial w_2}{\partial\rho} & \dfrac{1}{\rho}\dfrac{\partial w_2}{\partial\phi} & \dfrac{1}{\rho\sin\phi}\dfrac{\partial w_2}{\partial\theta} \\ \dfrac{\partial w_3}{\partial\rho} & \dfrac{1}{\rho}\dfrac{\partial w_3}{\partial\phi} & \dfrac{1}{\rho\sin\phi}\dfrac{\partial w_3}{\partial\theta} \end{bmatrix} \begin{bmatrix} d\rho \\ \rho\,d\phi \\ \rho\sin\phi\,d\theta \end{bmatrix}. \qquad (26)$$

Therefore, by (24),

$A\, dY(d\mathbf{w})$

$$
= \begin{bmatrix}
\dfrac{\partial w_1}{\partial \rho} & \dfrac{1}{\rho}\dfrac{\partial w_1}{\partial \phi} - \dfrac{w_2}{\rho} & \dfrac{1}{\rho \sin \phi}\dfrac{\partial w_1}{\partial \theta} - \dfrac{w_3}{\rho} \\[2mm]
\dfrac{\partial w_2}{\partial \rho} & \dfrac{1}{\rho}\dfrac{\partial w_2}{\partial \phi} + \dfrac{w_1}{\rho} & \dfrac{1}{\rho \sin \phi}\dfrac{\partial w_2}{\partial \theta} - \dfrac{w_3 \cot \phi}{\rho} \\[2mm]
\dfrac{\partial w_3}{\partial \rho} & \dfrac{1}{\rho}\dfrac{\partial w_3}{\partial \phi} & \dfrac{1}{\rho \sin \phi}\dfrac{\partial w_3}{\partial \theta} + \dfrac{w_1}{\rho} + \dfrac{w_2 \cot \phi}{\rho}
\end{bmatrix}
\begin{bmatrix}
d\rho \\[2mm]
\rho\, d\phi \\[2mm]
\rho \sin \phi\, d\theta
\end{bmatrix}.
$$

The square matrix here is the one for the linear transformation $D_\mathbf{r}\mathbf{w}$ with respect to the basis $\mathbf{b}_\rho$, $\mathbf{b}_\phi$, $\mathbf{b}_\theta$, and hence div $\mathbf{w}$ is its trace:

$$
\text{div } \mathbf{w} = \frac{\partial w_1}{\partial \rho} + \frac{1}{\rho}\frac{\partial w_2}{\partial \phi} + \frac{1}{\rho \sin \phi}\frac{\partial w_3}{\partial \theta} + \frac{2}{\rho} w_1 + \frac{\cot \phi}{\rho} w_2.
$$

To find **curl w**, get twice the skew-symmetric part of this matrix:

$$
\begin{bmatrix}
0 & \checkmark & \dfrac{1}{\rho \sin \phi}\dfrac{\partial w_1}{\partial \theta} - \dfrac{w_3}{\rho} - \dfrac{\partial w_3}{\partial \rho} \\[2mm]
\dfrac{\partial w_2}{\partial \rho} - \dfrac{1}{\rho}\dfrac{\partial w_1}{\partial \phi} + \dfrac{w_2}{\rho} & 0 & \checkmark \\[2mm]
\checkmark & \dfrac{1}{\rho}\dfrac{\partial w_3}{\partial \phi} - \dfrac{1}{\rho \sin \phi}\dfrac{\partial w_2}{\partial \theta} + \dfrac{\cot \phi}{\rho} w_3 & 0
\end{bmatrix}.
$$

The three entries written out are those from which curl is read, so

$$
\text{curl } w - \left(\frac{1}{\rho}\frac{\partial w_3}{\partial \phi} - \frac{1}{\rho \sin \phi}\frac{\partial w_2}{\partial \theta} + \frac{\cot \phi}{\rho} w_3 \right) \mathbf{b}_\rho
$$

$$
+ \left(\frac{1}{\rho \sin \phi}\frac{\partial w_1}{\partial \theta} - \frac{\partial w_3}{\partial \rho} - \frac{w_3}{\rho} \right) \mathbf{b}_\phi
$$

$$
+ \left(\frac{\partial w_2}{\partial \rho} - \frac{1}{\rho}\frac{\partial w_1}{\partial \phi} + \frac{w_2}{\rho} \right) \mathbf{b}_\theta.
$$

2. To find a gradient in spherical coordinates is much simpler:

$$
du = \frac{\partial u}{\partial \rho} d\rho + \frac{\partial u}{\partial \phi} d\phi + \frac{\partial u}{\partial \theta} d\theta
$$

$$
= \begin{bmatrix} \dfrac{\partial u}{\partial \rho} & \dfrac{\partial u}{\partial \phi} & \dfrac{\partial u}{\partial \theta} \end{bmatrix}
\begin{bmatrix} d\rho \\ d\phi \\ d\theta \end{bmatrix}
= \begin{bmatrix} \dfrac{\partial u}{\partial \rho} & \dfrac{1}{\rho}\dfrac{\partial u}{\partial \phi} & \dfrac{1}{\rho \sin \phi}\dfrac{\partial u}{\partial \theta} \end{bmatrix}
\begin{bmatrix} d\rho \\ \rho\, d\phi \\ \rho \sin \phi\, d\theta \end{bmatrix},
$$

and thus

$$\text{grad } u = \frac{\partial u}{\partial \rho}\, \mathbf{b}_\rho + \frac{1}{\rho}\frac{\partial u}{\partial \phi}\, \mathbf{b}_\phi + \frac{1}{\rho \sin \phi}\frac{\partial u}{\partial \theta}\, \mathbf{b}_\theta.$$

Combining this result with the formula for divergence in Example 1 yields the Laplacian in spherical coordinates:

$$\nabla^2 u = \text{div grad } u$$

$$= \frac{\partial^2 u}{\partial \rho^2} + \frac{1}{\rho^2}\frac{\partial^2 u}{\partial \phi^2} + \frac{1}{\rho^2 \sin^2 \theta}\frac{\partial^2 u}{\partial \theta^2} + \frac{2}{\rho}\frac{\partial u}{\partial \rho} + \frac{\cot \phi}{\rho^2}\frac{\partial u}{\partial \phi}.$$

3. One can also find a gradient geometrically. Recall (Section 7–2) that the component of **grad** u in any direction is the directional derivative of u in that direction. Now, in the spherical-coordinate directions, one has arc-length differentials and, hence, directional derivatives as follows:

$$\mathbf{b}_\rho: \quad ds = d\rho; \qquad \frac{du}{ds} = \frac{du}{d\rho} = \frac{\partial u}{\partial \rho},$$

$$\mathbf{b}_\phi: \quad ds = \rho\, d\phi; \qquad \frac{du}{ds} = \frac{du}{\rho\, d\phi} = \frac{1}{\rho}\frac{\partial u}{\partial \phi},$$

$$\mathbf{b}_\theta: \quad ds = \rho \sin \phi\, d\theta; \qquad \frac{du}{ds} = \frac{du}{\rho \sin \phi\, d\theta} = \frac{1}{\rho \sin \phi}\frac{\partial u}{\partial \theta}.$$

One might be tempted to define

$$\nabla = \mathbf{b}_\rho \frac{\partial}{\partial \rho} + \frac{1}{\rho}\mathbf{b}_\phi \frac{\partial}{\partial \phi} + \frac{1}{\rho \sin \phi}\mathbf{b}_\theta \frac{\partial}{\partial \theta}.$$

In terms of this operator,

$$\text{grad } u = \nabla u;$$

but here the similarity to the rectangular case ceases. For this "∇",

$$\text{div } \mathbf{w} = \nabla \cdot \mathbf{w} + \frac{2}{\rho} w_1 + \frac{\cot \phi}{\rho} w_2,$$

$$\text{curl } \mathbf{w} = \nabla \times \mathbf{w} + \frac{\cot \phi}{\rho} w_3 \mathbf{b}_\rho - \frac{1}{\rho} w_3 \mathbf{b}_\phi + \frac{1}{\rho} w_2 \mathbf{b}_\theta,$$

$$\text{div grad } u = \nabla \cdot \nabla u + \frac{2}{\rho}\frac{\partial u}{\partial \rho} + \frac{\cot \phi}{\rho}\frac{\partial u}{\partial \phi}.$$

The directional derivatives appear in the expected places, but in each case, there are perturbation terms due to the fact that the basis vectors are not constant. Note that the directional derivative terms come from $d(Z(\mathbf{w}))$, and the perturbation terms come from $d(R^{-1})RZ(\mathbf{w})$ [see (26) and (25) above].

EXERCISES

1. Derive the following formulas for the cylindrical coordinate system.

(a) $\text{grad } u = \dfrac{\partial u}{\partial r}\,\mathbf{b}_r + \dfrac{1}{r}\dfrac{\partial u}{\partial \theta}\,\mathbf{b}_\theta + \dfrac{\partial u}{\partial z}\,\mathbf{b}_z$

(b) If $\mathbf{w} = w_1\mathbf{b}_r + w_2\mathbf{b}_\theta + w_3\mathbf{b}_z$, then

$$\text{div } \mathbf{w} = \frac{\partial w_1}{\partial r} + \frac{1}{r}\frac{\partial w_2}{\partial \theta} + \frac{\partial w_3}{\partial z} + \frac{1}{r}w_1.$$

(c) With $\mathbf{w}$, as in part (b),

$$\text{curl } \mathbf{w} = \left(\frac{1}{r}\frac{\partial w_3}{\partial \theta} - \frac{\partial w_2}{\partial z}\right)\mathbf{b}_r + \left(\frac{\partial w_1}{\partial z} - \frac{\partial w_3}{\partial r}\right)\mathbf{b}_\theta$$

$$+ \left(\frac{\partial w_2}{\partial r} - \frac{1}{r}\frac{\partial w_1}{\partial \theta} + \frac{1}{r}w_2\right)\mathbf{b}_z.$$

(d) $\nabla^2 u = \dfrac{\partial^2 u}{\partial r^2} + \dfrac{1}{r^2}\dfrac{\partial^2 u}{\partial \theta^2} + \dfrac{\partial^2 u}{\partial z^2} + \dfrac{1}{r}\dfrac{\partial u}{\partial r}.$

2. Take the force vector of Example 1, Section 7–3.
   (a) Show that

$$\mathbf{w} = \frac{-x\mathbf{i} - y\mathbf{j} - z\mathbf{k}}{(x^2 + y^2 + z^2)^{3/2}} = \frac{-\mathbf{b}_\rho}{\rho^2} = \frac{-r\mathbf{b}_r - z\mathbf{b}_z}{(r^2 + z^2)^{3/2}}.$$

   (b) Compute the matrix for $D_r\mathbf{w}$ in both spherical and cylindrical coordinates, and verify from each that $\text{div } \mathbf{w} = 0$ and $\text{curl } \mathbf{w} = 0$.

3. Take the force vector of Example 2, Section 7–3.
   (a) Show that

$$\mathbf{w} = \frac{-x\mathbf{i} - y\mathbf{j}}{(x^2 + y^2)^{3/2}} = \frac{-\csc \phi\, \mathbf{b}_\rho - \csc \phi \cot \phi\, \mathbf{b}_\theta}{\rho^2} = \frac{-\mathbf{b}_r}{r^2}.$$

   (b) Compute the matrix for $D_r\mathbf{w}$ in both spherical and cylindrical coordinates, and verify from each that $\text{curl } \mathbf{w} = 0$ and $\text{div } \mathbf{w} = 1/r^3$.

4. Let $\mathbf{w} = r\mathbf{b}_\theta$.
   (a) Show that this is the velocity vector for a rigid body rotating with unit angular velocity about the z-axis.
   (b) Find $\mathbf{w}$ in rectangular and spherical coordinates.
   (c) Find $\text{div } \mathbf{w}$ and $\text{curl } \mathbf{w}$ in each of the three coordinate systems and compare.

5. Let s, t, z be coordinates related to the rectangular by

$$x = s^2 - t^2,$$

$$y = 2st,$$

$$z = z.$$

Find formulas in the (s, t, z)-system for

    (a) **grad** u (b) div **w**   (c) **curl w**   (d) $\nabla^2 u$

6. Let $\mathbf{b}_r$, $\mathbf{b}_\theta$, $\mathbf{b}_z$ be the basis for the cylindrical coordinate differentials.

    (a) Show that $d\mathbf{b}_r = \mathbf{b}_\theta \, d\theta$, $d\mathbf{b}_\theta = -\mathbf{b}_r \, d\theta$, $d\mathbf{b}_z = 0$.

    (b) Let $\mathbf{w} = w_1\mathbf{b}_r + w_2\mathbf{b}_\theta + w_3\mathbf{b}_z$; then

$$d\mathbf{w} = dw_1\mathbf{b}_r + dw_2\mathbf{b}_\theta + dw_3\mathbf{b}_z + w_1 \, d\mathbf{b}_r + w_2 \, d\mathbf{b}_\theta + w_3 \, d\mathbf{b}_z.$$

    Use part (a) to find $d\mathbf{w}$ in terms of $\mathbf{b}_r$, $\mathbf{b}_\theta$, and $\mathbf{b}_z$, and then solve Exercise 1 without the matrix formula (24).

7. Use the plan of Exercise 6 to rework Example 1 without the matrix formula (24). *Hint:* Show that

$$d\mathbf{b}_\rho = \mathbf{b}_\phi \, d\phi + \sin \phi \, \mathbf{b}_\theta \, d\theta,$$

$$d\mathbf{b}_\phi = -\mathbf{b}_\rho \, d\phi + \cos \phi \, \mathbf{b}_\theta \, d\theta,$$

$$d\mathbf{b}_\theta = (-\sin \phi \, \mathbf{b}_\rho - \cos \phi \, \mathbf{b}_\phi) \, d\theta.$$

CHAPTER 8

ITERATED INTEGRALS

8–1 Twofold iterated integrals. A good working formula for the volume of a solid is

$$\int_a^b \alpha(x)\, dx,$$

where $\alpha(x)$ is the area of the cross section at x. Problems encountered so far have been those in which the cross sections were simple, similar figures so that $\alpha(x)$ could be written out in terms of x by means of well-known formulas from elementary geometry.

More generally, consider the idea of finding $\alpha(x)$ by integration. Suppose that the solid is the one pictured in Fig. 8–1. The function to be integrated in order to find $\alpha(x)$ is the function on numbers generated by holding constant the first entry of f. The limits of integration in finding $\alpha(x)$ depend on x. Thus, one could write

$$\alpha(x) = \int_0^{g(x)} f(x, y)\, dy,$$

with the interpretation that the integration process indicated is the inverse of partial differentiation with respect to the second entry. That is, to compute $f(x, y)\, dy$, use the usual primitive formulas with x regarded as a

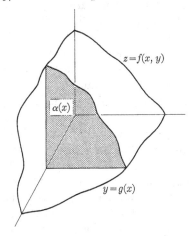

FIGURE 8–1

constant. In this notation the volume of the solid in Fig. 8–1 is

$$\int_0^a \left(\int_0^{g(x)} f(x, y) \, dy \right) dx. \tag{1}$$

This is usually written

$$\int_0^a \int_0^{g(x)} f(x, y) \, dy \, dx, \tag{2}$$

with the understanding that (at least for the present) the form (2) means what is described more explicitly by (1) and nothing more.

The form (2) is called an *iterated integral*. Many elementary texts mistakenly call it a double integral. The definition of a double integral will be presented in Chapter 9, and the distinction between this and an iterated integral will be discussed at some length there.

Embellishments of (2), introduced by moving the figure away from the axes and giving it a curved bottom and back side, can undoubtedly be supplied by the student.

The volume of the solid in Fig. 8–1 could also be found by considering cross sections perpendicular to the y-axis. This analysis would lead to an iterated integral of the form

$$\int_0^b \int_0^{\phi(y)} f(x, y) \, dx \, dy. \tag{3}$$

From geometric considerations, it is apparent that the iterated integrals (2) and (3) are equal, but there is no obvious algebraic manipulation to transform one into the other. They represent different analyses of the same problem. It should be borne in mind that the form (2) is merely a convenient abbreviation of the more explicit form (1). In (1) it is clear that the differentials belong in separate integration processes. Their juxtaposition in the abbreviated form (2) gives the unfortunate suggestion of ordinary multiplication, and this is quite misleading. If $dy \, dx$ in (2) meant dy times dx, then $dy \, dx = dx \, dy$ and all one would have to do would be to interchange them. However, this does not convert (2) into (3) at all. To reverse the order of integration, one must proceed from one iterated integral to the figure and from there to the other iterated integral. See Example 4 below.

A volume problem forms a natural informal introduction to the notion of a twofold iterated integral, but, in a way, this choice of an illustration is misleading. Recall the discussion in Section 5–1 to the effect that a problem based on a function on pairs is basically *two-dimensional*. Now twofold iterated integration is an operation on a function on pairs, so the domain of integration is two-dimensional. That is, each of the integrals (2) and (3) above is *over a region in the plane*. The application of the process to finding volumes is purely coincidental.

EXAMPLES

1. Evaluate

$$\int_0^1 \int_{y^2}^y (x^2 - y^2) \, dx \, dy.$$

Solution.

$$\int_0^1 \int_{y^2}^y (x^2 - y^2) \, dx \, dy = \int_0^1 \left(\frac{x^3}{3} - xy^2\right)\Big|_{y^2}^y dy$$

$$= \int_0^1 \left(\frac{y^3}{3} - y^3 - \frac{y^6}{3} + y^4\right) dy$$

$$= -\frac{y^4}{6} - \frac{y^7}{21} + \frac{y^5}{5}\Big|_0^1 = -\frac{1}{70}.$$

2. Set up an iterated integral for $f(x, y)$ over the figure bounded by the loci of

$$y = x^2 \qquad \text{and} \qquad y = 2x.$$

This figure is shown in Fig. 8–2(a). If the first integration is with respect to x, the limits of integration are the values of x in terms of y at the extremities of the horizontal line in Fig. 8–2(a). Thus, the required integral is

$$\int_0^4 \int_{y/2}^{\sqrt{y}} f(x, y) \, dx \, dy.$$

If an integral in the other order is desired, Fig. 8–2(b) is pertinent, and the result is

$$\int_0^2 \int_{x^2}^{2x} f(x, y) \, dy \, dx.$$

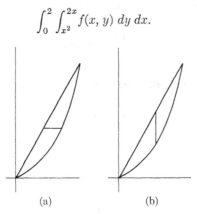

(a)   (b)

FIGURE 8–2

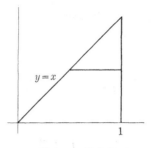

$y = x$

1

FIGURE 8–3

Note that the function f plays no role at all in the process of finding limits of integration. The integration is over the plane figure, and one can integrate all sorts of variables over it.

3. An alternative procedure in Example 2 is as follows. The plane region in question may itself be described as the locus of a system of inequalities. Specifically, the region involved in Example 2 is the locus of

$$\frac{y}{2} \leq x \leq \sqrt{y}, \qquad 0 \leq y \leq 4.$$

Or, it may be described as the locus of

$$x^2 \leq y \leq 2x,$$

$$0 \leq x \leq 2.$$

These descriptions of the plane region as a locus exhibit directly the limits of integration in iterated integrals over the region.

4. Reverse the order of integration in

$$\int_0^1 \int_0^x dy\, dx.$$

The region in question is described by

$$0 \leq y \leq x,$$

$$0 \leq x \leq 1.$$

The locus of these inequalities is the triangle shown in Fig. 8–3, and the other iterated integral is

$$\int_0^1 \int_y^1 dx\, dy.$$

Exercises

1. Evaluate each of the following iterated integrals.

(a) $\displaystyle\int_0^\pi \int_0^x x \sin y \, dy \, dx$

(b) $\displaystyle\int_0^1 \int_0^{\ln y} e^{x+y} \, dx \, dy$

(c) $\displaystyle\int_0^\pi \int_0^{\sin x} y \, dy \, dx$

(d) $\displaystyle\int_1^2 \int_y^{y^2} dx \, dy$

(e) $\displaystyle\int_0^1 \int_{\sqrt{x}}^x y^2 \, dy \, dx$

(f) $\displaystyle\int_0^1 \int_0^{x^3} e^{y/x} \, dy \, dx$

(g) $\displaystyle\int_{-1}^2 \int_{y^2}^{y+2} x^2 \, dx \, dy$

(h) $\displaystyle\int_0^2 \int_0^y xy \, dx \, dy$

(i) $\displaystyle\int_{-1}^1 \int_0^{\sqrt{1+x^2}} y^3 \, dy \, dx$

(j) $\displaystyle\int_0^{\pi/2} \int_0^{(\arcsin y)/y} y \cos xy \, dx \, dy$

(k) $\displaystyle\int_0^1 \int_{\sqrt{y}}^{y^2} (x^2 + y^2) \, dx \, dy$

(l) $\displaystyle\int_0^1 \int_{y^2}^y (x + y) \, dx \, dy$

(m) $\displaystyle\int_0^\infty \int_0^y xe^{-y^3} \, dx \, dy$

(n) $\displaystyle\int_0^\infty \int_{\sinh x}^{\cosh x} dy \, dx$

(o) $\displaystyle\int_0^\infty \int_0^{e-x} y \, dy \, dx$

(p) $\displaystyle\int_0^1 \int_{\ln x}^0 dy \, dx$

2. In each of the following integrals, reverse the order of integration.

(a) $\displaystyle\int_0^2 \int_1^{e^x} dy \, dx$

(b) $\displaystyle\int_0^1 \int_{\sqrt{y}}^1 dx \, dy$

(c) $\displaystyle\int_0^1 \int_{-\sqrt{1-y^2}}^{\sqrt{1-y^2}} dx \, dy$

(d) $\displaystyle\int_{-2}^1 \int_{x^2+4x}^{3x+2} dy \, dx$

(e) $\displaystyle\int_0^1 \int_0^{\sqrt{x-x^2}} dy \, dx$

(f) $\displaystyle\int_0^1 \int_y^{\sqrt{y}} dx \, dy$

(g) $\displaystyle\int_{-1}^2 \int_0^1 dx \, dy$

(h) $\displaystyle\int_0^{\ln 2} \int_1^{e^x} dy \, dx$

(i) $\displaystyle\int_0^{\pi/2} \int_0^{\sin x} dy \, dx$

(j) $\displaystyle\int_1^e \int_0^{\ln y} dx \, dy$

(k) $\displaystyle\int_0^1 \int_{y^2}^y dx\, dy$ (l) $\displaystyle\int_0^2 \int_{x^2}^{2x} dy\, dx$

8–2 Applications. In Example 2, Section 8–1, it was required to set up an iterated integral for $f(x, y)$ over a specified region in the plane. It was not specified what function f was; and indeed, this was not pertinent to the problem. This illustrates a key point in the application of iterated integrals to physical problems. Many quantities associated with a plane region can be found by iterated integration, and the solution to a problem of this type is very neatly divided into two independent parts:

(i) The limits of integration depend only on the region in question and not on the quantity to be found.

(ii) The expression behind the integral signs depends on the quantity to be found and not on the specific region involved.

Important quantities associated with a plane region are area and moments. Suppose that the area of a region is given by

$$\int_a^b [g(x) - f(x)]\, dx. \tag{4}$$

For any constant c and any positive n,

$$\int_a^b (x - c)^n [g(x) - f(x)]\, dx \tag{5}$$

is called the nth moment of x about c with respect to the area given by (4). This moment will be denoted by $\mu_n(x, c)$.

A word is in order about terminology. The preceding paragraph introduces a moment

of	*a variable*
about	*one of its values*
with respect to	*a weighting.*

This is essentially the language of probability theory in which the weighting is done by probabilities. (This application will be discussed later.) Weighting by physical measurements of figures is more typical of the applications to physics, and physicists use different terminology. They speak of a moment

of	*a weight*
about	*a locus.*

The connection is as follows. The nth moment of an area about the locus of $u = c$ means the nth moment of u about c with respect to the area. Note that in three-space the locus of $x = c$ is a plane; so one speaks of moments of volumes about planes.

TABLE 8–1

Quantity	Variable to integrate	
Area	$dx\,dy$ or	$dy\,dx$
$\mu_1(x, 0)$	$x\,dx\,dy$ or	$x\,dy\,dx$
$\mu_1(y, 0)$	$y\,dx\,dy$ or	$y\,dy\,dx$
$\mu_2(x, 0)$	$x^2\,dx\,dy$ or	$x^2\,dy\,dx$
$\mu_2(y, 0)$	$y^2\,dx\,dy$ or	$y^2\,dy\,dx$

With iterated integrals the question of what to integrate to find these quantities is reduced to pure routine and is summarized in Table 8–1.

Suppose that the region in question is the one shown in Fig. 8–5. Then, the rules summarized above yield

$$\alpha(A) = \int_a^b \int_{f(x)}^{g(x)} dy\,dx = \int_a^b y \Big|_{f(x)}^{g(x)} dx = \int_a^b [g(x) - f(x)]\,dx,$$

$$\mu_1(x, 0) = \int_a^b \int_{f(x)}^{g(x)} x\,dy\,dx = \int_a^b xy \Big|_{f(x)}^{g(x)} dx = \int_a^b x[g(x) - f(x)]\,dx,$$

$$\mu_2(x, 0) = \int_a^b \int_{f(x)}^{g(x)} x^2\,dy\,dx = \int_a^b x^2 y \Big|_{f(x)}^{g(x)} dx = \int_a^b x^2[g(x) - f(x)]\,dx.$$

The first of these is the integral (4) for area, and the other two are special cases of (5). Thus, certain entries in Table 8–1 are reducible to basic definitions.

However, the table indicates that, for the area of the region in Fig. 8–4,

$$\mu_2(y, 0) = \int_a^b \int_{f(x)}^{g(x)} y^2\,dy\,dx.$$

Performing the first integration here does not lead to anything like (5), but the other formula for $M_2(y, 0)$ can be reduced to the basic definition.

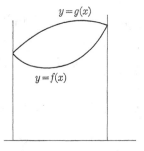

FIGURE 8–4

So, for each line in Table 8–1, one formula is provable by reference to (4) or (5), and the other depends on reversibility of order in an iterated integral. This reversibility will be taken for granted here. It is justified in all specific examples to be studied in this section, and a more general discussion of the question will appear in Chapter 9.

EXAMPLE

1. For the figure bounded by the graphs of $y = x^2$ and $y = x + 2$, use iterated integrals to find the area and the first and second moments of y about 0. This is a familiar example of a figure with a corner on it (see Fig. 8–5). In setting up iterated integrals, such corners govern the order of integration. In this case, integration with respect to y first is indicated because, if the first integration were with respect to x, the formula for the limits of integration would change at the corner:

$$\alpha(A) = \int_{-1}^{2} \int_{x^2}^{x+2} dy\, dx$$

$$= \int_{-1}^{2} (x + 2 - x^2)\, dx$$

$$= \left(\frac{x^2}{2} + 2x - \frac{x^3}{3} \right)\Bigg|_{-1}^{2} = \frac{9}{2}.$$

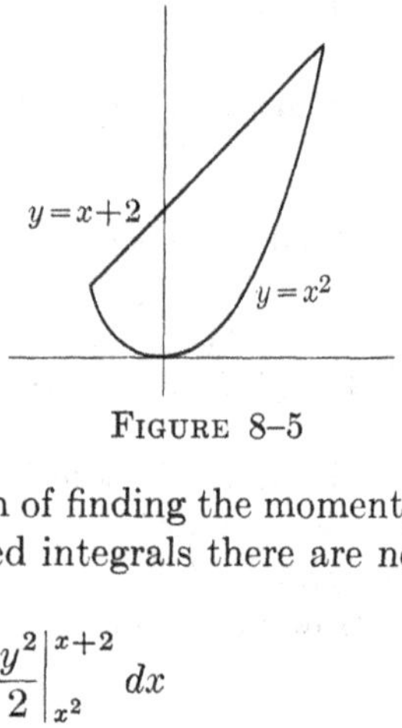

FIGURE 8–5

For this figure, the corner complicated the problem of finding the moments of y with one-dimensional integrals. With iterated integrals there are no such difficulties:

$$\mu_1(y, 0) = \int_{-1}^{2} \int_{x^2}^{x+2} y\, dy\, dx = \int_{-1}^{2} \frac{y^2}{2}\Bigg|_{x^2}^{x+2} dx$$

$$= \int_{-1}^{2} \tfrac{1}{2}(x^2 + 4x + 4 - x^4)\, dx$$

$$= \left(\frac{x^3}{6} + x^2 + 2x - \frac{x^5}{10} \right)\Bigg|_{-1}^{2} = \frac{36}{5},$$

$$\mu_2(y, 0) = \int_{-1}^{2} \int_{x^2}^{x+2} y^2\, dy\, dx = \int_{-1}^{2} \frac{y^3}{3}\Bigg|_{x^2}^{x+2} dx$$

$$= \int_{-1}^{2} \tfrac{1}{3}(x^3 + 6x^2 + 12x + 8 - x^6)\, dx$$

$$= \frac{x^4}{12} + \frac{2x^3}{3} + 2x^2 + \frac{8x}{3} - \frac{x^7}{21}\Bigg|_{-1}^{2} = \frac{423}{28}.$$

EXERCISES

1. In each of the following, the loci determine a bounded figure in the plane. For each of these figures, use twofold iterated integrals to find the area and the first two moments of each coordinate variable about zero.

(a) $y = x$, $y = x^2$ (b) $y = x^2$, $x = y^2$

(c) $x = y$, $y = x^2 - x$ (d) $y = x^2$, $8x = y^2$

(e) $x = y^2$, $x = 2y - y^2$ (f) $x^2 + y^2 = 2x$

(g) $y = \sqrt{1 - x^2}$, $x + y = 1$ (h) $\sqrt{x} + \sqrt{y} = 1$, $x + y = 1$

(i) $y = 4x^2$, $y = x^2 + 3$ (j) $y = \sin \pi x$, $y = 2x$

(k) $y = e^x$, $x = 0$, $y = e$ (l) $y = \ln x$, $x = 1$, $y = 1$

(m) $y = \cosh x$, $y = 5/4$ (n) $y = \arcsin x$, $y = 0$, $x = 1$

(o) $y = x$, $y = 2x$, $x = 1$ (p) $y = x^2$, $y = x^3$

(q) $xy = 1$, $x + y = 5/2$ (r) $y = e^x$, $y = (e - 1)x + 1$

(s) $y = x$, $y = 2x$, $y = 1$ (t) $xy = 1$, $x = 1$, $x = 2$, $y = 0$

8–3 Quadric surfaces. Just as in two dimensions it pays to recognize lines and conics, in three dimensions it also pays to recognize and sketch by mechanical rules the quadric surfaces; the more important ones are as follows. The *ellipsoid*, whose equation is

$$\frac{x^2}{a^2} + \frac{y^2}{b^2} + \frac{z^2}{c^2} = 1,$$

is shown in Fig. 8–6. The *elliptic paraboloid*, whose equation is

$$\frac{x^2}{a^2} + \frac{y^2}{b^2} = \frac{z}{c},$$

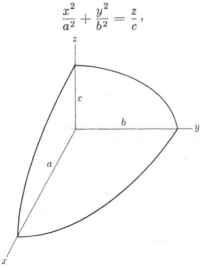

FIGURE 8–6

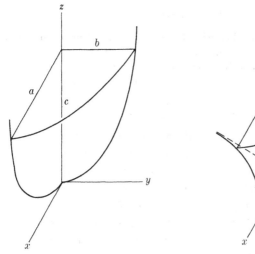

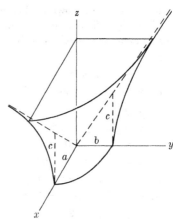

FIGURE 8–7　　　　　　　FIGURE 8–8

is shown in Fig. 8–7 for the case $c > 0$. For $c < 0$ it opens down. The *hyperboloid of one sheet*, whose equation is

$$\frac{x^2}{a^2} + \frac{y^2}{b^2} - \frac{z^2}{c^2} = 1,$$

is shown in Fig. 8–8. The *hyperboloid of two sheets*, whose equation is

$$\frac{x^2}{a^2} - \frac{y^2}{b^2} - \frac{z^2}{c^2} = 1,$$

is shown in Fig. 8–9. The *elliptic cone*, whose equation is

$$\frac{x^2}{a^2} + \frac{y^2}{b^2} = \frac{z^2}{c^2},$$

is shown in Fig. 8–10. Finally, note that an equation in two rectangular variables only is the equation of a cylindrical surface, with elements parallel to the axis of the missing variable.

Variants of the above forms are obtained by permuting the variables. For sketching purposes, one must first recognize the type of surface, and then determine the orientation of its axis. A good informal rule is that the axis is parallel to that of the "odd" variable. In the hyperboloids and cone, one variable is "odd" because of sign; on the paraboloid, one variable is "odd" because of power; and in the cylinder, one variable is "odd" because of its absence.

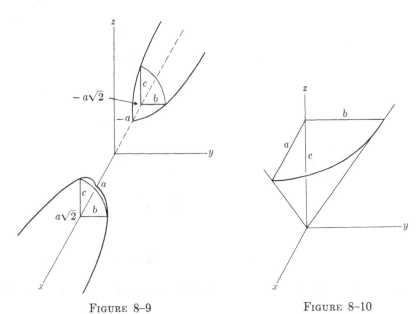

FIGURE 8–9   FIGURE 8–10

In the sketch of the hyperboloid of two sheets, a portion behind the yz-plane is shown in order to indicate the location of the second sheet. Otherwise, these sketches show only the first octant sections of the surfaces. In the case of the paraboloid, the first octant portion is one-fourth of the entire surface (reflect through the xz- and yz-planes in Fig. 8–7). In the other cases shown, the first octant portion is one-eighth of the entire surface (reflect through each of the three coordinate planes). As a rule, more intelligible sketches are obtained by limiting the sketch to the first octant, and this is recommended for the applications to triple integral problems in the next section. However, the symmetry of the figures should be clearly understood.

Finally, translations of these figures may be introduced in the standard manner. Uncalled for first power terms may be absorbed by completing the square, and the result reveals a translation.

EXAMPLES

1. Discuss and sketch the locus of

$$x^2 + y^2 = 1 - z.$$

This looks like a paraboloid except for the constant term. However, the

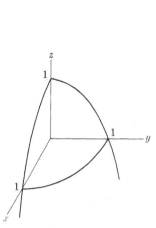

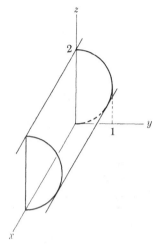

FIGURE 8–11   FIGURE 8–12

constant can be grouped with the linear term to indicate a translation as follows:

$$x^2 + y^2 = -(z - 1).$$

Now, a translation of the origin to $z = 1$ is indicated, and the form is a paraboloid that opens down. A sketch is shown in Fig. 8–11.

2. Discuss and sketch the locus of

$$y^2 + z^2 = 2z.$$

Here we have a cylindrical surface with elements parallel to the x-axis because there is no x in the equation. To get the section in the yz-plane (if it is not already recognizable), complete the square:

$$y^2 + z^2 - 2z + 1 = 1,$$

$$y^2 + (z - 1)^2 = 1.$$

This is a circle with radius 1 and center at $y = 0$, $z = 1$. A sketch of the cylinder is shown in Fig. 8–12.

EXERCISES

1. For each of the following equations name and sketch the locus.

(a) $x^2 + y^2 + z^2 = 1$ (b) $x^2 + y^2 - z^2 = 1$

(c) $x^2 - y^2 - z^2 = 1$ (d) $x^2 + y^2 - z^2 = -1$

(e) $x^2 - y^2 - z^2 = -1$ (f) $x^2 + y^2 - z = 1$

(g) $x^2 + y^2 + z = -1$ (h) $x^2 + y^2 - z^2 = 0$

(i) $x^2 - y^2 - z^2 = 0$ (j) $x^2 + y^2 - x = 0$

(k) $x^2 + y^2 + y = 0$ (l) $z^2 = x^2$

(m) $z^2 = x$ (n) $z^2 = x^2 + y^2$

(o) $z^2 = x - y^2$ (p) $z^2 = y - x^2$

(q) $z^2 = -y - x^2$ (r) $x^2 + y^2 + z^2 - 2x + 4y = 4$

(s) $x^2 - y^2 - z^2 - 2x + 4y = 4$

(t) $x^2 - y^2 - z^2 - 2x + 4y = 3$

(u) $x^2 - y^2 - x^2 - 2x + 4y = 2$

(v) $x^2 + y^2 - 2x + 4y - z = 4$

(w) $x^2 + y^2 - 2x + 4y + z = 4$

(x) $x^2 + y^2 - 2x + 4y = 4$

(y) $x^2 - y^2 - 2x + 4y = 4$ (z) $x^2 - y^2 - 2x + 4y = 2$

8-4 Threefold iterated integrals. The form

$$\int_a^b \int_{f(z)}^{g(z)} \int_{\phi(y,z)}^{\psi(y,z)} f(x, y, z) \, dx \, dy \, dz \tag{6}$$

is interpreted as indicating three successive integrations. More specifically, instructions for computing (6) are given by

$$\int_a^b \left\{ \int_{f(z)}^{g(z)} \left[\int_{\phi(y,z)}^{\psi(y,z)} f(x, y, z) \, dx \right] dy \right\} dz.$$

Geometrically, such a threefold iterated integral is taken over a region in three-space. For the form (6), the region in question is the locus of the simultaneous inequalities

$$\phi(y, z) \leq x \leq \psi(y, z), \qquad f(z) \leq y \leq g(z), \qquad a \leq z \leq b. \tag{7}$$

As with twofold iterated integrals, the applications determine the choice of $f(x, y, z)$; and, quite independently, the region in question determines the limits of integration. Applications to volumes and moments thereof are quickly summarized in a table like Table 8-1; such a table follows.

TABLE 8-2

Quantity	Variable to integrate
$Volume$	$dx \, dy \, dz$
$\mu_n(x, 0)$	$x^n \, dx \, dy \, dz$
$\mu_n(y, 0)$	$y^n \, dx \, dy \, dz$
$\mu_n(z, 0)$	$z^n \, dx \, dy \, dz$

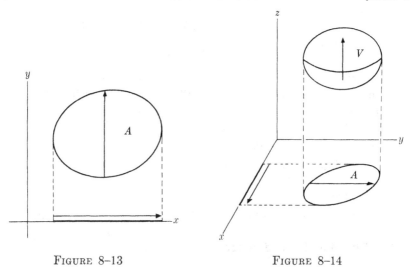

FIGURE 8–13   FIGURE 8–14

As in Table 8–1, different orders of integration may be used. There are six permutations of three things; so, for each entry in Table 8–2, there are five alternative forms, any one of which may be used if desired.

A common mistake in setting up iterated integrals is to retain too many variables in the limits of integration. Observe the correct pattern set by (6). Limits of integration for a given variable can involve only variables yet to be used as variables of integration. To see that this mechanical rule must be correct, think of the process of evaluating (6). The final result must be a number. Now x disappears after the first integration, but if it were to reappear in a subsequent limit of integration, there would be no way of getting rid of it, and it would appear in the answer. It is also well to bear in mind the geometric significance of the individual limits of integration.

First, consider a twofold integral over a plane figure A. Suppose the order of integration is $dy\,dx$. The first integration is from the bottom curve to the top curve bounding A, and the second is over the projection of A onto the x-axis. (See Fig. 8–13.) For an iterated integral over V in the order $dz\,dy\,dx$, the first integration is from the bottom surface to the top surface bounding V. The remaining twofold integration is over the projection A of V onto the xy-plane (Fig. 8–14) and proceeds according to plan for twofold iterated integrals.

An essential part of problem solving in connection with triple integrals is the sketching of three-dimensional figures. A general plan of attack is to set each variable equal to zero separately to find equations of the sections in the coordinate planes. If the sketches of these sections are insufficient for a picture of the figure, try other sections parallel to the coordinate planes; the equation is obtained by setting one variable equal to a constant, not zero. However, most of the exercises below are based on quadric surfaces. The outline presented in Section 8–3 is designed to handle most of the sketching problems in this section.

EXAMPLES

1. Evaluate

$$\int_0^1 \int_0^z \int_0^{yz} (x + y + z)\, dx\, dy\, dz.$$

Solution.

$$\int_0^1 \int_0^z \int_0^{yz} (x + y + z)\, dx\, dy\, dz$$

$$= \int_0^1 \int_0^z \left(\frac{x^2}{2} + xy + xz\right)\Big|_0^{yz} dy\, dz$$

$$= \int_0^1 \int_0^z \left(\frac{y^2 z^2}{2} + y^2 z + yz^2\right) dy\, dz$$

$$= \int_0^1 \left(\frac{y^3 z^2}{6} + \frac{y^3 z}{3} + \frac{y^2 z^2}{2}\right)\Big|_0^z dz$$

$$= \int_0^1 \left(\frac{z^5}{6} + \frac{z^4}{3} + \frac{z^4}{2}\right) dz = \left(\frac{z^6}{36} + \frac{z^5}{6}\right)\Big|_0^1 = \frac{7}{36}.$$

2. Set up six different iterated integrals for the volume of the figure bounded above by the plane on which $z = y$ and below by the paraboloid on which $z = x^2 + y^2$.

Integral 1. Integrate in the order $dz\, dx\, dy$. Integration with respect to z is from bottom to top surface; that is, from $x^2 + y^2$ to y. The projection of the figure onto the xy-plane is a circle (see Fig. 8–15). To obtain the

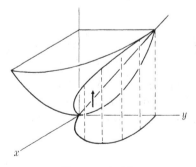

FIGURE 8–15

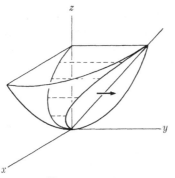

FIGURE 8–16   FIGURE 8–17

equation of this projection, eliminate z between the two equations given:

$$x^2 + y^2 = z = y, \qquad x^2 + y^2 = y.$$

In an integral over this circle in the order $dx\,dy$, x runs between $\pm\sqrt{y - y^2}$, and y runs from 0 to 1. So, the required threefold iterated integral is

$$\int_0^1 \int_{-\sqrt{y-y^2}}^{\sqrt{y-y^2}} \int_{x^2+y^2}^{y} dz\,dx\,dy.$$

Integral II. Integrate in the order $dz\,dy\,dx$. Limits for z are the same as above. The projection, that is the locus of

$$x^2 + y^2 = y,$$

is also the same. Solve this equation for y by the quadratic formula:

$$y = \frac{1 \pm \sqrt{1 - 4x^2}}{2}.$$

This formula gives the limits for y, and x runs from $-\frac{1}{2}$ to $\frac{1}{2}$ (Fig. 8–15). Thus, the integral is

$$\int_{-1/2}^{1/2} \int_{(1-\sqrt{1-4x^2})/2}^{(1+\sqrt{1-4x^2})/2} \int_{x^2+y^2}^{y} dz\,dy\,dx.$$

Integral III. Integrate in the order $dy\,dz\,dx$. Integration with respect to y is from the plane to the paraboloid (see Fig. 8–16); that is, from z to $\sqrt{z - x^2}$. The projection onto the xz-plane is another circle (Fig. 8–16) whose equation,

$$x^2 + z^2 = z,$$

is obtained by eliminating y between the two given equations. The integration over this circle in the order $dz\,dx$ follows the pattern of integral II above, and the required integral is

$$\int_{-1/2}^{1/2} \int_{(1-\sqrt{1-4x^2})/2}^{(1+\sqrt{1-4x^2})/2} \int_{z}^{\sqrt{z-x^2}} dy\,dz\,dx.$$

Integral IV. Integrate in the order $dy\,dx\,dz$. This merely calls for integration over the circle in Fig. 8–16 in the other order. The pattern for this was set in integral I. *Answer.*

$$\int_{0}^{1} \int_{-\sqrt{z-z^2}}^{\sqrt{z-z^2}} \int_{z}^{\sqrt{z-x^2}} dy\,dx\,dz.$$

Integral V. Integrate in the order $dx\,dz\,dy$. Integration with respect to x is from one branch of the paraboloid to the other; that is, between $\pm\sqrt{z-y^2}$. The projection of the figure onto the yz-plane is seen geometrically (Fig. 8–17) to be the plane figure bounded by the parabola on which $z = y^2$ and the line on which $z = y$. In an integration over this figure in the order $dz\,dy$, z runs from the parabola to the line, and y runs from 0 to 1. *Answer.*

$$\int_{0}^{1} \int_{y^2}^{y} \int_{-\sqrt{z-y^2}}^{\sqrt{z-y^2}} dx\,dz\,dy.$$

Integral VI. Integrate in the order $dx\,dy\,dz$. Limits for x are the same as above, and so is the projection. In an integral over this projection in the order $dy\,dz$, y runs from the line to the parabola, and z runs from 0 to 1. *Answer.*

$$\int_{0}^{1} \int_{z}^{\sqrt{z}} \int_{-\sqrt{z-y^2}}^{\sqrt{z-y^2}} dx\,dy\,dz.$$

EXERCISES

1. Evaluate each of the following integrals.

(a) $\displaystyle\int_{0}^{1} \int_{0}^{z} \int_{0}^{y} xy^2z^3\,dx\,dy\,dz$ (b) $\displaystyle\int_{0}^{1} \int_{0}^{1-y} \int_{0}^{1-x^2} z\,dz\,dx\,dy$

(c) $\displaystyle\int_{1}^{2} \int_{0}^{x} \int_{0}^{y\sqrt{3}} \frac{y}{y^2+z^2}\,dz\,dy\,dx$ (d) $\displaystyle\int_{0}^{1} \int_{0}^{\ln z} \int_{0}^{y+z} e^{x+y+z}\,dx\,dy\,dz$

(e) $\displaystyle\int_{0}^{1} \int_{z^2}^{1} \int_{0}^{1-z} z\,dy\,dz\,dx$ (f) $\displaystyle\int_{1}^{2} \int_{1}^{z} \int_{1/y}^{2} yz^2\,dx\,dy\,dz$

(g) $\displaystyle\int_{-1}^{0}\int_{0}^{y}\int_{1}^{x}(z^2 - y)\,dz\,dx\,dy$ (h) $\displaystyle\int_{0}^{1}\int_{0}^{\sqrt{1-z^2}}\int_{0}^{\sqrt{1-y^2-z^2}}yz\,dx\,dy\,dz$

2. In each of the following expressions a solid figure is described. Use a three-fold iterated integral in each case to find the volume of the figure.

    (a) Bounded above by the locus of $z = x^2 + y^2$, below by the locus of $z = 0$, and on the sides by that of $x^2 + y^2 = 1$

    (b) Bounded above by the locus of $z = x^2 + y^2$, below by the locus of $z = 0$, and on the sides by that of $x^2 + y^2 = 2y$

    (c) Bounded above by the locus of $z = y$, below by the locus of $z = 0$, and on the sides by the loci of $x = 4 - y^2$ and $x = 0$

    (d) Bounded by the loci of $z = x^2 + 4y^2$ and $z = 8 - x^2 - 4y^2$

    (e) Inside the loci of $x^2 + y^2 = 1$ and $x^2 + z^2 = 1$

    (f) Inside the loci of $x^2 + y^2 = 1$ and $x^2 + z^2 = 2z$

    (g) Bounded above by the locus of $z = x + 2$, below by the locus of $z = 0$, and on the sides by the locus of $x^2 + 4y^2 = 4$

    (h) Bounded above by the locus of $z = x + y$, below by the locus of $z = 0$, and on the sides by the loci of $x^2 + y^2 = 1$, $x = 1$ and $y = 1$

    (i) Bounded by the loci of $z = 2x + 3$ and $z = x^2 + y^2$

    (j) Bounded above by the locus of $z = 12 - x^2 - y^2$ and below by the top half of the locus of $z^2 = x^2 + y^2$

    (k) Bounded below by the locus of $z = 1$, above by the locus of $z = x$, and on the sides by the loci of $x = 2$, $y = 1$ and $y = x^2$

    (l) Bounded above by the locus of $z = 4 - y^2$, below by that of $z = 2 - y$, and on the sides by the loci of $x = 0$ and $x = 3$

    (m) Bounded by the locus of $(x/a) + (y/b) + (z/c) = 1$ and the coordinate planes

    (n) Bounded above by the locus of $z = 1 - y^2$, below by that of $z = 0$, and on the sides by the loci of $x = 0$ and $x + y = 1$

3. Find the centroid of each of the solid figures described in Exercise 2.

CHAPTER 9

MULTIPLE INTEGRALS

9–1 Oriented manifolds. Informally, a k-dimensional manifold is *oriented* if there is associated with each point in it an ordered k-tuple of vectors whose relative orientation does not change from point to point. In the plane, an ordered pair of vectors may be characterized as either clockwise or counterclockwise. In three-space an ordered triple of vectors is either right- or left-handed. Thus these same descriptions are applied to manifolds. However, even without getting into higher dimensions, there are situations in which this informal picture is not sufficiently precise, so a technical description proceeds along slightly different lines.

If M is a manifold, then ∂M is used to denote the boundary of M. Some manifolds have boundaries; others do not. For example, an arc of a curve has a two-point boundary (Fig. 9–1), a surface such as the one in Fig. 9–2 has a curve for a boundary, and a solid (Fig. 9–3) has a surface for a boundary. On the other hand, a simple closed curve has no boundary, nor does the surface of a sphere. However, a manifold with or without boundary may be partitioned into submanifolds with boundaries. Figure 9–4 shows the partitioning of a simple closed curve; Fig. 9–5 shows a surface without boundary, partitioned into pieces each of which has a boundary.

Figure 9–6 shows a counterclockwise figure in the plane with all sorts of things marked on it. It is given a counterclockwise boundary in an obvious sense—arrowheads marked 1, 2, 3. Several counterclockwise vector pairs are shown, and the one at p (a boundary point) consists of the exterior

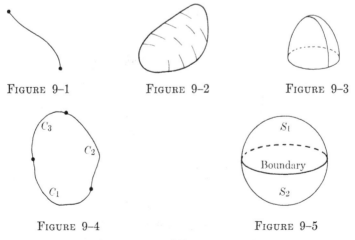

FIGURE 9–1   FIGURE 9–2   FIGURE 9–3

FIGURE 9–4   FIGURE 9–5

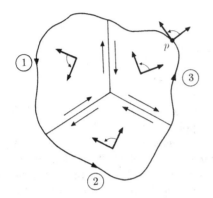

FIGURE 9-6

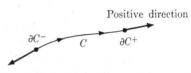

FIGURE 9-7

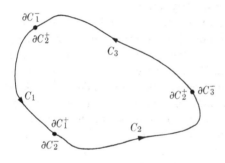

FIGURE 9-8

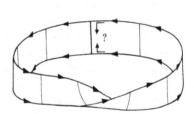

FIGURE 9-9

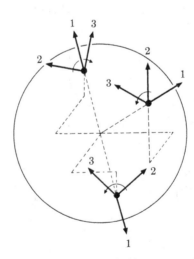

FIGURE 9-10

normal followed by the positive tangent. The figure is partitioned; and to give each of the two-dimensional submanifolds a counterclockwise boundary, one must orient each partition line one way as a boundary segment of one plane region and the other way as a boundary segment of the other.

It should be evident that to have everything counterclockwise, one must incorporate all the features shown in Fig. 9–6. In more general situations, these features may be used to define orientation. Specifically, oriented manifolds have two key properties.

(i) The orientation of M is always related to that of ∂M by the rule that, at a boundary point, an ordered set of vectors giving the orientation of M consists of the exterior normal followed by an ordered set giving the orientation of ∂M.

(ii) If an oriented manifold is partitioned, adjacent submanifolds induce opposite orientations on their common boundary.

One may now describe orientation by proceeding from ∂M to M, thus defining the notion in one higher dimension at each step. First, take an arc. Its two-point boundary may be oriented by labeling one point $(+)$ and the other $(-)$. In this case, (i) is interpreted to mean that the vector away from C at ∂C^+ gives the positive direction on C, and the vector away from C at ∂C^- gives the negative direction on C. This is shown in Fig. 9–7. To orient a simple closed curve, partition it; and as in Fig. 9–8, follow (ii) and make each partition point $(+)$ for one arc and $(-)$ for the other. If this is done consistently, it will orient each arc and hence the closed curve.

To orient a surface, orient its boundary (directions for doing this are in the preceding paragraph) and then use (i) above. The only tricky thing is that (ii) must be satisfied, and this is not always possible. That is, for surfaces in three-space, there is a question of *orientability*. The Mobius strip (Fig. 9–9) is a classical example of a nonorientable surface. The boundary is a simple closed curve and is easily oriented. However, a partition line across the strip cannot be oriented in such a way that (ii) is satisfied. Briefly, the situation is as follows: For $1 < k < n$, a k-dimensional manifold in n-space may or may not be orientable. The only case of this phenomenon to appear in the present discussion is $k = 2$, $n = 3$ (surfaces in three-space). Curves, solids, and plane figures are orientable.

To proceed then, knowing how to orient surfaces, one can orient solids by (i). Figure 9–10 shows a sphere with right-handed orientation.

Now a coordinate set on a manifold determines an ordered set of tangent vectors at each point. Say (u, v) is a coordinate set on a two-dimensional manifold M. Informally, for each point p of M, there are du and dv directions in the tangent plane T_p. More precisely, there are vectors $\mathbf{w}_u$ and $\mathbf{w}_v$ in T_p such that

$$du_p(\mathbf{w}_u) = 1, \quad du_p(\mathbf{w}_v) = 0, \quad dv_p(\mathbf{w}_u) = 0, \quad dv_p(\mathbf{w}_v) = 1.$$

The manifold M may be oriented quite independently of u and v; and if this is done, there is another ordered pair of tangent vectors at p. The pertinent question then is whether or not the two ordered pairs of vectors have the same or opposite orientations. The two possibilities are described by saying that the orientation of M is positive or negative with respect to (du, dv).

If M is oriented (satisfying (ii) above, in particular) and if (u, v) is a genuine coordinate set over all of M, then the relative orientation of M and (du, dv) does not change from point to point. However, one often uses variables that are coordinates except at exceptional points. In cases of practical importance, this causes very little trouble except for one thing: It can reverse relative orientation, and this must be watched for carefully.

EXAMPLES

1. For a closed surface S, such as that shown in Fig. 9–11, (x, y) is not a coordinate set over the entire manifold. The mapping $p \rightarrow \big(x(p), y(p)\big)$ is not one to one on S. If S is oriented, the orientations of the top and bottom halves must induce opposite orientations on the equator as shown in Fig. 9–11; thus the relative orientation of S and (dx, dy) will change from one portion of the figure to another. The orientation shown in Fig. 9–11 makes S the boundary of a right-handed solid, and in this case, the top half is positively oriented with respect to (dx, dy), while the relative orientation on the bottom half is negative. This example should be carefully noted, as it is often encountered in practice. One thinks of rectangular coordinates as being "completely well behaved." This is not quite true; the orientation problem for closed surfaces in three-space is a case in point.

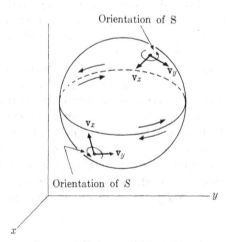

FIGURE 9–11

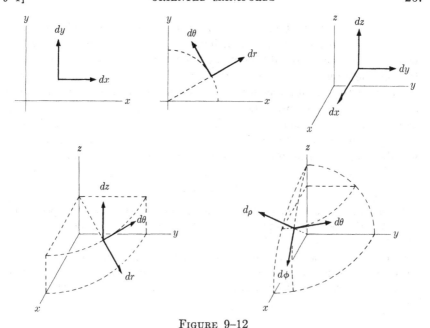

FIGURE 9–12

2. Let

$$X = \begin{bmatrix} x \\ y \end{bmatrix}, \qquad U = \begin{bmatrix} u \\ v \end{bmatrix}.$$

Suppose that X gives a coordinate set over M. The fundamental theorem equation

$$dU = D_X U \, dX$$

describes (among other things) a change of basis in the tangent spaces to M. Now, relative orientation of the basis vectors is preserved in a change of basis provided that the matrix for the transformation has a positive determinant; thus the relative orientation of (dx, dy) and (du, dv) is given by the sign of the Jacobian determinant

$$\det D_X U.$$

Since it is assumed that X gives a coordinate set, the relative orientation of M and dX is constant. The question of whether M and dU have constant relative orientation may then be determined analytically. It reduces to a question of constant sign for $\det D_X U$.

3. The coordinate systems in most common use are (x, y), (r, θ), (x, y, z), (r, θ, z), (ρ, ϕ, θ). Note that, when listed in the order indicated and drawn as they are consistently drawn in this book (Fig. 9–12), the two-dimen-

sional differential systems are both counterclockwise and the three-dimensional systems are all right-handed. (Note, however, Exercise 5 below.) The reader should be warned that there are many variations on this in the literature. Apparently, (x, y) and (r, θ) are always drawn counterclockwise, but a recent text lists polar coordinates (θ, r). The trend in recent years is to draw (x, y, z) right-handed, but most of the older analytic geometry and calculus books draw this system left-handed. Finally, some books list spherical coordinates (ρ, θ, ϕ), and no matter how you draw them this system is oppositely oriented from (x, y, z).

EXERCISES

1. Given that a one-dimensional manifold is oriented, show that specification of the positive tangential direction at one point determines uniquely all positive tangential directions.

2. Figure 9–13 shows two cubes, and in each case, the orientation of one edge with respect to one face is given. Show that the orientation of the cube is determined in each case. Which is right-handed and which is left-handed?

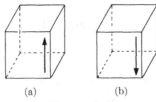

(a)   (b)

FIGURE 9–13

3. Show that no simple closed curve in the plane is oriented with respect to dx.

4. What convex simple closed curves in the plane are oriented with respect to $d\theta$?

5. What is the orientation of polar and cylindrical coordinate differentials when $r < 0$? Discuss the orientation of spherical coordinate differentials for $\rho < 0$, or $\phi > \pi$, or both. Do the signs of the coordinates, themselves, have any effect on the orientation of rectangular coordinate differentials? Discuss these questions both geometrically and analytically (see Example 2).

6. Draw a circular ring with counterclockwise orientation, with clockwise orientation.

7. Draw a torus which bounds a three-dimensional manifold with right-handed orientation, with left-handed orientation.

8. The parabola on which $y = x^2$ is orientable with respect to dx but not with respect to dy. Explain this situation geometrically, and also in terms of the sign of $D_x y$.

9. The unit circle is orientable with respect to $d\theta$, but not with respect to dx or dy. Compute $D_\theta x$ and $D_\theta y$ to show analytically where the relative orientation reverses for each of the rectangular coordinate differentials.

10. For $0 \leq \phi \leq \pi$, $0 \leq \theta \leq 2\pi$, the unit sphere is orientable with respect to $(d\phi, d\theta)$. Show analytically that the orientation with respect to (dx, dy) changes at $\phi = \pi/2$.

9–2 Exterior products. Let C be an oriented, one-dimensional manifold; the line integral

$$\int_C u \, ds$$

is taken with respect to a signed measure of distance on the tangent bundle for C. That is, ds measures lengths of oriented intervals on each tangent line, giving them positive or negative measures according as the orientation of C is positive or negative with respect to ds. One also has line integrals of the form

$$\int_C u \, dv.$$

Here dv is also a signed measure on oriented intervals on the tangent lines, but, in general, it does not measure lengths.

To define a multiple integral, one wishes to have analogous differential operators in two and more dimensions. First, consider the two-dimensional analogue to ds. Let the manifold in question be the plane so that the tangent spaces are translated planes. A signed measure of area on oriented plane figures in each tangent plane is required. This can be constructed from differentials in the following manner.

There is to be defined an *exterior product* of differentials, with the operation denoted by $\wedge$. This symbol is usually read "wedge." That is, $dx \wedge dy$ (dx wedge dy) will denote the exterior product of dx and dy. Each such exterior product will be a variable on oriented plane regions of the tangent planes. The variable on oriented regions of T_p will be written $(dx \wedge dy)_p$ just as the variable on T_p is written dx_p. If R is an oriented region of T_p bounded by level lines for a set of admissible coordinate variables, then

$$(dx \wedge dy)_p(R) = \pm \text{ area of } R, \tag{1}$$

with the sign giving the relative orientation of R and (dx, dy).

So far as $dx \wedge dy$ is concerned, (1) could be taken as a definition. This will be the geometric significance of $dx \wedge dy$; but it is best to arrive at this result in a slightly different way, because to use (1) as a definition would leave the following questions unanswered.

(i) For other differentials, du and dv, what is $(du \wedge dv)_p(R)$?

(ii) What are the algebraic properties of this $\wedge$ multiplication? For example, does

$$du \wedge (dv + dw) = (du \wedge dv) + (du \wedge dw)?$$

Why?

(iii) Is this actually an operation on differentials as they are defined in Section 5–4? For example, is it legitimate to substitute from the fundamental theorem:

$$du \wedge dv = du \wedge \left(\frac{\partial v}{\partial x} dx + \frac{\partial v}{\partial y} dy \right)?$$

Affirmative answers to (ii) and (iii) and a definitive answer to (i) may be obtained in the following way.

One way to define an area variable α over a large class of plane regions is to give the following postulates.

(a) The area of a rectangle is the base times the altitude.

(b) Area is *additive;* that is, if a region is subdivided into nonoverlapping subregions, then the area of the whole is the sum of the areas of the parts.

(c) Area is *monotone;* that is, if A is contained in B, then

$$\alpha(A) \leq \alpha(B).$$

Consider the area under a continuous curve. Form the upper and lower sums for the usual integral for this area. By postulate (a) each term in these sums is the area of a rectangle. By postulate (c) the area under the curve in each subdivision lies between that of the inside and outside rectangles. Thus, by postulate (b) the entire area under the curve lies between the upper and lower sums. The existence theorem for the integral guarantees that the upper and lower sums have a common limit; therefore the area under the curve must equal this limit.

A similar argument (details omitted) may be used to show that any variable on plane regions that is additive, monotone, and specified on rectangles is uniquely determined on all regions bounded by continuous curves.

To define $dx \wedge dy$ in this way it suffices to express the signed area of an oriented parallelogram in terms of dx and dy. To do this, recall that dx_p and dy_p are component variables on vectors in the tangent plane T_p. Let P be spanned by vectors $\mathbf{r}_1$ and $\mathbf{r}_2$ (Fig. 9–14). Then, the signed area of P is

$$\begin{vmatrix} dx_p(\mathbf{r}_1) & dx_p(\mathbf{r}_2) \\ dy_p(\mathbf{r}_1) & dy_p(\mathbf{r}_2) \end{vmatrix} \qquad (2)$$

(see Sections 4–6 and 4–7). So, take (2) as a definition of $(dx \wedge dy)_p(P)$; introduce additivity and monotonicity, and (1) is the result.

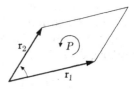

FIGURE 9–14

This sets an obvious pattern for a general definition of $\wedge$. Let du and dv be any differentials on the tangent bundle for the plane, and let P be the oriented parallelogram in T_p generated by $\mathbf{r}_1$ and $\mathbf{r}_2$ (Fig. 9–14). Define

$$(du \wedge dv)_p(P) = \begin{vmatrix} du_p(\mathbf{r}_1) & du_p(\mathbf{r}_2) \\ dv_p(\mathbf{r}_1) & dv_p(\mathbf{r}_2) \end{vmatrix}. \tag{3}$$

If P' is a translation of P, define

$$(du \wedge dv)_p(P') = (du \wedge dv)_p(P);$$

this defines $(du \wedge dv)_p$ on all the parallelograms in T_p. This variable probably does not give areas, but (3) is perfectly definite and tells what the variable does give on parallelograms. Finally, complete the definition by postulating that $(du \wedge dv)_p$ is *additive, monotone* and, *on parallelograms, is given by* (3).

These postulates should be checked for consistency. That is, $(du \wedge dv)_p$ is to be additive and monotone on all sorts of sets, but it is rigidly defined on parallelograms. Is it additive and monotone there? This check is left to the student (see Exercise 6 below for hints).

Questions (i) and (iii) above are now answered, and $du \wedge dv$ has been defined in terms of du and dv.

For the questions raised in (ii), algebraic properties of $\wedge$ are easily checked from formula (3). Actually, the computations that follow verify the key properties of exterior products on parallelograms. All these key properties can be extended to the complete domain of the exterior-product variables, but proofs of such extensions will not be given here.

Like matrix multiplication and vector multiplication in three-space, exterior multiplication of differentials is noncommutative. Specifically,

$$du \wedge dv = -dv \wedge du. \tag{4}$$

Proof. By reversing factors, rows are interchanged in the matrix whose determinant appears in (3).

Exterior multiplication is distributive over addition; that is,

$$du \wedge (dv + dw) = (du \wedge dv) + (du \wedge dw). \tag{5}$$

Proof. By (3), equation (5) follows from a simple identity for determinants:

$$\begin{vmatrix} a & b \\ c+d & e+f \end{vmatrix} = \begin{vmatrix} a & b \\ c & e \end{vmatrix} + \begin{vmatrix} a & b \\ d & f \end{vmatrix}.$$

Exterior multiplication on the tangent bundle commutes with multiplication of the differentials by variables on the base manifold. That is,

$$du \wedge (w\,dv) = (w\,du) \wedge dv = w(du \wedge dv). \qquad (6)$$

Proof. Result (6) also comes from a simple property of determinants:

$$\begin{vmatrix} a & b \\ kc & kd \end{vmatrix} = \begin{vmatrix} ka & kb \\ c & d \end{vmatrix} = k\begin{vmatrix} a & b \\ c & d \end{vmatrix}.$$

Finally, the exterior product of any differential with itself is zero:

$$du \wedge du = 0. \qquad (7)$$

Proof. $du \wedge du$ is given by the determinant of a matrix with identical rows.

Let $u_1, u_2, \ldots, u_n$ be a coordinate set in n-space. Let P be the oriented parallelepiped in a tangent n-space determined by vectors $\mathbf{r}_1, \mathbf{r}_2, \ldots, \mathbf{r}_n$. Define

$$du_1 \wedge du_2 \wedge \ldots \wedge du_n$$

by setting

$$(du_1 \wedge du_2 \wedge \ldots \wedge du_n)(P) = \det[du_i(\mathbf{r}_j)]$$

and postulating additivity and monotonicity.

Algebraic properties are similar to those for the two-dimensional case. The analogue to (4) is that an odd permutation of factors changes the sign of the exterior-product variable while an even permutation of factors leaves the variable unchanged. As in (5), if any factor is a sum, the result is a sum of products. As in (6), variables on the base manifold may be factored out. The analogue to (7) is that if any factor is repeated, the entire product is zero.

A general change of variable formula relating exterior products will appear in Section 9–4, but from the geometric significance of individual differentials, as studied in Section 5–5, one can discover signed area and volume variables in two- and three-space. The final result on determinant formulas for area and volume was that if $\mathbf{u}_1, \mathbf{u}_2, \ldots, \mathbf{u}_n$ is *any* orthonormal basis and $\mathbf{w}_1, \mathbf{w}_2, \ldots, \mathbf{w}_n$ spans a parallelepiped P, then the signed volume of P is $\det[u_i(\mathbf{w}_j)]$, where $u_i(\mathbf{w}_j)$ is the component of $\mathbf{w}_j$ with respect to $\mathbf{u}_i$ (see Section 4–7). Thus, to get a signed area or volume variable, take differentials (with multipliers, if necessary) that form an orthonormal component set. Their exterior product will be an area or volume variable, according to dimension. The technique for finding differentials that give orthonormal component systems was discussed in Section 5–5.

EXAMPLES

1. On the tangent bundle for the plane, the differential pair $(dr, r\, d\theta)$ generates orthonormal component variables. Therefore,

$$dr \wedge (r\, d\theta) = r(dr \wedge d\theta)$$

is a signed area variable. For $r > 0$, $(dr, d\theta)$ is a counterclockwise system; so, for $r > 0$, $r(dr \wedge d\theta)$ gives positive area on counterclockwise figures. For $r < 0$, $(dr, d\theta)$ is clockwise (Fig. 9–15); thus $dr \wedge d\theta$ is negative on counterclockwise figures, but $r(dr \wedge d\theta)$ is still positive.

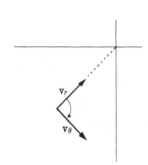

FIGURE 9–15	FIGURE 9–16

2. In Fig. 9–16, the rectangle P is generated by the vectors $\mathbf{r}_1$ and $\mathbf{r}_2$. Clearly,

$$dx(\mathbf{r}_1) = 2, \qquad dx(\mathbf{r}_2) = 0,$$

$$dy(\mathbf{r}_1) = 0, \qquad dy(\mathbf{r}_2) = 1;$$

so,

$$\text{area } (P) = (dx \wedge dy)(P) = \begin{vmatrix} 2 & 0 \\ 0 & 1 \end{vmatrix} = 2. \tag{8}$$

However (note dotted projection lines),

$$dr(\mathbf{r}_1) = \sqrt{2}, \qquad dr(\mathbf{r}_2) = 1/\sqrt{2},$$

$$r\, d\theta(\mathbf{r}_1) = -\sqrt{2}, \qquad r\, d\theta(\mathbf{r}_2) = 1/\sqrt{2};$$

hence,

$$\text{area } (P) = r\, (dr \wedge d\theta)(P) = \begin{vmatrix} \sqrt{2} & 1/\sqrt{2} \\ -\sqrt{2} & 1/\sqrt{2} \end{vmatrix} = 2. \tag{9}$$

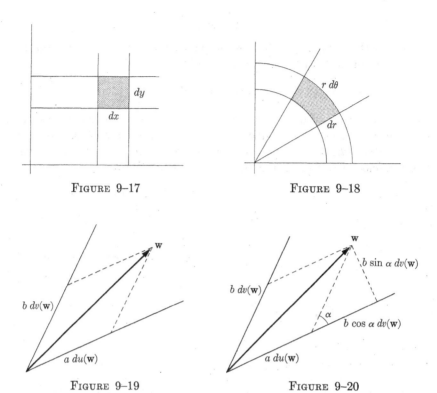

FIGURE 9–17   FIGURE 9–18

FIGURE 9–19   FIGURE 9–20

In many presentations of calculus "area variables," $dx\,dy$ and $r\,dr\,d\theta$ are derived from Figs. 9–17 and 9–18, this leads to at least two misconceptions. (i) One gets the idea that area can be given in terms of dx and dy only for rectangles like the shaded one in Fig. 9–17 and in terms of dr and $d\theta$ only for figures like the shaded one in Fig. 9–18. (ii) In the case of dr and $d\theta$, it appears that area can be obtained only for small figures and even then only approximately.

The trouble is that from these pictures one does not see a real product of differentials. The idea in Fig. 9–17 is that dx at one point times dy at another gives the area of the shaded rectangle. This is true, but an unfortunate coincidence obscures the real issue. One does not think of the determinant formula for area in connection with Fig. 9–17 because, as in (8), the matrix reduces to a diagonal one. Following the model of (9), one could equally well get the area of the shaded rectangle in Fig. 9–17 from $r\,dr \wedge d\theta$, but this is not apparent from the figure. As for the shaded region in Fig. 9–18, both $dx \wedge dy$ and $r\,dr \wedge d\theta$ give its area exactly, but since it is not a parallelogram, the determinant formula for these exterior products no longer applies.

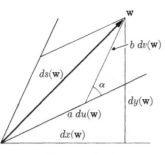

FIGURE 9–21

3. The area variable in polar coordinates is easy to see because the axes of dr and $d\theta$ are perpendicular, so only a multiplier is needed to convert these differentials into orthonormal component variables on the tangent planes. Once orthonormal component variables are found, their exterior product is an area variable. Now, suppose that du and dv are component variables with respect to a nonorthogonal basis, as shown in Fig. 9–19. Then (Fig. 9–20),

$$a\,du + b\cos\alpha\,dv, \qquad b\sin\alpha\,dv$$

are orthonormal component variables, and

$$(a\,du + b\cos\alpha\,dv) \wedge (b\sin\alpha\,dv)$$
$$= ab\sin\alpha\,du \wedge dv + b^2\cos\alpha\sin\alpha\,dv \wedge dv = ab\sin\alpha\,du \wedge dv$$

is a signed area variable.

If one is given x and y in terms of u and v over the plane, there is an easy way to find the multiplier $ab\sin\alpha$—the change of variable formula to appear in Section 9–4. However, the following technique finds $ab\sin\alpha$, yields a geometric analysis of the (du, dv)-coordinate system, and, unlike the change of variable formula, it can be generalized to the case in which the underlying manifold is no longer the plane but a surface in three-space.

Introduce ds, recall the law of cosines, and note from Fig. 9–21 that

$$a^2\,du^2 + b^2\,dv^2 + 2ab\cos\alpha\,du\,dv = ds^2 = dx^2 + dy^2. \qquad (10)$$

Now, use the fundamental theorem on differentials:

$$dx = \frac{\partial x}{\partial u}\,du + \frac{\partial x}{\partial v}\,dv, \qquad dy = \frac{\partial y}{\partial u}\,du + \frac{\partial y}{\partial v}\,dv.$$

Substitute in the right-hand member of (10); expand, and compare coefficients of du^2, dv^2, and $du\,dv$ to find a, b, and α. All these steps could be

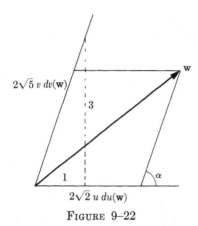

FIGURE 9–22

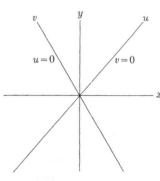

FIGURE 9–23

reduced to formulas, but that is hardly worthwhile. Consider a specific example. Let

$$x = u^2 - v^2,$$

$$y = u^2 + 2v^2,$$

$$dx = 2u\,du - 2v\,dv,$$ \hfill (11)

$$dy = 2u\,du + 4v\,dv.$$

Thus,

$$ds^2 = (2u\,du - 2v\,dv)^2 + (2u\,du + 4v\,dv)^2$$
$$= 8u^2\,du^2 + 20v^2\,dv^2 + 8uv\,du\,dv.$$

On comparing coefficients with the left-hand member of (10), one has

$$a = 2\sqrt{2}\,u, \qquad b = 2\sqrt{5}\,v;$$

so,

$$8uv = 2ab\cos\alpha = 8\sqrt{10}\,uv\cos\alpha, \qquad \cos\alpha = 1/\sqrt{10}.$$

Thus, $\sin\alpha = 3/\sqrt{10}$; so,

$$ab\sin\alpha\,du \wedge dv = 12uv\,du \wedge dv$$

is a signed area variable; and the (du, dv)-coordinate system is as shown in Fig. 9–22.

If the equations (11) are solved for u^2 and v^2, the result is

$$u^2 = \frac{y + 2x}{3}, \qquad v^2 = \frac{y - x}{3}.$$

So, the level curves for u and v are straight lines as shown in Fig. 9–23. From Fig. 9–23, it appears that Fig. 9–22 should be rotated by $\pi/4$ in order to conform to the usual direction conventions. An analysis of ds^2 yields a picture of the differential system, but it does not give its orientation with respect to the base manifold.

EXERCISES

1. For the parallelogram P of Fig. 9–24, compute area (P) from each of the following three exterior-product formulas.

    (a) $(dx \wedge dy)_{(1,2)}(P)$

    (b) $r(dr \wedge d\theta)_{(1,2)}(P)$

    (c) $12uv(du \wedge dv)_{(1,2)}(P)$, where u and v are the coordinates introduced in Example 3.

2. Each of the following exercises introduces coordinates u and v in the plane. In each case, sketch the (du, dv)-coordinate system, and find an area variable in terms of $du \wedge dv$.

    (a) $x = u^2 - v^2,\ y = 2uv$

    (b) $x = u \cosh v,\ y = u \sinh v$

    (c) $x = u + v,\ y = u - 2v$

3. Show that each of the following expressions is a volume variable because it is an exterior product of orthonormal component variables.

    (a) $dx \wedge dy \wedge dz$

    (b) $r\, dr \wedge d\theta \wedge dz$

    (c) $\rho^2 \sin\phi\, d\rho \wedge d\phi \wedge d\theta$

4. Let R be a region with right-handed orientation in a tangent three-space.

    (a) Show that $(dx \wedge dy \wedge dz)(R) > 0$.

    (b) Show that $(r\, dr \wedge d\theta \wedge dz)(R) > 0$ for $r > 0$ or $r < 0$.

    (c) Show that $(\rho^2 \sin\phi\, d\rho \wedge d\phi \wedge d\theta)(R) > 0$ for $\rho \neq 0$ and $\phi \neq n\pi$. Consider all cases: $\rho > 0,\ \rho < 0,\ 0 < \phi < \pi,\ \pi < \phi < 2\pi$.

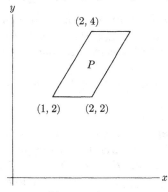

FIGURE 9–24

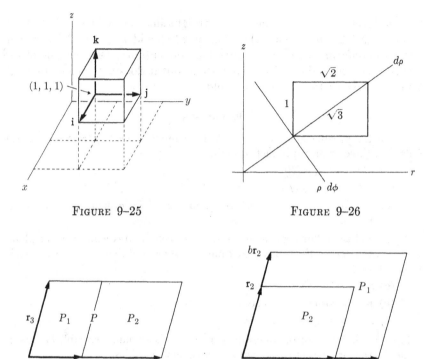

FIGURE 9–25   FIGURE 9–26

FIGURE 9–27   FIGURE 9–28

5. Let P be the oriented unit cube generated by $(\mathbf{i}, \mathbf{j}, \mathbf{k})$ at $(1, 1, 1)$ (see Fig. 9–25). Set up explicitly in determinant form and compute

    (a) $r(dr \wedge d\theta \wedge dz)_{(1,1,1)}(P)$

    (b) $\rho^2 \sin \phi (d\rho \wedge d\phi \wedge d\theta)_{(1,1,1)}(P)$.

[*Hint:* Figure 9–26 shows a cross section in the plane on which $\theta = \pi/4$.]

6. Verify that $du \wedge dv$ is additive and monotone on parallelograms.

    (a) In Fig. 9–27 one wishes that $(du \wedge dv)(P_1) + (du \wedge dv)(P_2) = (du \wedge dv)(P)$. Check this from the determinant formula.

    (b) In Fig. 9–28, P_1 contains P_2 if $a \geq 1$ and $b \geq 1$. Compare $(du \wedge dv)(P_1)$ and $(du \wedge dv)(P_2)$; discuss both cases of relative orientation.

9–3 Multiple and iterated integrals. Let R be an oriented region in the plane and let u, v, and w be variables on the plane with (u, v) forming an admissible coordinate set on R. Let R be partitioned into consistently oriented subregions $R_1, R_2, \ldots, R_n$. As in Fig. 9–29, the partitioning curves may be more or less arbitrary. In each subregion, R_i, choose a

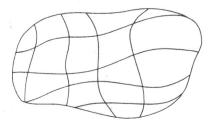

FIGURE 9–29

point p_i. Now, a *double integral* may be defined as follows:

$$\iint_R w\, du \wedge dv = \lim \sum_{i=1}^n w(p_i)(du \wedge dv)_{p_i}(R_i).$$

The limit here is to be taken over any sequence of partitions of R for which

$$\max_{1 \le i \le n} \text{diameter } (R_i) \to 0.$$

It can be shown that, for continuous w, this integral exists.

Note that in the definition of the double integral, there appears the expression $(du \wedge dv)_{p_i}(R_i)$. Now, strictly speaking, R_i is a set in the underlying manifold and $(du \wedge dv)_{p_i}$ operates on sets in T_{p_i}. However, in the present discussion the underlying manifold is the plane; so the tangent planes are superposed on it, and R_i determines, in an obvious way, a set in T_{p_i}.

In Chapter 10 an integral over a surface in three-space is to be defined, and in this project the principal problem is to associate a partition set on the surface with a set in a tangent plane. The solution is to reduce the problem through a basic mapping theorem back to the present one.

More important, for practical purposes, is a theorem relating the double integral to appropriate iterated integrals. This is analogous to the fundamental theorem of calculus in that a double integral is computed from the iterated integral forms. That is,

$$\iint_R w\, du \wedge dv = \int_a^b \int_{f(v)}^{g(v)} w\, du\, dv. \tag{12}$$

To be precise, however, one must discuss the relation between the *oriented* region R and the limits of integration in the iterated integral.

First, discussion must be limited to those regions R that can be described as on the right side of (12). That is, ∂R must consist of four connected pieces characterized by $v = a$, $v = b$, $u = f(v)$, and $u = g(v)$, respectively.

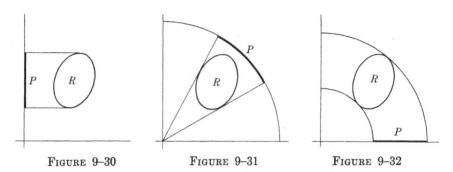

FIGURE 9–30   FIGURE 9–31   FIGURE 9–32

Let P be the (unoriented) projection of R by u onto a u-constant curve. For example, Fig. 9–30 shows a projection by x, Fig. 9–31 shows one by r, and Fig. 9–32 shows one by θ. Now, the basic formula may be written as follows:

$$\iint_R w \, du \wedge dv = \int_{\partial P^-}^{\partial P^+} \int_{\partial R^-}^{\partial R^+} w \, du \, dv. \tag{13}$$

The boundary of R is obviously divided into two parts, denoted here by ∂R^+ and ∂R^-. The projection P has two boundary points ∂P^- and ∂P^+. The basic rule in (13) is that the orientation of ∂R^+ is the same as that imposed on P by the choice of $(+)$ and $(-)$ boundary points.

Suppose the double integral is given. To set up the iterated integral, one would probably choose first the limits of integration with respect to u. The one that is put at the top automatically becomes ∂R^+; it has an orientation determined by that of R, and the orientation of P must be made to agree with it.

To go the other way, suppose the iterated integral is given. The entries at the top are automatically the $(+)$ boundaries. Now, start with P. Its orientation is obvious from the limits of integration. Then, ∂R^+ is identified by its position at the top of the integral sign, and its orientation is given by that of P. Once a part of ∂R is oriented, the orientation of R is determined by the "exterior normal followed by boundary" rule.

Realization that the limits of integration on the right side of (13) depend on the orientation of R clears up very quickly a point that is sometimes puzzling. It follows at once from the antisymmetry of exterior multiplication that

$$\iint_R w \, du \wedge dv = - \iint_R w \, dv \wedge du.$$

However, in Chapter 8 the order of integration in an iterated integral was changed without any change of signs. The explanation is that in Chapter 8 the partial integration was always taken in the positive direction for the

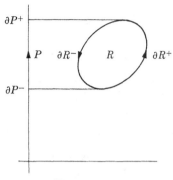
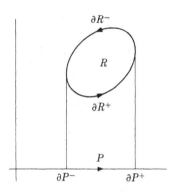

FIGURE 9–33   FIGURE 9–34

variable concerned, whereas a strict adherence to (13) produces this effect just half the time. To be specific, let $\int \uparrow$ denote partial integration in the direction of increase for the variable of integration, and let $\int \downarrow$ denote partial integration in the other direction. Then, whenever

$$\iint_R w \, du \wedge dv = \int \uparrow \int \uparrow w \, du \, dv,$$

it always follows that

$$\iint_R w \, dv \wedge du = \int \downarrow \int \uparrow w \, dv \, du;$$

so,

$$\int \uparrow \int \uparrow w \, dv \, du = - \iint_R w \, dv \wedge du$$

$$= \iint_R w \, du \wedge dv = \int \uparrow \int \uparrow w \, du \, dv,$$

in complete accordance with the practice employed in Chapter 8. To convince himself that this switch does occur, the student should study Figs. 9–33 and 9–34.

A *triple integral*,
$$\iiint_R w \, dt \wedge du \wedge dv,$$

is defined by a procedure completely analogous to that used above for the double integral. The details are left to the student.

The formula for reduction to an iterated integral follows the pattern of the two-dimensional case. Project R onto a two-dimensional manifold P_2; then project P_2 onto a one-dimensional manifold P_1. The formula is

$$\iiint_R w \, dt \wedge du \wedge dv = \int_{\partial P_1^-}^{\partial P_1^+} \int_{\partial P_2^-}^{\partial P_2^+} \int_{\partial R^-}^{\partial R^+} w \, dt \, du \, dv.$$

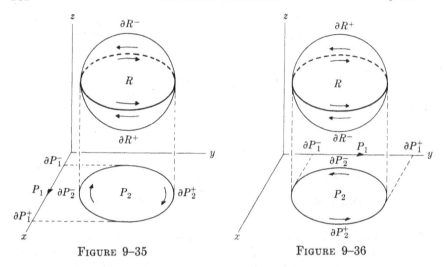

FIGURE 9–35   FIGURE 9–36

The interpretation is as follows. The form

$$\int_{\partial P_1^-}^{\partial P_1^+} \int_{\partial P_2^-}^{\partial P_2^+} \tag{14}$$

describes an orientation of P_2 in accordance with the rules for interpreting (13); then the orientation of ∂R^+ agrees with the orientation of P_2.

Suppose that R is given (oriented) and the iterated integral is to be set up. Find the two parts of ∂R and designate one of them as ∂R^+ and put it at the top of the integral sign. The orientation of R orients ∂R^+; so, follow the rules for the two-dimensional case in constructing the form (14) to give this same orientation for P_2.

Conversely, suppose the iterated integral is given. Start with P_1; its orientation is clear from the left-hand integral sign. The top limit on the middle integral gives ∂P_2^+; orient this to agree with P_1 and thereby determine the orientation of P_2. Finally, identify ∂R^+ by its position (top, right); orient it to agree with P_2, and find the orientation of R from this (exterior normal followed by boundary orientation).

Figure 9–35 shows an example in rectangular coordinates. Note that in Fig. 9–35, R is right-handed and the indicated order of integration is $dz\,dy\,dx$; so it is to be expected that the partial integrations come out as shown below,
$$\int \uparrow \int \downarrow \int \uparrow,$$
or something equivalent to this. The same integral in the order $dz\,dx\,dy$ is sketched in Fig. 9–36, leading to a form such as
$$\int \uparrow \int \uparrow \int \uparrow.$$

EXAMPLES

1. Let R be the right-handed region bounded by the paraboloid on which $z = x^2 + y^2$ and the plane on which $z = y$. Reduce to an iterated integral,

$$\iiint_R dz \wedge dx \wedge dy.$$

Suppose that the integral with respect to z is taken from paraboloid to plane:

$$\int_{x^2+y^2}^{y} dz.$$

This identifies $z = y$ as ∂R^+ and determines the orientation of the circle P_2 (Fig. 9–37). Figure 9–38 shows P_2 and its projection P_1. This indicates that the iterated integral is completed as follows:

$$\int_0^1 \int_{-\sqrt{y-y^2}}^{\sqrt{y-y^2}} \int_{x^2+y^2}^{y} dz\, dx\, dy.$$

2. Given that

$$\iiint_R dx \wedge dz \wedge dy = \int_0^1 \int_{y^2}^{y} \int_{-\sqrt{z-y^2}}^{\sqrt{z-y^2}} dx\, dz\, dy,$$

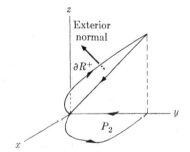

FIGURE 9–37

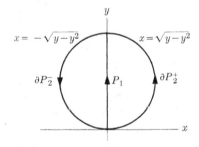

FIGURE 9–38

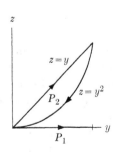

FIGURE 9–39

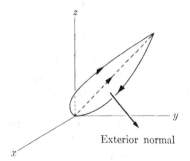

FIGURE 9–40

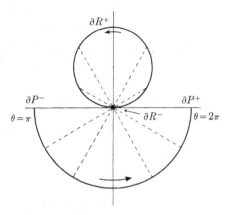

FIGURE 9–41

find the orientation of R. From the left and middle integral signs it appears that P_2 is as shown in Fig. 9–39. Thus, ∂R^+ is the $x \geq 0$ portion of the paraboloid on which $z = x^2 + y^2$, oriented as shown in Fig. 9–40, which indicates that R is the solid considered in Example 1, but this time with left-handed orientation.

3. Without computing the integral, show that

$$\int_{\pi}^{2\pi} \int_{0}^{\sin \theta} r \, dr \, d\theta > 0.$$

The region R in question is bounded by the circle in Fig. 9–41, but it is generated by negative values of r. Projection onto a level curve on which $r = 1$ is shown in Fig. 9–41. Integration with respect to θ from π to 2π orients this projection as shown. Now, the two "sections" of ∂R are the circle and the single point at which $r = 0$. Integration from 0 to $\sin \theta$ indicates that the circle must be ∂R^+; so its orientation must follow that of the projection. As shown in Fig. 9–41, this gives R counterclockwise orientation. Thus

$$\int_{\pi}^{2\pi} \int_{0}^{\sin \theta} r \, dr \, d\theta = \iint_{R_{\text{ccl}}} r \, dr \wedge d\theta.$$

However, as shown in Fig. 9–42, $(dr, d\theta)$ is a clockwise system for $r < 0$. Thus, the given integral represents the double integral of a negative variable r over a counterclockwise region with respect to a clockwise differential pair. Hence the result is positive. Direct computation shows that this integral does give the area of the circle.

The analysis could have been made from Fig. 9–43, which shows a projection onto a level curve on which $r = -1$. Values of θ have reversed geometric significance here, and the end result is the same.

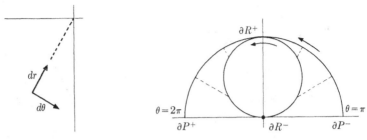

FIGURE 9–42   FIGURE 9–43

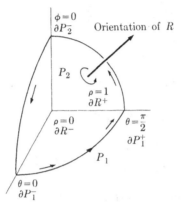

FIGURE 9–44

4. Without computing the integral, show that

$$\int_0^{\pi/2} \int_0^{\pi/2} \int_0^1 \rho^2 \sin\phi \, d\rho \, d\phi \, d\theta > 0.$$

The notations on Fig. 9–44 show that this represents a triple integral over a right-handed region. The solid is projected onto the spherical triangle P_2 which is projected onto the quarter circle P_1. Since $\theta = \pi/2$ at ∂P_1^+, P_1 is oriented as shown. Since $\phi = \pi/2$ on ∂P_2^+, it follows that ∂P_2^+ is the quarter circle P_1; and the orientation of P_2 is determined as shown. Finally, ∂R^+ must be the spherical triangle P_2 because $\rho = 1$ there; so the orientation of ∂R is determined, and that of R is found by drawing an exterior normal. As shown, the exterior normal followed by a positive rotation on ∂R is right-handed. Now, on this domain, $(d\rho, d\phi, d\theta)$ is right-handed and $\rho^2 \sin\phi > 0$; thus

$$\int_0^{\pi/2} \int_0^{\pi/2} \int_0^1 \rho^2 \sin\phi \, d\rho \, d\phi \, d\theta = \iiint_{R_{\mathrm{rh}}} \rho^2 \sin\phi \, d\rho \wedge d\phi \wedge d\theta > 0.$$

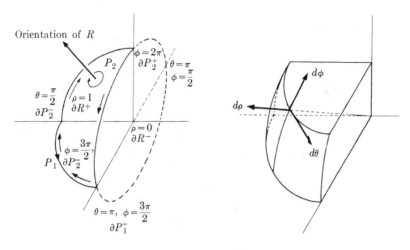

FIGURE 9–45   FIGURE 9–46

5. Without computing the integral, show that

$$\int_{\pi/2}^{\pi} \int_{3\pi/2}^{2\pi} \int_{0}^{1} \rho^2 \sin\phi \, d\rho \, d\phi \, d\theta < 0.$$

The orientation problem is shown in Fig. 9–45. The values of θ look out of place; but as shown for $\theta = \pi$, this is the effect of having $\pi \le \phi \le 2\pi$. In this case, the quarter circle in the horizontal plane is ∂P_2^-; so as a part of ∂P_2, it has the orientation opposite to that given to it as P_1. Finally, then, the exterior normal followed by positive rotation on ∂R is left-handed. However, it appears from Fig. 9–46 that on this domain, $(d\rho, d\phi, d\theta)$ is also left-handed. Finally, $\rho^2 \sin\phi < 0$; so,

$$\int_{\pi/2}^{\pi} \int_{3\pi/2}^{2\pi} \int_{0}^{1} \rho^2 \sin\phi \, d\rho \, d\phi \, d\theta = \iiint_{R_{1h}} \rho^2 \sin\phi \, d\rho \wedge d\phi \wedge d\theta < 0.$$

6. In the preceding examples a careful study of orientation was made in an effort to present concrete models of the basic ideas involved in a multiple integral. It is unnecessary to study the multiple integral in such detail simply to determine by inspection the sign of an iterated integral. In Example 3, for instance,

$$\pi \to 2\pi$$

indicates increasing θ, and on this domain,

$$0 \to \sin\theta$$

indicates decreasing r. Finally, $r < 0$; thus the schematic diagram

$$\int \uparrow \int \downarrow -$$

indicates a positive integral. Similar analyses of Examples 4 and 5 lead to the diagrams

$$\text{Example 4,} \qquad \int \uparrow \int \uparrow \int \uparrow +,$$

$$\text{Example 5,} \qquad \int \uparrow \int \uparrow \int \uparrow - \cdot$$

7. For the circle of Example 3 and the complete sphere, parts of which appeared in Examples 4 and 5, the natural iterated integrals for area and volume, respectively, are

$$\int_0^\pi \int_0^{\sin \theta} r \, dr \, d\theta, \qquad \int_0^{2\pi} \int_0^\pi \int_0^1 \rho^2 \sin \phi \, d\rho \, d\phi \, d\theta.$$

For the area of the circle

$$\int_\pi^{2\pi} \int_0^{\sin \theta} r \, dr \, d\theta, \qquad \text{or} \qquad \int_{-\pi/2}^{\pi/2} \int_0^{\sin \theta} r \, dr \, d\theta$$

will do just as well (compute these and see). Geometric reasons for this should be apparent from Example 3.

Sometimes there occurs the idea of getting the volume of the sphere from

$$\int_0^\pi \int_0^{2\pi} \int_0^1 \rho^2 \sin \phi \, d\rho \, d\phi \, d\theta.$$

Compute this integral; the value is zero. Geometric reasons for this should be apparent from Example 5.

EXERCISES

1. Without making any changes in order, transform each of the following iterated integrals into a multiple integral. Then, give the orientation of the region of integration, the orientation of the differential system, and the sign of the integrand. From the above data, determine the sign of the integral without evaluating it.

(a) $\displaystyle\int_0^1 \int_{y^2}^y x \, dx \, dy$

(b) $\displaystyle\int_{-\pi}^0 \int_0^{\sin x} y \, dy \, dx$

(c) $\displaystyle\int_0^1 \int_x^{\sqrt{x}} x \, dy \, dx$

(d) $\displaystyle\int_{\pi/2}^{3\pi/2} \int_0^{\cos \theta} r^2 \cos \theta \, dr \, d\theta$

(e) $\int_{-\pi/3}^{\pi/3} \int_{1-2\cos\theta}^{0} r\, dr\, d\theta$

(f) $\int_{\pi}^{3\pi/2} \int_{0}^{\sin 2\theta} r\, dr\, d\theta$

(g) $\int_{0}^{1} \int_{0}^{\arccos r} r\, d\theta\, dr$

(h) $\int_{-1}^{0} \int_{0}^{\arcsin r} r\, d\theta\, dr$

(i) $\int_{0}^{1} \int_{0}^{\sqrt{1-y^2}} \int_{0}^{\sqrt{1-y^2-z^2}} z\, dx\, dz\, dy$

(j) $\int_{0}^{1} \int_{0}^{\sqrt{1-y^2}} \int_{0}^{x^2+y^2-1} z\, dz\, dx\, dy$

(k) $\int_{-1}^{0} \int_{0}^{\sqrt{z+1}} \int_{0}^{\sqrt{z+1-y^2}} z\, dx\, dy\, dz$

(l) $\int_{0}^{2\pi} \int_{0}^{1} \int_{-r}^{r} r\, dz\, dr\, d\theta$

(m) $\int_{0}^{2\pi} \int_{1}^{0} \int_{z}^{0} r\, dr\, dz\, d\theta$

(n) $\int_{0}^{2\pi} \int_{-1}^{0} \int_{1}^{r^2} r\, dz\, dr\, d\theta$

(o) $\int_{0}^{1} \int_{r^2}^{r} \int_{0}^{2\pi} r\, d\theta\, dz\, dr$

(p) $\int_{0}^{1} \int_{0}^{2\pi} \int_{r}^{r^2} r\, dz\, d\theta\, dr$

(q) $\int_{0}^{2\pi} \int_{\pi/4}^{\pi/2} \int_{0}^{\csc\phi} \rho^2 \sin\phi\, d\rho\, d\phi\, d\theta$

(r) $\int_{0}^{2\pi} \int_{0}^{1} \int_{0}^{\operatorname{arccsc}\rho} \rho^2 \sin\phi\, d\phi\, d\rho\, d\theta$

(s) $\int_{-1}^{0} \int_{0}^{2\pi} \int_{0}^{\pi/4} \rho^2 \sin\phi\, d\phi\, d\theta\, d\rho$

(t) $\int_{-1}^{0} \int_{0}^{2} \int_{\pi}^{5\pi/4} \rho^2 \sin\phi\, d\phi\, d\theta\, d\rho$

(u) $\int_{0}^{\pi} \int_{3\pi/4}^{\pi} \int_{0}^{1} \rho^2 \sin\phi\, d\rho\, d\phi\, d\theta$

(v) $\int_{0}^{\pi/2} \int_{0}^{2\pi} \int_{0}^{\cos\phi} \rho^2 \sin\phi\, d\rho\, d\theta\, d\phi$

2. In each of the following exercises complete the iterated integral form.

(a) $\iint_{R} dx \wedge dy = \iint^{y} dx\, dy$;

R is clockwise and bounded by the loci of $y = x$ and $y = x^2$.

(b) $\iint_{R} dy \wedge dx = \iint^{x} dy\, dx$;

R is the same as in part (a).

(c) $\iint_{R} dx \wedge dy = \iint^{1} dx\, dy$;

R is counterclockwise and bounded by the loci of $y = x$, $x = 1$, and $y = 0$.

(d) $\iint_{R} dy \wedge dx = \iint^{x} dy\, dx$;

R is the same as in part (c).

(e) $\iint_R r\,dr \wedge d\theta = \iint^{\sin\theta} r\,dr\,d\theta$;

R is counterclockwise and bounded on the right by the locus of $r = \sin\theta$ and, on the left, by the locus $\theta = \pi/2$.

(f) $\iint_R r\,d\theta \wedge dr = \iint^{\arcsin r} r\,d\theta\,dr$;

R is the same as in part (e).

(g) $\iint_R r\,dr \wedge d\theta = \iint^1 r\,dr\,d\theta$;

R is the clockwise quarter circle bounded by the loci of $r = 1$, $\theta = 0$, and $\theta = \pi/2$.

(h) $\iint_R r\,d\theta \wedge dr = \iint^{\pi/2} r\,d\theta\,dr$;

R is the same as in part (g).

(i) $\iiint_R dx \wedge dy \wedge dz = \iiint^{1-y-z} dx\,dy\,dz$;

R is right-handed and bounded by the loci of $x + y + z = 1$, $x = 0$, $y = 0$, and $z = 0$.

(j) $\iiint_R dz \wedge dy \wedge dx = \iiint^{1-z-y} dz\,dy\,dx$;

R is the same as in part (i).

(k) $\iiint_R dz \wedge dy \wedge dx = \iiint^1 dz\,dy\,dx$;

R is left-handed and bounded by the loci of $z = x^2 + y^2$ and $z = 1$.

(l) $\iiint_R dy \wedge dz \wedge dx = \iiint^{\sqrt{z-x^2}} dy\,dz\,dx$;

R is the same as in part (k).

(m) $\iiint_R r\,dr \wedge d\theta \wedge dz = \iiint^z r\,dr\,d\theta\,dz$;

R is right-handed and bounded by the loci of $r = z$ and $z = 1$.

(n) $\iiint_R r\,dz \wedge dr \wedge d\theta = \iiint^1 r\,dz\,dr\,d\theta$;

R is the same as in part (m).

(o) $\iiint_R r\, dr \wedge dz \wedge d\theta = \iiint^1 r\, dr\, dz\, d\theta$;

R is left-handed and bounded by the loci of $z = r^2$, $r = 1$, and $z = 0$.

(p) $\iiint_R r\, dz \wedge dr \wedge d\theta = \iiint^{r^2} r\, dz\, dr\, d\theta$;

R is the same as in part (o).

(q) $\iiint_R \rho^2 \sin\phi\, d\rho \wedge d\phi \wedge d\theta = \iiint^1 \rho^2 \sin\phi\, d\rho\, d\phi\, d\theta$;

R is right-handed and bounded by the loci of $\rho = 1, \phi = \pi/4, \phi = \pi/2$.

(r) $\iiint_R \rho^2 \sin\phi\, d\phi \wedge d\rho \wedge d\theta = \iiint^{\pi/2} \rho^2 \sin\phi\, d\phi\, d\rho\, d\theta$;

R is the same as in part (q).

(s) $\iiint_R \rho^2 \sin\phi\, d\phi \wedge d\rho \wedge d\theta = \iiint^{\text{arccos}\,\rho} \rho^2 \sin\phi\, d\phi\, d\rho\, d\theta$;

R is left-handed and bounded by the locus of $\rho = \cos\phi$.

(t) $\iiint_R \rho^2 \sin\phi\, d\rho \wedge d\phi \wedge d\theta = \iiint^{\cos\phi} \rho^2 \sin\phi\, d\rho\, d\phi\, d\theta$;

R is the same as in part (s).

3. By using six different orders of integration in each of the three major coordinate systems, set up 18 iterated integrals for the (positive) volume of the first octant of the sphere with radius 1 and center at the origin. Write each integral as a triple integral and determine the orientation imposed on the figure.

9–4 Change of variable. Many changes of variable in multiple integrals can be accomplished by recalling from geometric considerations what the area or volume differential is in the coordinate systems in question. For example, given

$$\iiint_R \rho^3 \sin 2\phi\, d\rho \wedge d\phi \wedge d\theta,$$

transform to cylindrical coordinates. Note that

$$\rho^3 \sin 2\phi\, d\rho \wedge d\phi \wedge d\theta = 2\rho^3 \sin\phi \cos\phi\, d\rho \wedge d\phi \wedge d\theta$$

$$= (2\rho \cos\phi)(\rho^2 \sin\phi\, d\rho \wedge d\phi \wedge d\theta).$$

Now,

$$\rho^2 \sin \phi \, d\rho \wedge d\phi \wedge d\theta = r \, dr \wedge d\theta \wedge dz$$

because these are the volume differentials, and

$$2\rho \cos \phi = 2z$$

by direct substitution; so the cylindrical-coordinate integral is

$$\iiint_R 2zr \, dr \wedge d\theta \wedge dz.$$

There is, however, a simple substitution formula that yields all such results by straightforward computation and also furnishes a direct generalization of the substitution formula for line integrals (Section 2–5). This is obtained from the fundamental theorem on differentials and the rules of exterior algebra as follows:

$$
\begin{aligned}
du \wedge dv &= \left(\frac{\partial u}{\partial s} ds + \frac{\partial u}{\partial t} dt \right) \wedge \left(\frac{\partial v}{\partial s} ds + \frac{\partial v}{\partial t} dt \right) \\[2mm]
&= \frac{\partial u}{\partial s} \frac{\partial v}{\partial s} ds \wedge ds + \frac{\partial u}{\partial s} \frac{\partial v}{\partial t} ds \wedge dt \\[2mm]
&\quad + \frac{\partial u}{\partial t} \frac{\partial v}{\partial s} dt \wedge ds + \frac{\partial u}{\partial t} \frac{\partial v}{\partial t} dt \wedge dt \\[2mm]
&= \frac{\partial u}{\partial s} \frac{\partial v}{\partial t} ds \wedge dt - \frac{\partial u}{\partial t} \frac{\partial v}{\partial s} ds \wedge dt \\[2mm]
&= \begin{vmatrix} \dfrac{\partial u}{\partial s} & \dfrac{\partial u}{\partial t} \\[3mm] \dfrac{\partial v}{\partial s} & \dfrac{\partial v}{\partial t} \end{vmatrix} ds \wedge dt.
\end{aligned}
$$

This determinant is the Jacobian of u and v with respect to s and t (see Section 5–6). In matrix notation this reads as follows. Let

$$U = \begin{bmatrix} u \\ v \end{bmatrix}, \qquad S = \begin{bmatrix} s \\ t \end{bmatrix}; \tag{15}$$

then

$$du \wedge dv = \det D_S U \, ds \wedge dt. \tag{16}$$

THEOREM. *Change of variable for double integrals.* If each of the matrices U and S in (15) gives an admissible coordinate set on an oriented plane region R, and if w is continuous on R, then

$$\iint_R w \, du \wedge dv = \iint_R w \det D_S U \, ds \wedge dt.$$

Proof. Partition R into subregions $R_1, R_2, \ldots, R_n$ and choose points p_i in R_i. Since (16) is an identity,

$$\sum_{i=1}^n w(p_i)(du \wedge dv)_{p_i}(R_i) = \sum_{i=1}^n w(p_i)(\det D_S U)_{p_i}(ds \wedge dt)_{p_i}(R_i).$$

That is, for any partition of R, the approximating sums for the two integrals are equal; hence the integrals are equal.

Like the proof of the substitution theorem for line integrals (Section 2–5), this proof depends on the fact that, in the definition of the integral, the partitioning process is independent of the coordinate system.

An analogous result holds in any number of dimensions. That is, let

$$U = \begin{bmatrix} u_1 \\ u_2 \\ \vdots \\ u_n \end{bmatrix}, \qquad V = \begin{bmatrix} v_1 \\ v_2 \\ \vdots \\ v_n \end{bmatrix};$$

then

$$du_1 \wedge du_2 \wedge \cdots \wedge du_n = \det D_V U \, dv_1 \wedge dv_2 \wedge \cdots \wedge dv_n, \qquad (17)$$

and this substitution may be made in a multiple integral. For a complete proof of the identity (17), one should write

$$du_1 \wedge \cdots \wedge du_n = \left(\sum_{i=1}^n \frac{\partial u_1}{\partial v_i} \, dv_i \right) \wedge \cdots \wedge \left(\sum_{i=1}^n \frac{\partial u_n}{\partial v_i} \, dv_i \right) \qquad (18)$$

and proceed as in the proof of (16). Actually, (17) may be derived from (18) by noting which terms in the expansion of the right-hand side of (18) are different from zero, studying permutations of subscripts in these terms, and recalling the definition of a determinant. If one is content to prove (17) on parallelepipeds, the determinant representation of exterior products yields an easy matrix proof. The (i, j)-entry in the matrix product,

$$\left[\frac{\partial u_i}{\partial v_j} \right] [dv_i(\mathbf{r}_j)],$$

is

$$\sum_{k=1}^n \frac{\partial u_i}{\partial v_k} \, dv_k(\mathbf{r}_j) = du_i(\mathbf{r}_j)$$

by the fundamental theorem on differentials; so,

$$\det\,[du_i(\mathbf{r}_j)] \;=\; \det\left[\frac{\partial u_i}{\partial v_j}\right]\det\,[dv_i(\mathbf{r}_j)],$$

and this expression is (17) for parallelepipeds.

Given an iterated integral

$$\int_a^b \int_{f_1(v)}^{f_2(v)} F(u,\,v)\;du\;dv, \tag{19}$$

let

$$u = \phi(s,\,t), \qquad v = \psi(s,\,t) \tag{20}$$

define a change of coordinate variables. The end result of transforming the integral (19) will be in the form

$$\int_c^d \int_{g_1(t)}^{g_2(t)} F[\phi(s,\,t),\,\psi(s,\,t)] \begin{vmatrix} \dfrac{\partial u}{\partial s} & \dfrac{\partial u}{\partial t} \\[2ex] \dfrac{\partial v}{\partial s} & \dfrac{\partial v}{\partial t} \end{vmatrix} ds\;dt. \tag{21}$$

This much comes directly from the theorem for change of variable. If

$$w = F(u,\,v),$$

then, given (20),

$$w = F[\phi(s,\,t),\,\psi(s,\,t)].$$

The integral (19) is equal to

$$\iint_R w\;du \wedge dv, \tag{22}$$

and, by the theorem for change of variable, (22) is equal to

$$\iint_R w \begin{vmatrix} \dfrac{\partial u}{\partial s} & \dfrac{\partial u}{\partial t} \\[2ex] \dfrac{\partial v}{\partial s} & \dfrac{\partial v}{\partial t} \end{vmatrix} ds \wedge dt. \tag{23}$$

Integral (23) in turn is equal to (21). The only problem is that of changing the limits of integration from (19) to (21). This must be done geometrically. The oriented region R is the same in (22) and (23); so, the procedure is to determine from (19) what the region of integration is and how it is oriented, and then to describe that region and orientation in the new coordinates to obtain (21).

EXAMPLES

1. Transform

$$\int_0^{2\pi} \int_0^{\pi/4} \int_0^1 \rho^3 \sin 2\phi \, d\rho \, d\phi \, d\theta \qquad (24)$$

into an iterated integral in cylindrical coordinates. As noted in the opening paragraph of this section,

$$\rho^2 \sin \phi \, d\rho \, d\phi \, d\theta \rightarrow r \, dr \, d\theta \, dz,$$

and the remainder is

$$2\rho \cos \phi = 2z;$$

so

$$\rho^3 \sin 2\phi \, d\rho \, d\phi \, d\theta \rightarrow 2zr \, dr \, d\theta \, dz.$$

To obtain the same result by the formula for change of variable, let

$$S = \begin{bmatrix} \rho \\ \phi \\ \theta \end{bmatrix}, \qquad C = \begin{bmatrix} r \\ \theta \\ z \end{bmatrix}.$$

Recall that

$$r = \rho \sin \phi, \qquad \theta = \theta, \qquad z = \rho \cos \phi,$$

so

$$D_S C = \begin{bmatrix} \sin \phi & \rho \cos \phi & 0 \\ 0 & 0 & 1 \\ \cos \phi & -\rho \sin \phi & 0 \end{bmatrix}.$$

Thus,

$$\det D_C S = \det (D_S C)^{-1} = \frac{1}{\det D_S C} = \frac{1}{\rho},$$

and

$$\rho^3 \sin 2\phi \, d\rho \wedge d\phi \wedge d\theta = 2\rho^3 \sin \phi \cos \phi \, (1/\rho) \, dr \wedge d\theta \wedge dz$$
$$= 2(\rho \sin \phi)(\rho \cos \phi) \, dr \wedge d\theta \wedge dz \qquad (25)$$
$$= 2rz \, dr \wedge d\theta \wedge dz.$$

Now, the region of integration is the "ice-cream cone" shown in Fig. 9–47 and, if the region is right-handed, the given iterated integral is equal to a triple integral over this cone with respect to $d\rho \wedge d\phi \wedge d\theta$. Therefore,

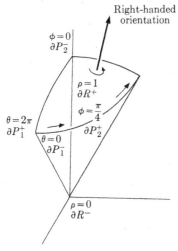

FIGURE 9–47

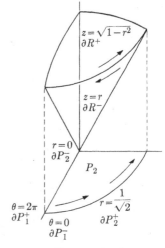

FIGURE 9–48

by (25) it goes into an integral with respect to $dr \wedge d\theta \wedge dz$ over this right-handed region. However, because of "corner trouble," one would prefer the order $dz\, dr\, d\theta$ in cylindrical coordinates.

Now,

$$dr \wedge d\theta \wedge dz = dz \wedge dr \wedge d\theta$$

because a cyclic permutation of three terms is even; so one still wants a right-handed region for integrating with respect to $dz \wedge dr \wedge d\theta$. As shown in Fig. 9–48, this objective calls for the following limits of integration:

$$\int_0^{2\pi} \int_0^{1/\sqrt{2}} \int_r^{\sqrt{1-r^2}} 2rz\, dz\, dr\, d\theta. \tag{26}$$

Though one must go from one integral to the picture to the other integral, often a detailed analysis of orientation can be bypassed merely by checking signs in the end result. In (24) the sign of $\rho^3 \sin 2\phi$ and the directions of integration are indicated by

$$\int \uparrow \int \uparrow \int \uparrow +. \tag{27}$$

A similar analysis of (26) yields the same schematic diagram, so the integrals are equal. Note that

$$\int_0^{1/\sqrt{2}} \int_0^{2\pi} \int_r^{\sqrt{1-r^2}} 2rz\, dz\, d\theta\, dr \tag{28}$$

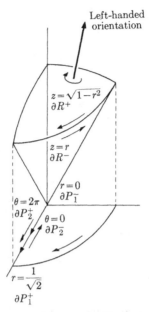

Left-handed
orientation

$z = \sqrt{1 - r^2}$
∂R^+

$z = r$
∂R^-

$r = 0$
∂P_1^-

$\theta = 2\pi$
∂P_2^+

$\theta = 0$
∂P_2^-

$r = \dfrac{1}{\sqrt{2}}$
∂P_1^+

FIGURE 9–49

will do as well as (26), because a sign check here also leads to (27). However, $(dz, d\theta, dr)$ is a left-handed system, and as Fig. 9–49 shows, integration over a left-handed region leads to (28).

2. Transform

$$\int_0^1 \int_0^1 (x^2 + y^2)\sqrt[3]{x^2 - y^2}\, dx\, dy$$

into an iterated integral with respect to u and v where

$$u = 2xy, \qquad v = x^2 - y^2.$$

Let

$$X = \begin{bmatrix} x \\ y \end{bmatrix}, \qquad U = \begin{bmatrix} u \\ v \end{bmatrix};$$

then

$$D_X U = \begin{bmatrix} 2y & 2x \\ 2x & -2y \end{bmatrix}.$$

Thus,

$$\det D_U X = \frac{1}{\det D_X U} = \frac{1}{-4(x^2 + y^2)},$$

and

$$(x^2 + y^2)\sqrt[3]{x^2 - y^2}\, dx \wedge dy = \sqrt[3]{v}\, \frac{(x^2 + y^2)}{-4(x^2 + y^2)}\, du \wedge dv$$

$$= \tfrac{1}{4}\sqrt[3]{v}\, dv \wedge du.$$

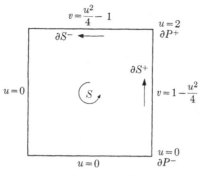

$$v = \frac{u^2}{4} - 1$$

$u = 2$
∂P^+

∂S^- ⟵

∂S^+

$u = 0$

S

$v = 1 - \frac{u^2}{4}$

$u = 0$
∂P^-

$u = 0$

FIGURE 9–50

So, the double-integral transformation is

$$\iint_S (x^2 + y^2)\sqrt[3]{x^2 - y^2}\, dx \wedge dy = \iint_S (\sqrt[3]{v}/4)\, dv \wedge du,$$

where S is the unit square with counterclockwise orientation.

To get limits of integration for an iterated integral, turn to Fig. 9–46. The left and bottom sides of the square have the equation $u = 0$. The right side has the equation $x = 1$; substitution into the transformation equations yields $u = 2y$, and $v = 1 - y^2$, which are easily solved:

$$v = 1 - \frac{u^2}{4}$$

or

$$u = 2\sqrt{1 - v}.$$

Similarly, the top of the square has equations

$$v = \frac{u^2}{4} - 1$$

or

$$u = 2\sqrt{1 + v}.$$

Integration with respect to u first would run from bottom to right side or from left to top side, depending on the value of v. Thus, the formula for the upper limit of integration would change at $v = 0$. For this reason (and not because $dv \wedge du$ appeared in the double integral), it is desirable to integrate with respect to v first. In this process, the first integration runs from top to right side of the square. The final result is

$$\int_0^1 \int_0^1 (x^2 + y^2)\sqrt[3]{x^2 - y^2}\, dx\, dy = \int_0^2 \int_{(u^2/4)-1}^{1-(u^2/4)} (\sqrt[3]{v}/4)\, dv\, du.$$

An analysis of orientation for the new integral is shown in Fig. 9–50.

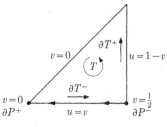

FIGURE 9–51

3. Transform

$$\int_0^1 \int_y^1 \sqrt{x^2 - y^2}\, dx\, dy$$

into an iterated integral with respect to u and v where

$$x = u + v, \qquad y = u - v.$$

Let

$$X = \begin{bmatrix} x \\ y \end{bmatrix}, \qquad U = \begin{bmatrix} u \\ v \end{bmatrix};$$

then

$$\det D_U X = \begin{vmatrix} 1 & 1 \\ 1 & -1 \end{vmatrix} = -2;$$

so, again (dv, du) is the counterclockwise pair. The transformation equations may be solved for u and v:

$$u = \frac{x + y}{2},$$

$$v = \frac{x - y}{2}.$$

From this it appears that

$$\sqrt{x^2 - y^2} = 2\sqrt{uv};$$

so the double integral transformation will be

$$\iint_T \sqrt{x^2 - y^2}\, dx \wedge dy = \iint_T 4\sqrt{uv}\, dv \wedge du,$$

where T is the triangle of Fig. 9–51 with counterclockwise orientation. For this triangle the hypotenuse has the equation

$$v = 0;$$

the bottom side has the equation

$$y = 0 \quad \text{or} \quad u = v;$$

the right side has the equation

$$x = 1 \quad \text{or} \quad u + v = 1.$$

Thus, an iterated integral in the order $du \, dv$ is indicated, and

$$\int_0^1 \int_y^1 \sqrt{x^2 - y^2} \, dx \, dy = \int_0^{1/2} \int_v^{1-v} 4\sqrt{uv} \, du \, dv.$$

The student may check that these are both positive. The orientation analysis for the new integral is shown in Fig. 9–51. From this analysis one gets

$$\int_0^1 \int_y^1 \sqrt{x^2 - y^2} \, dx \, dy = \iint_T \sqrt{x^2 - y^2} \, dx \wedge dy$$

$$= \iint_T 4\sqrt{uv} \, dv \wedge du$$

$$= -\iint_T 4\sqrt{uv} \, du \wedge dv$$

$$= -\int_{1/2}^0 \int_v^{1-v} 4\sqrt{uv} \, du \, dv$$

$$= \int_0^{1/2} \int_v^{1-v} 4\sqrt{uv} \, du \, dv.$$

4. Find the centroid of the "ice-cream cone" studied in Example 1. The volume is

$$\int_0^{2\pi} \int_0^{\pi/4} \int_0^1 \rho^2 \sin \phi \, d\rho \, d\phi \, d\theta = \int_0^{2\pi} \int_0^{\pi/4} \tfrac{1}{3} \sin \phi \, d\phi \, d\theta$$

$$= \int_0^{2\pi} \tfrac{1}{3} \left(1 - \frac{1}{\sqrt{2}} \right) d\theta$$

$$= \frac{\pi}{3} (2 - \sqrt{2}).$$

Now,

$$\mu_1(z, 0) = \iiint_R z \, dx \wedge dy \wedge dz,$$

but the computations are easier if one substitutes

$$z = \rho \cos \phi, \qquad dx \wedge dy \wedge dz = \rho^2 \sin \phi \, d\rho \wedge d\phi \wedge d\theta;$$

then

$$\mu_1(z, 0) = \int_0^{2\pi} \int_0^{\pi/4} \int_0^1 \rho^3 \cos \phi \sin \phi \, d\rho \, d\phi \, d\theta$$

$$= \int_0^{2\pi} \int_0^{\pi/4} \tfrac{1}{4} \cos \phi \sin \phi \, d\phi \, d\theta = \int_0^{2\pi} \frac{1}{8}\left(\frac{1}{\sqrt{2}}\right)^2 d\theta = \frac{\pi}{8}.$$

Thus,

$$\bar{z} = \frac{\mu_1(z, 0)}{\text{vol}} = \frac{\pi/8}{\pi(2 - \sqrt{2})/3} = \frac{3}{16 - 8\sqrt{2}}.$$

EXERCISES

1. For each of the following double integrals make the indicated coordinate transformation, reduce to an iterated integral, and evaluate. The domains of integration are shown in Fig. 9–52.

(a) $\iint_A (1 - x^2 - y^2) \, dx \wedge dy$; $x = r \cos \theta, y = r \sin \theta$

(b) $\iint_B (x - y)^2 \sin^2 (x + y) \, dx \wedge dy$; $u = x - y, v = x + y$

(c) $\iint_C (x^2 + y^2) \, dy \wedge dx$; $u = 2xy, v = x^2 - y^2$

(d) $\iint_D \sqrt{x^2 + y^2} \, dy \wedge dx$; $x = 2uv, y = u^2 - v^2$

2. Transform each of the following iterated integrals into an iterated integral in the indicated new coordinates. Do not evaluate.

(a) $\int_0^1 \int_0^x \sqrt{1 + x^2 + y^2} \, dy \, dx$; $x = u + v, y = u - v$

(b) $\int_0^1 \int_0^x \sqrt{x^2 + y^2} \, dy \, dx$; $u = 2xy, v = x^2 - y^2$

(c) $\int_0^1 \int_{1-x}^{1+x} \sqrt{x^2 + y^2} \, dy \, dx$; $x = u, y = u + v$

(d) $\int_0^1 \int_{-1}^1 \int_{-\sqrt{1-y^2}}^{\sqrt{1-y^2}} x^2 \sqrt{x^2 + y^2} \, dx \, dy \, dz$; cylindrical coordinates

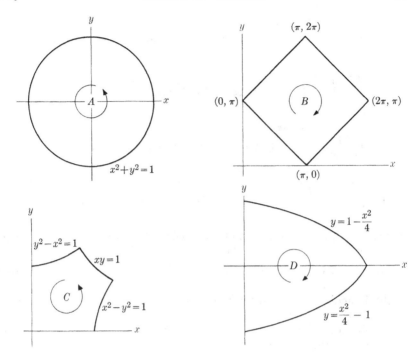

FIGURE 9–52

(e) $\int_0^1 \int_0^{\sqrt{1-x^2}} \int_0^{1+x+y} xyz\, dz\, dy\, dx$; cylindrical coordinates

(f) $\int_{-1}^1 \int_{-\sqrt{1-y^2}}^{\sqrt{1-y^2}} \int_{-\sqrt{4-x^2-y^2}}^{\sqrt{4-x^2-y^2}} z\, dz\, dx\, dy$; cylindrical coordinates

(g) $\int_{-1}^1 \int_{-\sqrt{1-z^2}}^{\sqrt{1-z^2}} \int_{-\sqrt{1-y^2-z^2}}^{\sqrt{1-y^2-z^2}} \sqrt{x^2+y^2+z^2}\, dx\, dy\, dz$;

spherical coordinates

(h) $\int_{-1}^1 \int_{-\sqrt{1-x^2}}^{\sqrt{1-x^2}} \int_{\sqrt{x^2+y^2}}^1 z\, dz\, dy\, dx$; spherical coordinates

(i) $\int_{-\sqrt{2}}^{\sqrt{2}} \int_{-\sqrt{2-y^2}}^{\sqrt{2-y^2}} \int_{\sqrt{x^2+y^2}}^{\sqrt{4-x^2-y^2}} z^2\, dz\, dx\, dy$; spherical coordinates

(j) $\int_0^{2\pi} \int_{-1/\sqrt{2}}^{1/\sqrt{2}} \int_z^{\sqrt{1-z^2}} r^3\, dr\, dz\, d\theta$; spherical coordinates

(k) $\displaystyle\int_0^{2\pi}\int_0^{1/\sqrt{2}}\int_r^{\sqrt{1-r^2}}$ $rz\,dz\,dr\,d\theta$; spherical coordinates

(l) $\displaystyle\int_0^{2\pi}\int_0^{\pi/4}\int_0^{\sec\phi}$ $\rho^3\sin 2\phi\,d\rho\,d\phi\,d\theta$; cylindrical coordinates

(m) $\displaystyle\int_0^{\pi}\int_0^{\sin\theta}\int_0^{r\sin\theta}$ $r^2\sin\theta\,dz\,dr\,d\theta$; rectangular coordinates

(n) $\displaystyle\int_0^1\int_0^{\pi/4}\int_0^{2\pi}$ $\rho^3\sin 2\phi\,d\theta\,d\phi\,d\rho$; rectangular coordinates

3. For each of the following solid figures choose the most favorable coordinate system, and find the volume and the centroid.

  (a) Bounded above by the locus of $x^2 + y^2 + z^2 = 1$ and below by that of $z = 0$
  (b) Bounded above by the locus of $z = 1$ and below by the top half of the locus of $z^2 = x^2 + y^2$
  (c) Bounded above by the locus of $z = 1$, below by the locus of $z = 0$, and on the sides by that of $x^2 + y^2 = 1$
  (d) Bounded above and below by the locus of $z^2 = x^2 + y^2$ and on the sides by the locus of $x^2 + y^2 = 1$
  (e) Bounded above and below by the locus of $z^2 = x^2 + y^2$ and on the sides by the locus of $x^2 + y^2 + z^2 = 1$
  (f) The spherical shell between the loci of $x^2 + y^2 + z^2 = 1$ and $x^2 + y^2 + z^2 = 4$
  (g) Bounded above and below by the locus of $z^2 = x^2 + y^2$ and on the sides by that of $x^2 + y^2 = x$
  (h) Bounded above by the locus of $z^2 = x^2 + y^2$ and below by that of $x^2 + y^2 + z^2 = 2z$
  (i) Bounded above by the locus of $z = x$ and below by that of $z = x^2 + y^2$
  (j) Bounded above by the locus of $z = x^2 + y^2$, below by the locus of $z = 0$, and on the sides by that of $x^2 + y^2 = 1$
  (k) Bounded above by the locus of $z^2 = x^2 + y^2$ and below by the locus of $z = x^2 + y^2$
  (l) Bounded above by the locus of $z = x^2 + y^2$, below by the locus of $z = 0$, and on the sides by that of $x^2 + y^2 = 1$. Compare Exercise 2(a), Section 8–4.
  (m) Bounded above by the locus of $z = x^2 + y^2$, below by the locus of $z = 0$, and on the sides by the locus of $x^2 + y^2 = 2y$. Compare Exercise 2(b), Section 8–4.
  (n) Bounded above by the locus of $z = 12 - x^2 - y^2$ and below by the top half of the locus of $z^2 = x^2 + y^2$. Compare Exercise 2(j), Section 8–4.

(o) Bounded above and below by the locus of $z^2 = x^2 + y^2$, and on the sides by the locus of $x^2 + y^2 = 2x$

(p) Bounded above by the locus of $z^2 = x^2 + y^2$, and below by that of $z = x^2 + y^2$

(q) Bounded above by the locus of $x^2 + y^2 + z^2 = 1$, and below by the top half of the locus of $z^2 = x^2 + y^2$

(r) Bounded above by the locus of $x^2 + y^2 + z^2 = 2$, and below by the locus of $z = x^2 + y^2$

(s) Inside the locus of $x^2 + y^2 + z^2 = 4$ and the locus of $x^2 + y^2 = 1$

(t) Bounded by the loci of $z = x^2 + 4y^2$ and $z = 8 - x^2 - 4y^2$

9–5 Mass. Suppose that the plane is given counterclockwise orientation, and denote by $d\alpha$ the positive area variable over the tangent bundle. This "area differential" has many different representations in terms of exterior products. Specifically, if $(a\,du,\,b\,dv)$ is any counterclockwise system of orthonormal component variables, then

$$d\alpha = ab\,du \wedge dv.$$

The most frequently used representations are

$$d\alpha = dx \wedge dy = r\,dr \wedge d\theta.$$

Now, suppose that there is a mass distribution in the plane, and let

$$m(A)$$

denote the mass contained in A for each counterclockwise region A. For certain mass distributions, there is a variable u on the points of the plane such that

$$m(A) = \iint_A u\,d\alpha$$

for each A. It can be shown (the proof is omitted) that if this is the case, then

$$u(p) = \lim_{A \to p} \frac{m(A)}{\alpha(A)}. \tag{29}$$

The notation "$A \to p$" means, roughly, that the region A "shrinks to the point p." Certain technical restrictions must be placed on this limit process, but these are not pertinent to the present informal discussion.

In the light of (29), it is natural to call u the *mass-density variable* and denote it by

$$D_\alpha m.$$

That is, $D_\alpha m$ measures the limiting value of mass per unit area.

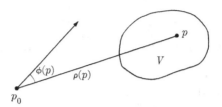

FIGURE 9–53

Now, look at it the other way. The area differential $d\alpha$ can always be expressed in terms of coordinates. Suppose that $D_\alpha m$ is given in terms of coordinates; then

$$m(A) = \iint_A D_\alpha m \, d\alpha$$

gives the mass of each set A.

Similarly, let

$$d\tau = dx \wedge dy \wedge dz = r \, dr \wedge d\theta \wedge dz = \rho^2 \sin\phi \, d\rho \wedge d\phi \wedge d\theta = \text{etc.}$$

be the volume variable on the tangent bundle to three-space. A mass density will be written $D_\tau m$, and

$$m(V) = \iiint_V D_\tau m \, d\tau$$

will give the mass of each solid V.

In either case it is natural to write

$$dm = D_\alpha m \, d\alpha$$

or

$$dm = D_\tau m \, d\tau,$$

with the understanding that the dimension of dm must be clear from the context.

Note that mass is intrinsically positive; thus, when dm is written as an exterior product and used in a multiple integral, the orientation of the coordinate system must be the same as that of the domain of integration. In the next step, when the multiple integral is reduced to an iterated integral, the limits of integration must be chosen so that the integral of dm itself will be positive, even though (see Section 9–7) there may be other factors present that lead to a negative answer to the problem at hand.

Consider a three-dimensional manifold V with mass distributed in it, and let p_0 be some fixed point of three-space, either inside or outside V; it makes no difference. Define ρ over V by letting $\rho(p)$ be the distance between p and p_0. Take a fixed direction through p_0, and define ϕ over V

by letting $\phi(p)$ be the angle from the given fixed direction to the direction from p_0 to p (see Fig. 9-53). The inverse-square gravitation law may be defined as saying that for a unit mass at p_0, the component of force in the given direction due to the mass in V is

$$\iiint_V \frac{\cos \phi \, dm}{\rho^2}.$$

Given this same setup, the gravitational potential at p_0 for the mass in V is defined as

$$\iiint_V \frac{dm}{\rho}.$$

EXAMPLES

1. Find the mass inside the circle $r = 2a \cos \theta$, given that $D_\alpha m = r$. In this case,

$$dm = r \, d\alpha = r^2 \, dr \wedge d\theta;$$

so, the required total mass is

$$\int_{-\pi/2}^{\pi/2} \int_0^{2a \cos \theta} r^2 \, dr \, d\theta = \int_{-\pi/2}^{\pi/2} \frac{r^3}{3} \bigg|_0^{2a \cos \theta} d\theta$$

$$= \int_{-\pi/2}^{\pi/2} \frac{8a^3}{3} \cos^3 \theta \, d\theta$$

$$- \int_{-\pi/2}^{\pi/2} \frac{8a^3}{3} (1 - \sin^2 \theta) \cos \theta \, d\theta$$

$$= \frac{8a^3}{3} \left(\sin \theta + \frac{\sin^3 \theta}{3} \right) \bigg|_{-\pi/2}^{\pi/2} = \frac{64a^3}{9}.$$

2. Find the gravitational potential at a point p_0 due to the mass in a spherical shell with inner radius a and outer radius b; assume that $D_\tau m = 1$ over the shell. Place the center of the shell at the origin; place p_0 on the z-axis; and introduce spherical coordinates (see Fig. 9-54). It appears from this figure that the distance between p_0 and p is

$$\sqrt{\rho^2 \sin^2 \phi + (c - \rho \cos \phi)^2} = \sqrt{\rho^2 - 2c\rho \cos \phi + c^2}.$$

Since $D_\tau m = 1$,

$$dm = d\tau = \rho^2 \sin \phi \, d\rho \wedge d\phi \wedge d\theta.$$

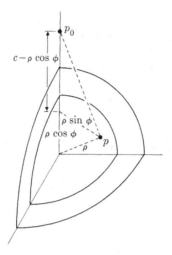

FIGURE 9–54

Thus, the required potential is

$$\int_0^{2\pi} \int_a^b \int_0^\pi \frac{\rho^2 \sin \phi}{\sqrt{\rho^2 - 2c\rho \cos \phi + c^2}} \, d\phi \, d\rho \, d\theta.$$

This order of integration is chosen because the partial integral with respect to ϕ is easily computed:

$$\int_0^\pi \frac{\rho^2 \sin \phi}{\sqrt{\rho^2 - 2c\rho \cos + c^2}} \, d\phi = \frac{\rho}{c} \sqrt{\rho^2 - 2c\rho \cos \phi + c^2} \Big|_0^\pi$$

$$= \frac{\rho}{c} (\sqrt{\rho^2 + 2c\rho + c^2} - \sqrt{\rho^2 - 2c\rho + c^2}).$$

There are now two cases to consider. The point p_0 was drawn outside the shell in Fig. 9–54, but the analysis down to this stage is the same if p_0 is inside the shell. However, if the situation is as pictured ($c > b$), then $c > \rho$ for all applicable values of ρ, and

$$\sqrt{\rho^2 - 2c\rho + c^2} = c - \rho.$$

So, if the point is outside the shell, the potential is

$$\int_0^{2\pi} \int_a^b \frac{2\rho^2}{c} \, d\rho \, d\theta = \frac{4\pi}{3c} (b^3 - a^3).$$

This is the volume of the shell divided by c; that is, at points outside, the

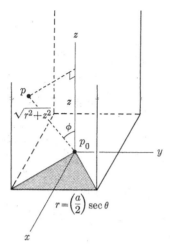

FIGURE 9–55

potential is the same as if the entire mass were concentrated at the center of the shell. Now, if p_0 is inside the shell ($c < a$), then $c < \rho$ for all applicable values of ρ, and

$$\sqrt{\rho^2 - 2c\rho + c^2} = \rho - c.$$

In this case the potential is

$$\int_0^{2\pi} \int_a^b 2\rho \, d\rho \, d\theta = 2\pi(b^2 - a^2).$$

The interesting thing about this result is that it does not depend on c. The potential is the same at all points inside the shell.

3. A solid bar with mass density 1 has a square face of side a and extends to infinity behind that face (Fig. 9–55). Find the gravitational force on a unit mass at the center of this face. The direction of attraction is down the axis of the bar; so, the component of force in that direction is required. Place the bar as shown in Fig. 9–55, and introduce cylindrical coordinates. With this setup,

$$\rho = \sqrt{r^2 + z^2},$$

$$\cos \phi = z/\sqrt{r^2 + z^2},$$

$$dm = d\tau = r \, dr \wedge d\theta \wedge dz.$$

The symmetry of the figure permits us to take the integral over the solid above the shaded area in Fig. 9–55 and then multiply it by 4. Therefore, the attractive force is

$$4 \int_{-\pi/4}^{\pi/4} \int_0^{(a/2)\sec\theta} \int_0^\infty \frac{zr}{(r^2+z^2)^{3/2}}\, dz\, dr\, d\theta$$

$$= 4 \int_{-\pi/4}^{\pi/4} \int_0^{(a/2)\sec\theta} \left. \frac{-r}{\sqrt{r^2+z^2}} \right|_0^\infty dr\, d\theta$$

$$= 4 \int_{-\pi/4}^{\pi/4} \int_0^{(a/2)\sec\theta} dr\, d\theta$$

$$= 2a \int_{-\pi/4}^{\pi/4} \sec\theta\, d\theta = 2a \ln \frac{\sqrt{2}+1}{\sqrt{2}-1}.$$

EXERCISES

1. Find the area and mass of each of the following plane figures.
   (a) Bounded by the loci of $y = \sin x$, $y = 0$ $(0 \le x \le \pi)$; $D_\alpha m = y$
   (b) Bounded by the loci of $y = x^2$, $y = x$; $D_\alpha m = y$
   (c) Bounded by the loci of $x = 0$, $y = 0$, $x = 3$, $y = 2$; $D_\alpha m = 2x$
   (d) The same figure in part (c); $D_\alpha m = x^2 + y^2$
   (e) Bounded by the loci of $y = 0$, $x = a$, $y = bx/a$; $D_\alpha m = x^2 + y^2$
   (f) Bounded by the loci of $y = 1 - x^2$, $y = 0$; $D_\alpha m = y$
   (g) Semicircle of radius a; $D_\alpha m =$ distance from the center of the diameter
   (h) Bounded by the loci of $r = \theta$, $\theta = \pi$ $(0 \le \theta \le \pi)$; $D_\alpha m = r$
   (i) Circle of radius a; $D_\alpha m =$ square of the distance from the center
   (j) Inside the cardioid on which $r = 1 - \cos\theta$; $D_\alpha m = r$
   (k) Inside one loop of the locus of $r = \cos 2\theta$; $D_\alpha m = r^3$
   (l) Inside the lemniscate on which $r^2 = \sin 2\theta$; $D_\alpha m = r^2$

2. Find the volume and mass of each of the following solid figures.
   (a) Bounded by the loci of $y = x^2$, $x = y^2$, $z = 0$, $z = 1 + x + y$; $D_\tau m = 2y$
   (b) Bounded by the loci of $x^2 + z^2 = 4$, $y = x^2$, $y = 2x$; $D_\tau m = |z|$
   (c) Cube of side a; $D_\tau m =$ sum of the distances from three adjacent faces
   (d) Bounded by the loci of $y = \sqrt{1 - x^2}$, $z = y$, $z = 2y$; $D_\tau m = z^2$
   (e) Cube of side a; $D_\tau m =$ square of distance from one corner
   (f) Bounded by the loci of $z = \sqrt{1 - x^2 - y^2}$, $z = 0$; $D_\tau m = |xyz|$
   (g) Right circular cylinder of radius a and height h; $D_\tau m =$ distance from base
   (h) Inside the locus of $r^2 + z^2 = 4$, outside that of $r = 1$; $D_\tau m = |z|$
   (i) Bounded by the loci of $z = r^2$, $z = 2r$; $D_\tau m = r$
   (j) Above the cone on which $z = r$, inside the sphere on which $\rho = 1$; $D_\tau m = z$

(k) Outside the cone on which $z = r$, inside the sphere on which $\rho = 1$; $D_\tau m = r$

(l) Sphere of radius a; $D_\tau m$ = distance from a fixed point on the surface. [*Hint:* Put this fixed point at the origin; put the center of the sphere on the z-axis; use spherical coordinates.]

3. In each of the following statements there is given a figure and a point. Assume that the mass density is constant over the figure, and find the attractive force on a unit mass at the given point.

(a) Cylindrical shell of inner radius a, outer radius b, and height h. The point is at the center of one end.

(b) Solid right circular cylinder of radius a and height h. The point is on the axis of the cylinder a distance m from one base—outside the figure.

(c) Solid right circular cone of base radius a and height h. The point is at the vertex.

(d) Solid sphere of radius a. The point is at a distance b ($b > a$) from the center.

(e) Thin circular plate of radius a. The point is at a distance b from the center on a line perpendicular to the plate.

4. In parts (c) and (e) of Exercise 3 find the gravitational potential due to the given mass at the given point.

9–6 Probability. The mathematical model for the theory of probability presents the notion of probability as something quite analogous to mass. The first thing to have in mind is the model that is used to represent a probability problem.

Roughly speaking, probability is a "likelihood measure" for "experimental results." In order to study such an idea mathematically, however, it helps to represent the results by points. If the set of all possible experimental results is represented as a set of points, this master point set is called the *event space* for the problem. Then, an *event* is defined as a set in the event space. The connection between this definition and the usual usage of the word "event" is that an event (point set) E represents the "event" (à la Webster) that the experimental result is one of those represented by a point of E. Probability is to be associated with events; so, in the event-space model this amounts to a variable P that associates each event E with a number $P(E)$ called the *probability of E*. Probability is a variable on point sets, just as mass is.

In terms of this model, probability can be defined by postulates. Roughly speaking, any variable that is additive, monotone, and has the value 1 on the entire event space is called a probability variable. This section and the following section will present a few basic manipulations with certain types of event spaces and probabilities. First, however, it is important to note a few things that the discussion does not cover.

(i) Problems will be posed by specifying an event space and probability variable. In other words, analysis of physical experiments and determination of basic probabilities will not be discussed here. This is similar to the point of view of Section 9–5. In that section, one merely said, "Let $D_\alpha m$ be as follows." No question was raised concerning how mass might get distributed in this way.

(ii) The purpose of this discussion is to give more practice with integrals; so, our attention will be focused on those cases in which probabilities are given by integrals. Many standard probability problems (dice, cards, etc.) do not fit this model.

Given an event space S and a probability P on sets in S, a variable x on the points of S is called a *random variable*. If x is a random variable and a is a number, then the inequality

$$x \leq a$$

has a locus which is an event (set in S). Now, P is applied to events; so, P applies to this locus. However, the usual convention is to write

$$P(x \leq a)$$

meaning probability of the locus of $x \leq a$.

Just as with mass, probability may often be determined by a *probability density* variable

$$D_\alpha P.$$

Given such a probability density, one has

$$P(E) = \int_E D_\alpha P \, d\alpha.$$

It is also convenient to introduce a differential of probability,

$$dP = D_\alpha P \, d\alpha.$$

Suppose that $D_\alpha P$ is given over the plane or a portion thereof. The coordinate variables x and y on the plane are then random variables, and

$$\phi(x, y)$$

is another random variable. For any number a,

$$P[\phi(x, y) \leq a] = \iint_E D_\alpha P \, d\alpha,$$

where E is the locus of $\phi(x, y) \leq a$.

Examples

1. Let the unit square be an event space with $D_\alpha P = 2y$. Find the probability that $x + y \le a$. Figure 9–56 shows two cases: (a) $0 \le a \le 1$, and (b) $1 \le a \le 2$. In each case, the probability of the shaded portion is required. Now,

$$dP = 2y\, d\alpha = 2y\, dx \wedge dy;$$

so, for $0 \le a \le 1$,

$$P(x + y \le a) = \int_0^a \int_0^{a-y} 2y\, dx\, dy = \int_0^a 2y(a - y)\, dy$$

$$= \left. (ay^2 - \frac{2}{3}y^3) \right|_0^a = \frac{a^3}{3}.$$

For $1 \le a \le 2$, it is easier to integrate over the unshaded triangle and subtract from 1:

$$P(x + y \le a) = 1 - \int_{a-1}^1 \int_{a-y}^1 2y\, dx\, dy = 1 - \int_{a-1}^1 2y(1 - a + y)\, dy$$

$$= 1 - \left. (y^2 - ay^2 + \tfrac{2}{3}y^3) \right|_{a-1}^1 = \tfrac{1}{3}[3a - 2 - (a - 1)^3].$$

2. If the event space is the real line with x as the usual coordinate, a probability density assumes the form $D_x P$, and

$$P(E) = \int_E D_x P\, dx.$$

An important example of such a one-dimensional event space is the one on which

$$D_x P = \frac{1}{\sqrt{2\pi}} e^{-x^2/2}. \tag{30}$$

For reasons that will not be discussed here, many problems involving

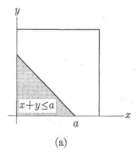
(a)

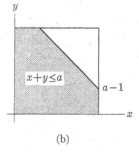
(b)

Figure 9–56

chance errors may be cast in this model. That is, given the real line with probability density (30), the coordinate variable x often has the physical significance that it measures errors in numerical measurements of some sort. The variable (30) is called the *normal* probability density.

Not every variable can be a probability density. For the real line, one must have

$$P(S) = \int_{-\infty}^{\infty} D_x P \, dx = 1.$$

The following shows that (30) qualifies in this respect:

$$\left[\int_{-\infty}^{\infty} \frac{1}{\sqrt{2\pi}} e^{-x^2/2} \, dx \right]^2 = \frac{1}{2\pi} \int_{-\infty}^{\infty} e^{-x^2/2} \, dx \int_{-\infty}^{\infty} e^{-y^2/2} \, dy$$

$$= \frac{1}{2\pi} \int_{-\infty}^{\infty} \int_{-\infty}^{\infty} e^{-(x^2+y^2)/2} \, dx \, dy$$

$$= \frac{1}{2\pi} \int_{0}^{2\pi} \int_{0}^{\infty} e^{-r^2/2} \, r \, dr \, d\theta$$

$$= \frac{1}{2\pi} \int_{0}^{2\pi} -e^{-r^2/2} \Big|_{0}^{\infty} \, d\theta$$

$$= \frac{1}{2\pi} \int_{0}^{2\pi} d\theta = 1.$$

3. The primitive of the normal density,

$$\int \frac{1}{\sqrt{2\pi}} e^{-x^2/2} \, dx,$$

is a nonelementary function compounded with x. Despite this seeming obstacle to computation, many integrals involving the normal density may be computed by using the result of Example 2.

For example, let

$$D_\alpha P = \frac{1}{2\pi} e^{-(x^2+y^2)/2}$$

over the plane; find

$$D_\alpha P(x + y \leq a).$$

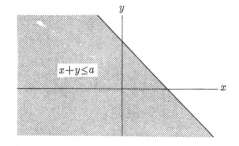

FIGURE 9–57

The locus of $x + y \le a$ is the shaded region of Fig. 9–57; so,

$$P(x + y \le a) = \int_{-\infty}^{\infty} \int_{-\infty}^{a-y} \frac{1}{2\pi} e^{-(x^2+y^2)/2} \, dx \, dy$$

$$= \int_{-\infty}^{\infty} \frac{1}{2\pi} e^{-y^2/2} \int_{-\infty}^{a-y} e^{-x^2/2} \, dx \, dy$$

$$= \int_{-\infty}^{\infty} \frac{1}{2\pi} e^{-y^2/2} F(a - y) \, dy,$$

where F is a primitive of $e^{-l^2/2}$. So,

$$D_a F(a - y) = e^{-(a-y)^2/2},$$

and it can be shown that

$$D_a P(x + y \le a) = \int_{-\infty}^{\infty} \frac{1}{2\pi} e^{-y^2/2} D_a F(a - y) \, dy$$

$$= \int_{-\infty}^{\infty} \frac{1}{2\pi} e^{-y^2/2} e^{-(a-y)^2/2} \, dy.$$

Now, the exponent in this last integral is

$$-\tfrac{1}{2}[y^2 + (a - y)^2] = -\tfrac{1}{2}(2y^2 - 2ay + a^2)$$

$$= -\left(y^2 - ay + \frac{a^2}{4} + \frac{a^2}{4}\right)$$

$$= -\left(y - \frac{a}{2}\right)^2 - \frac{a^2}{4}.$$

Thus,

$$D_a P(x + y \le a)$$

$$= \int_{-\infty}^{\infty} \frac{1}{\sqrt{2\pi}} e^{-a^2/4} \frac{1}{\sqrt{2\pi}} e^{-[y-(a/2)]^2} \, dy$$

$$= \frac{1}{2\sqrt{\pi}} e^{-a^2/4} \int_{-\infty}^{\infty} \frac{1}{\sqrt{2\pi}} e^{-[y\sqrt{2}-(a/\sqrt{2})]^2/2} \sqrt{2} \, dy$$

$$= \frac{1}{2\sqrt{\pi}} e^{-a^2/4} \int_{-\infty}^{\infty} \frac{1}{\sqrt{2\pi}} e^{-[y\sqrt{2}-(a/\sqrt{2})]^2/2} \, d\left(y\sqrt{2} - \frac{a}{\sqrt{2}}\right)$$

$$= \frac{1}{2\sqrt{\pi}} e^{-a^2/4},$$

because, by Example 2, the last integral has the value 1.

EXERCISES

1. In each of the following exercises, a probability density is specified and an inequality describing a portion of the plane is given. In each case, find the probability of the given inequality.

(a) $D_a P = x + y$ for $0 \leq x \leq 1$, $0 \leq y \leq 1$, $D_a P = 0$ otherwise; $x + y \leq u$

(b) $D_a P$ as in part (a); $xy \leq u$

(c) $D_a P = 1$ for $0 \leq x \leq 1$, $0 \leq y \leq 1$, $D_a P = 0$ otherwise; $xy \leq u$

(d) $D_a P$ as in part (c); $x + y \leq u$

(e) $D_a P$ as in part (c); $\max (x, y) \leq u$

(f) $D_a P$ as in part (c); $\min (x, y) \leq u$

(g) $D_a P = abe^{-ax-by}$ for $x \geq 0$, $y \geq 0$ $(a > 0, b > 0, a \neq b)$; $x + y \leq u$

(h) Same as part (g) with $a = b$

2. Let $D_a P = (1/2\pi)e^{-(x^2+y^2)/2}$ over the plane. Find

(a) $D_a P(x \leq a)$ (b) $D_a P(y \leq a)$

(c) $P(\sqrt{x^2 + y^2} \leq a)$. [*Hint:* Use polar coordinates.]

(d) $P(x^2 + y^2 \leq a)$

9–7 Moments. The notion of moment is associated with a variable on the points of a space and a weighting of sets in the space. In Sections 8–2 and 8–4 the space was a region A of the plane or three-space, and the weight of a set was its area or volume, as the case might be. Mass and probability furnish equally good (perhaps even better) examples of weights.

For purposes of a general summary, let $W(E)$ be the weight of E. In special cases, W may be α (area), τ (volume), m (mass), or P (probability). Let u be a variable on the points of the space, and define a function F_u by setting

$$F_u(t) = W(u \leq t).$$

As in Section 9–6, the "weight of an inequality" means the weight of the locus of the inequality. The function F_u is called the *distribution function* for u. By definition,

$$\mu_n(u, a) = \int_{-\infty}^{\infty} (t - a)^n \, dF_u(t). \tag{31}$$

Often this reduces to an integral from c to d because $dF_u(t) = 0$ for $t < c$ and for $t > d$.

If the space extends from $u = c$ to $u = d$, and if weight is area, then (31) becomes

$$\mu_n(u, a) = \int_c^d (t - a)^n \, D_t \alpha(u \leq t) \, dt,$$

and this is the formula for moments used as a definition in Section 8–2.

In Section 8–2, moments were computed from iterated integrals. The double integrals from which these iterated integrals come would assume the form

$$\iint_A (u - a)^n \, d\alpha,$$

where A is the space.

Under quite general conditions the quantity defined in terms of the distribution function is given by the appropriate multiple integral. The following computations do not constitute a proof, but they should serve to make plausible this basic theorem about moments. To compare the two integrals for a moment of u, assume that dW can be expressed in terms of an exterior product involving du:

$$dW = f(u, v) \, du \wedge dv.$$

Then,

$$W(u \leq t) = \iint_{u \leq t} f(u, v) \, du \wedge dv = \int_{-\infty}^{\infty} \int_{-\infty}^{t} f(u, v) \, du \, dv;$$

so,

$$D_t W(u \leq t) = \int_{-\infty}^{\infty} f(t, v) \, dv.$$

Thus, starting from (31), one has

$$\mu_n(u, a) = \int_{-\infty}^{\infty} (t - a)^n \, D_t W(u \leq t) \, dt = \int_{-\infty}^{\infty} \int_{-\infty}^{\infty} (t - a)^n \, f(t, v) \, dv \, dt$$

$$= \int_{-\infty}^{\infty} \int_{-\infty}^{\infty} (u - a)^n f(u, v) \, du \, dv = \iint_A (u - a)^n \, dW. \qquad (32)$$

Thus, although (31) is the definition, (32) is often a better working formula and will be employed here.

Terminology with respect to moments was discussed in Section 8–2. The following is a brief summary of this, together with the required multiple integral formulas.

1. *Moments with respect to mass* (the language of physics). The nth moment of x about 0 is called the nth moment of the mass about the y-axis. The nth moment of y about 0 is called the nth moment of the mass about the x-axis. In particular, second moments are called *moments of inertia*. The point with coordinates $(\bar{x}, \bar{y})$ where

$$\bar{x} = \frac{1}{m(A)} \iint_A x \, dm, \qquad \bar{y} = \frac{1}{m(A)} \iint_A y \, dm$$

is called the centroid or *center of mass* of A.

2. *Moments with respect to probability* (the language of statistics). The first moment of any variable u about 0 is called the *expectation* of u. It is denoted by $E(u)$ or by $\bar{u}$:

$$\bar{u} = E(u) = \iint_S u\, dP,$$

where S is the event space. The second moment of u about $\bar{u}$ is called the *variance* of u:

$$\text{var}\,(u) = \iint_S (u - \bar{u})^2\, dP.$$

In probability theory, there are several important consequences of the fact that moments may be given by multiple integrals without a return to the distribution function every time.

(i) A random variable represents a numerical measurement of experimental results. However, the event-space model for a given experiment is by no means unique. Thus, the given measurement may have many random-variable representations. One can speak of the expectation of the measurement, because all random variables representing the measurement have the same distribution function (probability that the measurement is less than or equal to t does not depend on the representation). Therefore, these random variables all have the same expectation.

(ii) The *addition theorem* for expectation is trivial when expectations are written as multiple integrals:

$$E(u + v) = \iint_S (u + v)\, dP = \iint_S u\, dP + \iint_S v\, dP = E(u) + E(v).$$

(iii) Any moment may be written as an expectation:

$$\mu_n(u, a) = \iint_S (u - a)^n\, dP = E[(u - a)^n].$$

EXAMPLES

1. Find the center of mass of the unit square with $D_\alpha m = xy$. By the fundamental theorem on moments,

$$\mu_1(x, 0) = \int_0^1 \int_0^1 x^2 y\, dy\, dx = \int_0^1 (x^2/2)\, dx = \tfrac{1}{6},$$

$$\mu_1(y, 0) = \int_0^1 \int_0^1 xy^2\, dx\, dy = \int_0^1 (y^2/2)\, dy = \tfrac{1}{6},$$

$$m(A) = \int_0^1 \int_0^1 xy\, dx\, dy = \int_0^1 (y/2)\, dy = \tfrac{1}{4}.$$

Therefore,

$$\bar{x} = \bar{y} = \frac{1/6}{1/4} = \frac{2}{3}.$$

2. The results of Example 1, Section 9–6, may be interpreted as follows. If $D_\alpha P = 2y$ on the unit square and if the random variable u is defined on this square by $u = x + y$, then

$$F_u(t) = \begin{cases} t^3/3 & \text{for} \quad 0 \le t \le 1, \\ \frac{1}{3}[3t - 2 - (t-1)^3] & \text{for} \quad 1 \le t \le 2. \end{cases}$$

Thus, by definition,

$$E(u) = \int_{t=0}^{t=1} t\,d(t^3/3) + \int_{t=1}^{t=2} t\,d\{\tfrac{1}{3}[3t - 2 - (t-1)^3]\}$$

$$= \int_0^1 t^3\,dt + \int_1^2 t[1 - (t-1)^2]\,dt$$

$$= \int_0^1 t^3\,dt + \int_1^2 (2t^2 - t^3)\,dt = \frac{t^4}{4}\Big|_0^1 + \left(\frac{2}{3}t^3 - \frac{t^4}{4}\right)\Big|_1^2 = \frac{7}{6}.$$

By the fundamental theorem on moments,

$$E(u) = \int_0^1 \int_0^1 (x + y)2y\,dx\,dy = \int_0^1 \int_0^1 (2xy + 2y^2)\,dx\,dy$$

$$= \int_0^1 (y + 2y^2)\,dy = \frac{y}{2} + \frac{2y^3}{3}\Big|_0^1 = \frac{7}{6}.$$

For this distribution,

$$F_x(t) = P(x \le t) = \int_0^1 \int_0^t 2y\,dx\,dy = \int_0^1 2ty\,dy = t;$$

therefore, by definition,

$$E(x) = \int_0^1 t\,dt = \tfrac{1}{2}.$$

Similarly,

$$F_y(t) = P(y \le t) = \int_0^t \int_0^1 2y\,dx\,dy = \int_0^t 2y\,dy = t^2,$$

$$E(y) = \int_{t=0}^{t=1} t\,d(t^2) = \int_0^1 2t^2\,dt = \tfrac{2}{3}.$$

These results illustrate that for this particular example, $E(x + y)$ is the same by definition and by the fundamental theorem; also

$$E(x + y) = E(x) + E(y).$$

3. Show that the variance of a random variable x [by definition, $\mu_2(x, \bar{x})$] is given by

$$E(x^2) - \bar{x}^2.$$

As noted in item (iii) above, in probability theory all moments may be regarded as expectations; so

$$\text{var } (x) = \mu_2(x, \bar{x}) = E[(x - \bar{x})^2] = E(x^2 - 2x\bar{x} + \bar{x}^2).$$

Thus, by the addition theorem for expectations,

$$\text{var } (x) = E(x^2) - 2\bar{x}E(x) + \bar{x}^2 = E(x^2) - \bar{x}^2$$

because $\bar{x} = E(x)$ by definition.

EXERCISES

1. Find the center of mass of each of the figures in Exercises 1 (a through j) and 2 (d through l), Section 9–5.

2. The polar moment of inertia of a mass distribution in the plane is defined as $\mu_2(r, 0)$. Prove that the polar moment of inertia is the sum of the moments of inertia about the two rectangular coordinate axes.

3. Each part of Exercise 1, Section 9–6, may be interpreted (see Example 2 above) as finding a distribution function for a certain random variable. Take each of these results and compute the expectation of the indicated variable from the definition. Then, compute each of these expectations from the two-dimensional distribution by means of the fundamental theorem on moments, and check.

4. Parts (b) and (d) of Exercise 1, Section 9–6, deal with the random variable xy. Show in part (d) that $E(xy) = E(x)E(y)$, and show in part (b) that this does not hold.

5. Let x be a random variable with probability density

$$F_x'(t) = \frac{1}{\sigma\sqrt{2\pi}} e^{-t^2/2\sigma^2},$$

where σ is a constant. Then, x is called *normally distributed*. (In connection with this density function, see Example 2, Section 9–6.) Show that

(a) $\mu_1(x, 0) = 0$ \hspace{2cm} (b) $\mu_2(x, 0) = \sigma^2$

6. Consider a normal probability distribution in the plane defined by

$$D_\alpha P = \frac{1}{2\pi\sigma^2} e^{-(x^2+y^2)/2\sigma^2}.$$

On this event space find

  (a) $\mu_1(r, 0)$ [*Hint:* Use polar coordinates, and note Exercise 5(b).]
  (b) $\mu_2(x, 0)$
  (c) $\mu_2(r, 0)$ [*Hint:* Use part (b) and the addition theorem for expectations.]
  (d) $\mu_2(r, \bar{r})$ [*Hint:* Use (iii) in this section.]

CHAPTER 10

LINE AND SURFACE INTEGRALS

10–1 Line integrals, recapitulation. Chapter 9 has presented double integrals,

$$\iint_A w \, du \wedge dv, \tag{1}$$

where A is an oriented two-dimensional submanifold of the plane. The present project is to study surface integrals,

$$\iint_S w \, du \wedge dv, \tag{2}$$

where S is an oriented surface in three-space.

By way of introduction, review the problem in one lower dimension. If J is a straight-line interval, then

$$\int_J u \, dv \tag{3}$$

may be defined in the following manner. Partition J by points

$$p_0, p_1, \ldots, p_n,$$

and set

$$\int_J u \, dv = \lim \sum_{i=0}^{n-1} u(p_i) \, dv_{p_i}(p_{i+1}).$$

The notation $dv_{p_i}(p_{i+1})$ is permissible because the tangent line T_{p_i} coincides with J; so, p_{i+1} coincides with a point on T_{p_i} at which dv_{p_i} is defined.

The same simplification is used in defining (1) (Section 9–3). Briefly,

$$\iint_A w \, du \wedge dv = \lim \sum_{i=1}^{n} w(p_i)(du \wedge dv)_{p_i}(A_i).$$

The A_i are really subsets of A, but each coincides with a subset of T_{p_i}; hence it is permissible to write $(du \wedge dv)_{p_i}(A_i)$. That is, (1) has been defined, and the question is how to get (2). The analogous question in one dimension is, Given (3), how do you get

$$\int_C u \, dv, \tag{4}$$

where C is a curve in the plane? The following questions and answers indicate the nature and solution of the major problems encountered in developing (4) from (3).

320

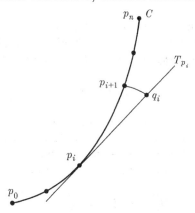

FIGURE 10–1

(i) What is the formal definition of (4)? Partition C by points $p_0, p_1, \ldots, p_n$. At each p_i, draw the tangent plane T_{p_i}. On T_{p_i} mark off an interval $[p_i, q_i]$ whose length is equal to the length of the arc from p_i to p_{i+1} on C (see Fig. 10–1). Then,

$$\int_C u \, dv = \lim \sum_{i=1}^{n-1} u(p_i) \, dv_{p_i}(q_i).$$

This method is used in Section 2–5.

(ii) What do we mean by length of an arc on C? We answer this question in the following way. Assume that x is a coordinate over the whole of C. In general, if q is on T_p, $dx_p(q)$ does not give the distance from p to q. However, there is a multiplier t such that

$$\text{dist } (p, q) = t(p) \, dx_p(q)$$

for each p on C and q on T_p. Specifically,

$$t = [1 + (D_x y)^2]^{1/2},$$

but this is immaterial; there is such a variable t. Now, project C onto an interval J on the x-axis. Let $p \to p'$ in this projection process; x and t are defined on C. Now define x' and t' on J by setting

$$x'(p') = x(p), \qquad t'(p') = t(p).$$

Then, define

$$\text{length of } C = \int_J t' \, dx'.$$

This defines arc length on C in terms of the form (3).

(iii) What happens in (ii) if x is not a coordinate over the whole of C? Find some variable w that is. Again find a multiplier t such that $t\,dw$ measures distance on tangent lines. (For example, if w is the polar coordinate θ, then

$$t = [(D_\theta r)^2 + r^2]^{1/2}.$$

Instead of the projection process in (ii), introduce the w-map of C. That is, by definition, w maps C one to one onto an interval of real numbers, and there is a natural map of this onto a straight line interval J. This J is the w-map of C. As we did in (ii), transfer w and t on C to w' and t' on J, and define

$$\text{length of } C = \int_J t'\,dw'.$$

(iv) All this defines (4), but how does one compute it? There is a theorem to the effect that if J is any coordinate map of C, if p on C corresponds to p' on J, if $u'(p') = u(p)$, if $v'(p') = v(p)$, and if v' is a coordinate on J, then

$$\int_C u\,dv = \int_J u'\,dv'.$$

Thus, the evaluation of (4) is reduced to the evaluation of (3).

Examples

1. Let C be the semicircle defined by

$$x = \cos\theta, \qquad y = \sin\theta \qquad (0 \le \theta \le \pi).$$

Evaluate

$$\int_C (y^2\,dx + x^2\,dy).$$

The parameter θ is a coordinate on C; so, a transfer to the θ-map yields

$$\int_C (y^2\,dx + x^2\,dy) = \int_0^\pi [\sin^2\theta(-\sin\theta\,d\theta) + \cos^2\theta(\cos\theta\,d\theta)]$$

$$= \int_0^\pi (\cos^3\theta - \sin^3\theta)\,d\theta$$

$$= \int_0^\pi [(1 - \sin^2\theta)\cos\theta - (1 - \cos^2\theta)\sin\theta]\,d\theta$$

$$= (\sin\theta - \tfrac{1}{3}\sin^3\theta + \cos\theta - \tfrac{1}{3}\cos^3\theta)\Big|_0^\pi = -\tfrac{4}{3}.$$

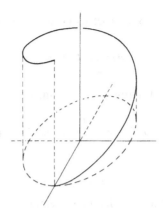

FIGURE 10–2

2. Let C be the circular helix defined by

$$x = \cos t, \qquad y = \sin t, \qquad z = t \qquad (0 \le t \le 2\pi)$$

(see Fig. 10–2). Evaluate

$$\int_C (y\,dx + z\,dy + x\,dz).$$

Make substitutions, and transfer to the t-map:

$$\int_C (y\,dx + z\,dy + x\,dz) = \int_0^{2\pi} (-\sin^2 t + t \cos t + \cos t)\,dt$$

$$= \left(-\frac{t}{2} + \frac{1}{4}\sin 2t + t \sin t + \cos t + \sin t\right)\Bigg|_0^{2\pi} = -\pi.$$

3. The notation

$$\int_{(0,0)}^{(1,1)} (x\,dy - y\,dx)$$

is not well defined because there are many paths from $(0, 0)$ to $(1, 1)$ and the value depends on the choice of path. For example, let C be the locus of $y = x$ $(0 \le x \le 1)$; then

$$\int_C (x\,dy - y\,dx) = \int_0^1 (x\,dx - x\,dx) = 0.$$

If C is the locus of $y = x^2$ $(0 \le x \le 1)$, then

$$\int_C (x\,dy - y\,dx) = \int_0^1 [x(2x\,dx) - x^2\,dx] = \int_0^1 x^2\,dx = \tfrac{1}{3}.$$

If C is the broken line from $(0,0)$ to $(1,0)$ to $(1,1)$, then

$$\int_C (x\,dy - y\,dx) = \int_0^1 0\,dx + \int_0^1 1\,dy = 1.$$

Let C be the other rectangular path, from $(0,0)$ to $(0,1)$ to $(1,1)$; then

$$\int_C (x\,dy - y\,dx) = \int_0^1 0\,dy + \int_0^1 -1\,dx = -1.$$

EXERCISES

1. Evaluate each of the following line integrals.

   (a) $\int_C (y\,dx + x\,dy)$ where C is the locus of $y = x^2$ $(0 \le x \le 2)$
   (b) $\int_C (y^2\,dx + xy\,dy)$ where C is the square with vertices $(\pm 1, \pm 1)$, traversed counterclockwise
   (c) $\int_C (y\,dx - x\,dy)$ where C is the unit circle, traversed counterclockwise
   (d) $\int_C (x^2y^2\,dx - xy^3\,dy)$ where C is the triangle with vertices $(0,0)$, $(1,0)$, $(1,1)$, traversed counterclockwise
   (e) $\int_C (x^2\,dx - xz\,dy + y^2\,dz)$ where C is the straight line segment from $(1,0,1)$ to $(2,3,2)$
   (f) $\int_C (z\,dx + x\,dy + y\,dz)$ where C is the locus of $x = \cos t$, $y = \cos t$, $z = \sqrt{2}\sin t$ $(0 \le t \le \pi/2)$

2. Evaluate

$$\int_{(0,0)}^{(1,1)} (x\,dy + y\,dx)$$

over each of the four paths considered in Example 3.

10–2 Surface integrals. Let S be an oriented, two-dimensional manifold in three-space. Let s and t be coordinates on S such that the axes of ds and dt are perpendicular on each tangent plane. Such variables s and t will be called *orthogonal coordinates* on S; for example, on the unit sphere where $\rho = 1$, ϕ and θ are orthogonal coordinates. The question of finding orthogonal coordinates will be discussed at greater length in Section 10–3. It is sufficient to say for the present that they do exist on any two-dimensional manifold (proof is omitted).

If s and t are orthogonal coordinates on S, there is a multiplier r such that

$$r\,ds \wedge dt$$

measures signed areas on the tangent planes to S. Arrange things so that $r > 0$ and (ds, dt) has the same orientation as S, otherwise interchange s

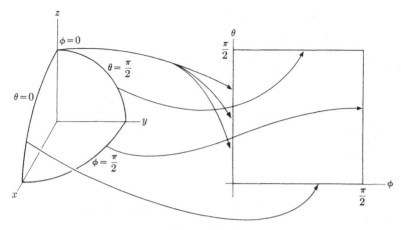

FIGURE 10–3

and t. Now, let A be the (s, t)-map of S. That is, by definition of coordinate set, the mapping

$$p \to \big(s(p), t(p)\big)$$

carries S one to one onto a set of number pairs; and since each coordinate has an interval range, this set of number pairs corresponds to a rectangle A in the plane. For example, the (ϕ, θ)-map of the unit sphere is shown in Fig. 10–3.

Transfer the variables from S to A. That is, if p on S is mapped into p' in A, define r', s', and t' on A by

$$r'(p') = r(p), \qquad s'(p') = s(p), \qquad t'(p') = t(p).$$

Now, define the *surface-area variable* σ on subsets S_0 of S by

$$\sigma(S_0) = \iint_{A_0} r' \, ds' \wedge dt', \tag{5}$$

where A_0 is the (s, t)-map of S_0. Computation of surface area will be discussed in Section 10–3. The only point here is that (5) defines σ in terms of a double integral of the type defined in Section 9–3. In terms of the one-dimensional analogy in Section 10–1, items (ii) and/or (iii) have been covered so far.

The *surface integral*,

$$\iint_S w \, du \wedge dv \tag{6}$$

may now be defined as follows. Partition S into two-dimensional submanifolds $S_1, S_2, \ldots, S_n$, and in each S_i choose a point p_i (Fig. 10–4). On

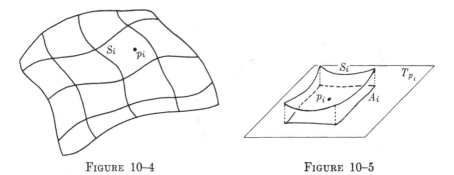

FIGURE 10–4   FIGURE 10–5

each tangent plane T_{p_i} construct a set A_i whose orientation agrees with that of S and for which

$$\sigma(S_i) = \alpha(A_i).$$

Note: σ is defined by (5), and α means plane area. Figure 10–5 shows the general idea, but note that there is no general geometric method for constructing A_i. As a rule, there is no area-preserving "unrolling" process as in the case of arc length. However, if $\alpha(A_i)$ is given, then $(du \wedge dv)_{p_i}(A_i)$ is determined; hence define

$$\iint_S w \, du \wedge dv = \lim \sum_{i=L}^{n} w(p_i)(du \wedge dv)_{p_i}(A_i).$$

The limiting process is the usual one; the maximum diameter of the S_i must tend to zero.

To evaluate a surface integral, transfer it to a coordinate map. For the integral (6), one would probably choose the (u, v)-map, but this is not essential. In any case, let M be the map of S and define the primed variables on M as usual. If (u', v') is an admissible coordinate set on M, then (the proof is omitted)

$$\iint_S w \, du \wedge dv = \iint_M w' \, du' \wedge dv'. \tag{7}$$

The integral on the right is a double integral of the type studied in Chapter 9. It may be reduced to an iterated integral and evaluated.

Note that although surface area must be defined before the surface integral is, a given surface integral may have nothing to do with surface area. Furthermore, surface areas are not needed to compute a surface integral. This is quite analogous to the one-dimensional situation. Arc length is needed to define a line integral, but line integrals give many things other than arc length, and one does not need to know the length of a curve to evaluate an integral over it.

Finally, the formula for change of variable is the same as for double integrals in the plane. Since

$$du \wedge dv = \det D_{[^r_s]}\begin{bmatrix} u \\ v \end{bmatrix} dr \wedge ds$$

over the tangent bundle for S, it follows that

$$\iint_S w \, du \wedge dv = \iint_S w \det D_{[^r_s]}\begin{bmatrix} u \\ v \end{bmatrix} dr \wedge ds \tag{8}$$

because the approximating sums are not changed by this substitution.

EXAMPLES

1. Evaluate

$$\iint_S (x \, dy \wedge dz + y \, dz \wedge dx + z \, dx \wedge dy),$$

where S is the hemisphere on which $z = \sqrt{1 - x^2 - y^2}$ with positive orientation with respect to (dx, dy). The (x, y)-map of S is the unit circle with positive orientation; thus the last term is computed at once by (3):

$$\iint_S z \, dx \wedge dy = \int_{-1}^{1} \int_{-\sqrt{1-y^2}}^{\sqrt{1-y^2}} \sqrt{1 - x^2 - y^2} \, dx \, dy$$

$$= \int_0^{2\pi} \int_0^1 r\sqrt{1 - r^2} \, dr \, d\theta$$

$$= \int_0^{2\pi} -\tfrac{1}{3}(1 - r^2)^{3/2} \Big|_0^1 \, d\theta = \frac{2\pi}{3}.$$

In projecting onto the yz-plane one must divide S into two parts. Let S_1 be the $x \geq 0$ side of S; this projects onto the yz-plane into the oriented-plane semicircle M_1 shown in Fig. 10–6. Thus,

$$\iint_{S_1} x \, dy \wedge dz = \iint_{M_1} \sqrt{1 - y^2 - z^2} \, dy \wedge dz$$

$$= \int_0^1 \int_{-\sqrt{1-z^2}}^{\sqrt{1-z^2}} \sqrt{1 - y^2 - z^2} \, dy \, dz = \frac{\pi}{3}.$$

The integral is evaluated by noting that it is half the one computed above. Now S_2, the $x \leq 0$ side of S, projects onto the yz-plane into the oriented-

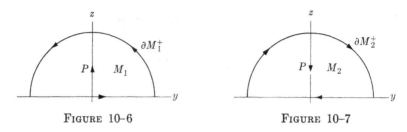

FIGURE 10-6   FIGURE 10-7

plane semicircle M_2 shown in Fig. 10-7. Thus,

$$\iint_{S_2} x \, dy \wedge dz = \iint_{M_2} -\sqrt{1 - y^2 - z^2} \, dy \wedge dz$$

$$= \int_1^0 \int_{-\sqrt{1-z^2}}^{\sqrt{1-z^2}} \sqrt{1 - y^2 - z^2} \, dy \, dz = \frac{\pi}{3},$$

and

$$\iint_S x \, dy \wedge dz = \frac{2\pi}{3}.$$

The middle term leads to the same computations; hence

$$\iint_S (x \, dy \wedge dz + y \, dz \wedge dx + z \, dx \wedge dy) = 2\pi.$$

2. The hemisphere S of Example 1 may be described parametrically with the spherical-coordinate angle variables as parameters:

$$x = \sin \phi \cos \theta, \qquad y = \sin \phi \sin \theta, \qquad z = \cos \phi,$$

$$0 \le \phi \le \pi/2, \qquad 0 \le \theta \le 2\pi.$$

The (ϕ, θ)-map of S is a rectangle; so the integral in Example 1 may be reduced to an iterated integral over this rectangle. Note that

$$\det D_{\left[\begin{smallmatrix}\phi\\\theta\end{smallmatrix}\right]} \begin{bmatrix} x \\ y \end{bmatrix} = \begin{vmatrix} \cos \phi \cos \theta & -\sin \phi \sin \theta \\ \cos \phi \sin \theta & \sin \phi \cos \theta \end{vmatrix} = \cos \phi \sin \phi,$$

$$\det D_{\left[\begin{smallmatrix}\phi\\\theta\end{smallmatrix}\right]} \begin{bmatrix} z \\ x \end{bmatrix} = \begin{vmatrix} -\sin \phi & 0 \\ \cos \phi \cos \theta & -\sin \phi \sin \theta \end{vmatrix} = \sin^2 \phi \sin \theta,$$

$$\det D_{\left[\begin{smallmatrix}\phi\\\theta\end{smallmatrix}\right]} \begin{bmatrix} y \\ z \end{bmatrix} = \begin{vmatrix} \cos \phi \sin \theta & \sin \phi \cos \theta \\ -\sin \phi & 0 \end{vmatrix} = \sin^2 \phi \cos \theta;$$

thus

$$x\, dy \wedge dz = \sin \phi \cos \theta \sin^2 \phi \cos \theta\, d\phi \wedge d\theta = \sin^3 \phi \cos^2 \theta\, d\phi \wedge d\theta,$$

$$y\, dz \wedge dx = \sin \phi \sin \theta \sin^2 \phi \sin \theta\, d\phi \wedge d\theta = \sin^3 \phi \sin^2 \theta\, d\phi \wedge d\theta,$$

$$z\, dx \wedge dy = \cos \phi \cos \phi \sin \phi\, d\phi \wedge d\theta = \cos^2 \phi \sin \phi\, d\phi \wedge d\theta.$$

Adding these, one obtains for the coefficient of $d\phi \wedge d\theta$,

$$\sin \phi \sin^2 \phi \,(\cos^2 \theta + \sin^2 \theta) + \sin \phi \cos^2 \phi = \sin \phi;$$

hence by (8), the integral in Example 1 becomes

$$\iint_S \sin \phi\, d\phi \wedge d\theta = \int_0^{2\pi} \int_0^{\pi/2} \sin \phi\, d\phi\, d\theta = 2\pi.$$

3. In Example 1,

$$\iint_S x\, dy \wedge dz$$

could be computed by integrating over the front half of the hemisphere and doubling for reasons of symmetry. Note that for this same surface,

$$\iint_S x^2\, dy \wedge dz = 0,$$

because the sign of x^2 does not change from front to back, but the orientation of the projection does. Symmetry is a useful tool in the computation of surface integrals, but habits formed by using symmetry in double integrals in the plane may lead to errors. Note that the integral of the odd power doubles, whereas the integral of even power cancels.

EXERCISES

1. Compute each of the following surface integrals.
   (a) $\iint_S (dy \wedge dz + dz \wedge dx + dx \wedge dy)$, where S is the hemisphere of Example 1
   (b) $\iint_S (x\, dy \wedge dz + y\, dz \wedge dx + z\, dx \wedge dy)$, where S is the triangle with corners at $(1, 0, 0)$, $(0, 1, 0)$, and $(0, 0, 1)$ oriented so that on the boundary these three points come in the indicated order
   (c) $\iint_S xz\, dy \wedge dz$ where, S is the cylindrical surface on which $x^2 + y^2 = 1$, $0 \le z \le 1$; and the $z = 0$ boundary has counterclockwise orientation
   (d) $\iint_S (x\, dy \wedge dz + y\, dz \wedge dx + z\, dx \wedge dy)$, where S is the conical surface on which $z^2 = x^2 + y^2$, $0 \le z \le 1$, positively oriented with respect to (dx, dy)

2. Each of the following suggests a parametric representation for the surface in the corresponding part of Exercise 1. Set up each of the integrals in Exercise 1 in terms of the indicated parameter variables; compute and check.

(a) $x = \sin\phi\cos\theta,\ y = \sin\phi\sin\theta,\ z = \cos\phi$
(b) $x = u + v,\ y = u - v,\ z = 1 - 2u$
(c) $x = \cos\theta,\ y = \sin\theta,\ z = z$
(d) $x = r\cos\theta,\ y = r\sin\theta,\ z = r$

3. Let S be the hemisphere of Example 1. Give a symmetry argument (do not integrate) to show that each of the following integrals is zero.

(a) $\iint_S dz \wedge dx$　　　　　　　(b) $\iint_S y^2\, dz \wedge dx$
(c) $\iint_S xyz\, dx \wedge dy$　　　　　(d) $\iint_S x^2 z\, dy \wedge dz$
(e) $\iint_S xz^2\, dx \wedge dy$　　　　　(f) $\iint_S x^2 y\, dy \wedge dx$

10–3 Surface area. The definition of surface area appears in Section 10–2 as an essential part of the definition of the surface integral. However, many surface integrals are not directly connected with surface area. The present section is concerned with specific determination of surface areas.

One could write as general formulas for arc length and plane area

$$s(C) = \int_C ds, \qquad \alpha(A) = \iint_A d\alpha.$$

At first glance these expressions do not seem very informative, but they do serve to isolate the problem. Given a coordinate u on C, find a multiplier t such that $ds = t\, du$, and then you can find arc length. Given coordinates u and v in the plane, find a multiplier t such that $d\alpha = t\, du \wedge dv$, and then you can find plane areas. In each case, it is a problem in geometric significance of differentials.

Similarly,

$$\sigma(S) = \iint_S d\sigma,$$

and the problem is to find exterior-product representations for $d\sigma$. The definition of surface area yields the most direct solution to this problem. If u and v are orthogonal coordinates on S, find multipliers r and s such that $r\, du$ and $s\, dv$ give orthonormal component variables on the planes tangent to S; then

$$d\sigma = rs\, du \wedge dv.$$

Frequently this can be set up by inspection (Example 2 below). However, the following questions seem pertinent.

(i) How can one tell when coordinates are orthogonal?
(ii) If the multipliers are not obvious geometrically, how can one find them analytically?
(iii) How can one get surface area from nonorthogonal coordinates?

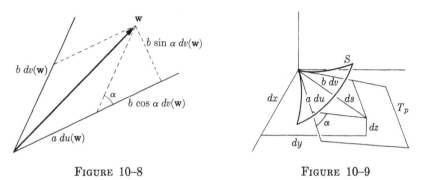

FIGURE 10–8   FIGURE 10–9

There are several ways to answer each of these questions; but one systematic answer to all three is contained in the following development, which is patterned after Example 3, Section 9–2.

Suppose that Fig. 10–8 shows differential component variables on a plane tangent to S. As shown there, orthonormal component variables are given by

$$a\,du + b \cos \alpha\, dv, \qquad b \sin \alpha\, dv;$$

hence by definition

$$d\sigma = (a\,du + b \cos \alpha\, dv) \wedge (b \sin \alpha\, dv)$$

$$= ab \sin \alpha\, du \wedge dv. \tag{9}$$

From Fig. 10–9 and the law of cosines, one also has

$$ds^2 = a^2\,du^2 + 2\,ab \cos \alpha\, du\, dv + b^2\,dv^2. \tag{10}$$

However, on any tangent plane in three-space,

$$ds^2 = dx^2 + dy^2 + dz^2. \tag{11}$$

Now, if u and v are coordinates on S, then u and v together determine points of S uniquely and thus determine x, y, and z. That is,

$$x = f(u, v), \qquad y = g(u, v), \qquad z = \phi(u, v) \tag{12}$$

on S. In other words, (12) is a set of parametric equations for S—implicit function theorem: three equations in five variables; locus two-dimensional. From (12), one can compute

$$dx = \frac{\partial x}{\partial u}\,du + \frac{\partial x}{\partial v}\,dv, \qquad dy = \frac{\partial y}{\partial u}\,du + \frac{\partial y}{\partial v}\,dv, \qquad dz = \frac{\partial z}{\partial u}\,du + \frac{\partial z}{\partial v}\,dv.$$

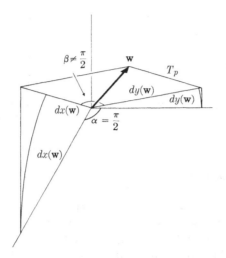

FIGURE 10–10

If these are substituted into (11) and the binomial squares are expanded, the result is

$$ds^2 = E\,du^2 + 2F\,du\,dv + G\,dv^2, \qquad (13)$$

where

$$E = \left(\frac{\partial x}{\partial u}\right)^2 + \left(\frac{\partial y}{\partial u}\right)^2 + \left(\frac{\partial z}{\partial u}\right)^2,$$

$$F = \frac{\partial x}{\partial u}\frac{\partial x}{\partial v} + \frac{\partial y}{\partial u}\frac{\partial y}{\partial v} + \frac{\partial z}{\partial u}\frac{\partial z}{\partial v},$$

$$G = \left(\frac{\partial x}{\partial v}\right)^2 + \left(\frac{\partial y}{\partial v}\right)^2 + \left(\frac{\partial z}{\partial v}\right)^2.$$

It is hardly necessary to remember these formulas for E, F, and G. In any specific example, convert (11) into (13) by substitution, and E, $2F$, and G are the coefficients.

Comparison of (10) and (13) shows the geometric significance of E, F, and G:

$$E = a^2, \qquad G = b^2, \qquad F = ab\cos\alpha. \qquad (14)$$

From this result it appears that the factor $ab\sin\alpha$ in (9) is given by

$$ab\sin\alpha = \sqrt{a^2b^2 - a^2b^2\cos^2\alpha} = \sqrt{EG - F^2}. \qquad (15)$$

The three questions raised above are now easily answered. Remember that E, F, and G are computable from (12); thus answers in terms of E, F, and G constitute an analytic solution to the problem.

(i) The coordinates u and v are orthogonal when $\cos \alpha = 0$. By (14), this is equivalent to $F = 0$.

(ii) If u and v are orthogonal, the multipliers $a = \sqrt{E}$ and $b = \sqrt{G}$ give orthonormal differential component variables.

(iii) It follows from (15) that in all cases

$$d\sigma = \sqrt{EG - F^2}\, du \wedge dv. \qquad (16)$$

Finally, consider the following question. Suppose that S is the locus of

$$z = f(x, y);$$

what is $d\sigma$ in terms of $dx \wedge dy$? Many discussions of surface area begin with this problem. This approach is basically backward, because (dx, dy) is an unnatural component system on the tangent bundle for S. On slanting tangent planes, dx and dy are never orthonormal (Fig. 10–10). However, the problem is easily solved from (16). Parametric equations for S are

$$x = x, \qquad y = y, \qquad z = f(x, y);$$

hence

$$ds^2 = dx^2 + dy^2 + dz^2 = dx^2 + dy^2 + \left(\frac{\partial z}{\partial x} dx + \frac{\partial z}{\partial y} dy\right)^2$$

$$= \left[1 + \left(\frac{\partial z}{\partial x}\right)^2\right] dx^2 + 2\frac{\partial z}{\partial x}\frac{\partial z}{\partial y} dx\, dy + \left[1 + \left(\frac{\partial z}{\partial y}\right)^2\right] dy^2.$$

Thus,

$$EG - F^2 = \left[1 + \left(\frac{\partial z}{\partial x}\right)^2\right]\left[1 + \left(\frac{\partial z}{\partial y}\right)^2\right] - \left(\frac{\partial z}{\partial x}\frac{\partial z}{\partial y}\right)^2$$

$$= 1 + \left(\frac{\partial z}{\partial x}\right)^2 + \left(\frac{\partial z}{\partial y}\right)^2,$$

and

$$d\sigma = \sqrt{1 + \left(\frac{\partial z}{\partial x}\right)^2 + \left(\frac{\partial z}{\partial y}\right)^2}\, dx \wedge dy.$$

Examples

1. Find the surface area of a hemisphere with radius a. The equation of the surface is

$$z = \sqrt{a^2 - x^2 - y^2};$$

so

$$\left(\frac{\partial z}{\partial x}\right)_y = \frac{-x}{\sqrt{a^2 - x^2 - y^2}}, \qquad \left(\frac{\partial z}{\partial y}\right)_x = \frac{-y}{\sqrt{a^2 - x^2 - y^2}},$$

and the rectangular coordinate formula for surface area becomes

$$\sigma(S) = \iint_S \sqrt{1 + \frac{x^2 + y^2}{a^2 - x^2 - y^2}}\, dx \wedge dy$$

$$= \int_{-a}^{a} \int_{-\sqrt{a^2-y^2}}^{\sqrt{a^2-y^2}} \frac{a}{\sqrt{a^2 - x^2 - y^2}}\, dx\, dy$$

$$= \int_0^{2\pi} \int_0^{a} \frac{ar}{\sqrt{1 - r^2}}\, dr\, d\theta = 2\pi a^2.$$

2. The hemisphere of Example 1 may be described parametrically by the equations

$$x = a \sin \phi \cos \theta, \qquad y = a \sin \phi \sin \theta, \qquad z = a \cos \phi,$$

$$0 \le \phi \le \pi/2, \qquad 0 \le \theta \le 2\pi.$$

To set up an integral for surface area in terms of these parameters one needs only to note that on the tangent bundle for the sphere,

$$a\, d\phi \qquad \text{and} \qquad a \sin \phi\, d\theta$$

yield orthonormal coordinate variables; hence by definition,

$$d\sigma = a^2 \sin \phi\, d\phi \wedge d\theta,$$

and

$$\sigma(S) = \iint_S a^2 \sin \phi\, d\phi \wedge d\theta = \int_0^{2\pi} \int_0^{\pi/2} a^2 \sin \phi\, d\phi\, d\theta = 2\pi a^2.$$

3. Find the centroid of a hemispherical surface with radius a. The centroid has coordinates $(0, 0, \bar{z})$, where

$$\bar{z} = \frac{1}{\sigma(S)} \iint_S z\, d\sigma.$$

Introduce the spherical coordinate parameters as in Example 2 so that

$$d\sigma = a^2 \sin \phi\, d\phi \wedge d\theta;$$

then

$$\bar{z} = \frac{1}{2\pi a^2} \int_0^{2\pi} \int_0^{\pi/2} a^3 \sin \phi \cos \phi\, d\phi\, d\theta = \frac{\pi a^3}{2\pi a^2} = \frac{a}{2}.$$

EXERCISES

1. Find the surface area and the centroid of each of the following surfaces.
   (a) The part of the sphere on which $x^2 + y^2 + z^2 = a^2$ above the cone on which $z^2 = x^2 + y^2$
   (b) The part of the cylinder on which $x^2 + y^2 = ax$ above the locus of $z = 0$ and inside the sphere on which $x^2 + y^2 + z^2 = a^2$
   (c) The part of the cylinder on which $x^2 + y^2 = a^2$ lying between the planes on which $y = \pm a/2$ and $z = \pm a/2$
   (d) The part of the cone on which $z^2 = x^2 + y^2$ above the locus of $z = 0$ and inside the cylinder on which $x^2 + y^2 = 2ax$

2. Each of the following suggests coordinates for a parametric representation of the surface in the corresponding part of Exercise 1. In each case, determine the surface area and centroid from this parametric representation.
   (a) Spherical coordinates ϕ, θ
   (b) Cylindrical coordinates θ, z. [*Hint:* Show that $ds^2 = a^2\,d\theta^2 + dz^2$ on this surface.]
   (c) Cylindrical coordinates θ, z
   (d) Polar coordinates r, θ. [*Hint:* Show that $ds^2 = 2\,dr^2 + r^2\,d\theta^2$ on the cone.]

3. Consider each of the following surfaces as a thin shell with mass equal to surface area, and find the force on a unit mass and the gravitational potential at the point indicated.
   (a) Hemisphere of radius a; center of sphere
   (b) Right circular cylinder with radius a, height h; point on axis at one end

4. Let S be given by the equation

$$f(x, y, z) = 0;$$

show that

$$d\sigma = \frac{\sqrt{[f_1(x, y, z)]^2 + [f_2(x, y, z)]^2 + [f_3(x, y, z)]^2}}{\pm f_3(x, y, z)}\,dx \wedge dy,$$

and discuss the question of signs.

5. Let S be given by the equation

$$z = f(r, \theta).$$

   (a) Find $d\sigma$ in terms of $dr \wedge d\theta$.
   (b) Apply the result of part (a) to the problem of Exercise 2(d) above, and compare.

10–4 Stokes type theorems. It has been pointed out (Section 3–3) that if C is a sufficiently smooth, counterclockwise, simple closed curve in the plane, then each of the line integrals

$$\int_C x\,dy \qquad \text{and} \qquad \int_C -y\,dx$$

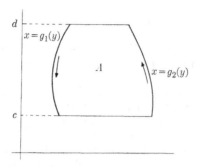

FIGURE 10–11

gives the area of the region enclosed by C. Thus, if $C = \partial A$, then

$$\alpha(A) = \int_{\partial A} x \, dy = \iint_A dx \wedge dy, \qquad (17)$$

and

$$-\alpha(A) = \int_{\partial A} y \, dx = \iint_A dy \wedge dx.$$

From these relations a pattern emerges that could be described by saying that an integral over the boundary of an oriented manifold is equal to an integral over the interior in which the integrand variable has been replaced by its differential. The expected generalized formula would be

$$\int \cdots \int_{\partial M} u \, dv_1 \wedge \cdots \wedge dv_k = \iint \cdots \int_M du \wedge dv_1 \wedge \cdots \wedge dv_k, \qquad (18)$$

where M is an oriented $(k+1)$-dimensional manifold.

The chief difficulty in proving formulas of this type lies in the fact that the basic first step requires quite drastic restrictions on the shape of the base manifold. Thus, one considers a special case first and then generalizes in successive steps. In this respect (18) is no more difficult to prove than (17); the higher-dimensional cases offer no new complications. However, to see the difficulties geometrically, it is well to consider an analytic proof of (17) first.

Case 1. The manifold A is defined by

$$c \le y \le d, \qquad g_1(y) \le x \le g_2(y)$$

(see Fig. 10–11). In this case,

$$\iint_A dx \wedge dy = \int_c^d \int_{g_1(y)}^{g_2(y)} dx \, dy = \int_c^d [g_2(y) - g_1(y)] \, dy$$

$$= \int_c^d g_2(y) \, dy + \int_d^c g_1(y) \, dy = \int_{\partial A} x \, dy$$

inasmuch as $dy = 0$ on each horizontal segment of ∂A.

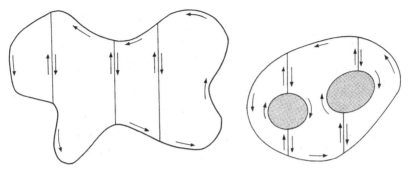

FIGURE 10–12   FIGURE 10–13

Case 2. The manifold A can be partitioned into a finite number of two-dimensional submanifolds, each of which satisfies the conditions for Case 1 (see Fig. 10–12). Apply the result of Case 1 to each of the submanifolds and add. The double integrals clearly add to that over A. Since common boundary sections for adjacent submanifolds have opposite orientations, the line integrals cancel except over ∂A, and the result follows. Note that multiply connected regions (regions with holes in them; see Fig. 10–13) are admitted under this case.

Case 3. The boundary of A has a tangent line at all but a finite number of points. (The proof is omitted.) This is the difficult extension. The method is to approximate the general figure by figures satisfying Case 2. It is easy to show that the double integrals over the simple figures converge to that over the general one, but the corresponding limit theorem for the line integrals is far from trivial.

The condition in Case 3 is frequently stated, "∂A is piecewise smooth." In this form the statement is easily generalized. If M is a $(k + 1)$-dimensional manifold in Euclidean n-space ($n \geq k + 1$), then ∂M is *piecewise smooth* provided that it has a unique tangent k-space everywhere except on a $(k - 1)$-dimensional submanifold. The complete theorem involving (18) now is as follows.

GENERALIZED STOKES THEOREM. Let M be a $(k + 1)$-dimensional manifold with a piecewise smooth boundary, and let u, v_1, v_2, $\ldots$, v_k be an admissible coordinate set on M; then

$$\int \cdots \int_{\partial M} u \, dv_1 \wedge \cdots \wedge dv_k = \iint \cdots \int_M du \wedge dv_1 \wedge \cdots \wedge dv_k.$$

For Case 1 of this proof, M is the locus of

$$\phi(v_1, v_2, \ldots, v_k) \leq 0,$$

$$f_1(v_1, v_2, \ldots, v_k) \leq u \leq f_2(v_1, v_2, \ldots, v_k).$$

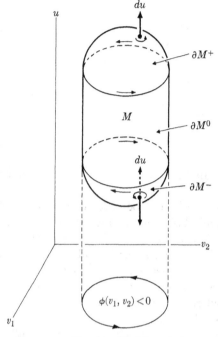

FIGURE 10–14

Intuitively, M is a portion of a cylinder with elements running in the u-direction (see Fig. 10–14 for $k = 2$). Here M has a "top,"

$$\partial M^+\colon\ u = f_2(v_1, v_2, \ldots, v_k), \qquad \phi(v_1, v_2, \ldots, v_k) < 0,$$

a "bottom,"

$$\partial M^-\colon\ u = f_1(v_1, v_2, \ldots, v_k), \qquad \phi(v_1, v_2, \ldots, v_k) < 0,$$

and "sides,"

$$\partial M^0\colon\ \phi(v_1, v_2, \ldots, v_k) = 0, \qquad f_1(v_1, v_2, \ldots, v_k) \le u \le f_2(v_1, v_2, \ldots, v_k).$$

The integral over M reduces to an iterated integral in the usual way:

$$\iint \cdots \int_M du \wedge dv_1 \wedge \cdots \wedge dv_k$$

$$= \int \cdots \int_{\phi(v_1, v_2, \ldots, v_k) \le 0} \left[\int_{\partial M^-}^{\partial M^+} du \right] dv_1 \wedge \cdots \wedge dv_k$$

$$= \int \cdots \int_{\partial M^+} u\, dv_1 \wedge \cdots \wedge dv_k + \int \cdots \int_{\partial M^-} u\, dv_1 \wedge \cdots \wedge dv_k.$$

There is a plus instead of a minus between these last two integrals because of orientation, and this is the crux of the whole matter. Suppose that the orientation of M is that of $(du, dv_1, \ldots, dv_k)$. Then, on ∂M^+, du gives the exterior normal, and $(dv_1, \ldots, dv_k)$ gives the orientation of ∂M^+. However, on ∂M^-, du gives the interior normal; hence $(dv_1, \ldots, dv_k)$ has orientation opposite to that of ∂M^-, and the sign must be changed. Finally,

$$\int \cdots \int_{\partial M^0} u \, dv_1 \wedge \cdots \wedge dv_k = 0$$

because $\phi(v_1, v_2, \ldots, v_k) = 0$ on ∂M^0; thus by the implicit function theorem, some dv_i is a linear combination of the others, and $dv_1 \wedge \cdots \wedge dv_k$ contains a repeated factor. Extension to more general manifolds will not be discussed here.

Appropriate specializations of the generalized Stokes theorem yield three important theorems of classical analysis. These are as follows.

GREEN'S THEOREM. Let A be an oriented, two-dimensional manifold in the plane with a piecewise smooth boundary, and let u and v be admissible coordinates on A; then

$$\int_{\partial A} (u \, dx + v \, dy) = \iint_A \left(\frac{\partial v}{\partial x} - \frac{\partial u}{\partial y} \right) dx \wedge dy. \tag{19}$$

Proof. Applying the generalized Stokes theorem to each term in the line integral separately, one has

$$\int_{\partial A} (u \, dx + v \, dy) = \iint_A (du \wedge dx + dv \wedge dy). \tag{20}$$

By the fundamental theorem on differentials and the rules of exterior multiplication,

$$du \wedge dx = \left(\frac{\partial u}{\partial x} dx + \frac{\partial u}{\partial y} dy \right) \wedge dx = \frac{\partial u}{\partial y} dy \wedge dx = -\frac{\partial u}{\partial y} dx \wedge dy,$$

$$dv \wedge dy = \left(\frac{\partial v}{\partial x} dx + \frac{\partial v}{\partial y} dy \right) \wedge dy = \frac{\partial v}{\partial x} dx \wedge dy;$$

hence (20) reduces to (19).

CLASSICAL STOKES THEOREM. Let S be an oriented, two-dimensional manifold in three-space with a piecewise smooth boundary, and let $u, v,$ and w be admissible coordinates on S; then

$$\int_{\partial S} (u \, dx + v \, dy + w \, dz) =$$

$$\iint_S \left[\left(\frac{\partial w}{\partial y} - \frac{\partial v}{\partial z} \right) dy \wedge dz + \left(\frac{\partial u}{\partial z} - \frac{\partial w}{\partial x} \right) dz \wedge dx + \left(\frac{\partial v}{\partial x} - \frac{\partial u}{\partial y} \right) dx \wedge dy \right].$$

Proof. Again, the generalized Stokes theorem may be applied to each term in the line integral:

$$\int_{\partial S} (u\,dx + v\,dy + w\,dz) = \iint_S (du \wedge dx + dv \wedge dy + dw \wedge dz).$$

By the fundamental theorem,

$$du \wedge dx = \left(\frac{\partial u}{\partial x} dx + \frac{\partial u}{\partial y} dy + \frac{\partial u}{\partial z} dz\right) \wedge dx$$

$$= \frac{\partial u}{\partial y} dy \wedge dx + \frac{\partial u}{\partial z} dz \wedge dx$$

$$= -\frac{\partial u}{\partial y} dx \wedge dy + \frac{\partial u}{\partial z} dz \wedge dx.$$

Thus, the first term in the line integral yields the third and sixth terms in the surface integral. Others are paired off in a similar manner.

DIVERGENCE THEOREM. Let V be an oriented, three-dimensional manifold in three-space with a piecewise smooth boundary, and let u, v, and w be admissible coordinates on V; then

$$\iint_{\partial V} (u\,dy \wedge dz + v\,dz \wedge dx + w\,dx \wedge dy)$$

$$= \iiint_V \left(\frac{\partial u}{\partial x} + \frac{\partial v}{\partial y} + \frac{\partial w}{\partial z}\right) dx \wedge dy \wedge dz.$$

Proof. The generalized Stokes theorem yields the triple integral

$$\iiint_V (du \wedge dy \wedge dz + dv \wedge dz \wedge dx + dw \wedge dx \wedge dy).$$

This reduces to the required form by the fundamental theorem. Take the middle term, for example:

$$dv \wedge dz \wedge dx = \left(\frac{\partial v}{\partial x} dx + \frac{\partial v}{\partial y} dy + \frac{\partial v}{\partial z} dz\right) \wedge dz \wedge dx$$

$$= \frac{\partial v}{\partial y} dy \wedge dz \wedge dx = \frac{\partial v}{\partial y} dx \wedge dy \wedge dz.$$

Other terms are checked in a similar manner.

The reason for the name "divergence theorem" should be apparent. Since

$$\text{div} (u\mathbf{i} + v\mathbf{j} + w\mathbf{k}) = \frac{\partial u}{\partial x} + \frac{\partial v}{\partial y} + \frac{\partial w}{\partial z},$$

the triple integral is that of the divergence of a vector variable. Further comments on the vector significance of these theorems will appear in Section 10–5.

In each theorem of this section it is required that the variables that appear be admissible coordinates on the manifold in question. This places two important restrictions on them.

(i) They must have continuous partial derivatives.

(ii) An appropriate set of them must map the manifold one to one into ordered sets of numbers.

Similar restrictions have been imposed before, and minor infractions of the rules have caused little trouble. For example,

$$\iint_A r \, dr \wedge d\theta$$

is presumably defined only if (r, θ) is an admissible coordinate set on A; but this integral gives $\alpha(A)$ quite satisfactorily when A contains the origin even though r and θ are not coordinates there. On the other hand, Stokes type theorems are peculiarly sensitive to multivalued mappings and discontinuities; thus the reader is warned to be scrupulous in checking hypotheses as he uses these theorems. Examples 2 and 3 below illustrate some of the pitfalls.

EXAMPLES

1. Evaluate

$$\int_C [(2x^3 - y^3) \, dx + (x^3 + y^3) \, dy],$$

where C is the unit circle on which $x^2 + y^2 = 1$, taken in a counterclockwise direction. By Green's theorem,

$$\int_C [(2x^3 - y^3) \, dx + (x^3 + y^3) \, dy] = \iint_A (3x^2 + 3y^2) \, dx \wedge dy$$

$$= \iint_A 3r^3 \, dr \wedge d\theta = \int_0^{2\pi} \int_0^1 3r^3 \, dr \, d\theta = \frac{3\pi}{2}.$$

2. By exactly the same argument used in the proof of Green's theorem,

$$\int_{\partial A} (u \, dr + v \, d\theta) = \iint_A \left[\left(\frac{\partial v}{\partial r} \right)_\theta - \left(\frac{\partial u}{\partial \theta} \right)_r \right] dr \wedge d\theta \qquad (21)$$

(note that in the double integral here there appears only $dr \wedge d\theta$, not $r \, dr \wedge d\theta = d\alpha$) provided that A is the locus of

$$f_1(\theta) \leq r \leq f_2(0), \qquad a \leq \theta \leq b.$$

Of course, it is also possible to piece together regions of this type, but it must be borne in mind that θ is not a genuine coordinate in the plane. The loci of $\theta = 0$ and $\theta = 2\pi$ are the same line.

For example,

$$\frac{\partial}{\partial r}\left(\frac{1}{2}r^2\right) = r, \qquad \frac{\partial}{\partial \theta}(r\theta) = r;$$

hence by (21),

$$\int_{\partial A} \tfrac{1}{2}r^2 \, d\theta = \iint_A r \, dr \wedge d\theta = \alpha(A), \qquad (22)$$

$$\int_{\partial A} -r\theta \, dr = \iint_A r \, dr \wedge d\theta = \alpha(A), \qquad (23)$$

wherever the formula applies. A circle with radial lines in and out, as in Fig. 10–15, is an $r\theta$-rectangle; thus (21) does apply to it, but the line integrals along the radii do not cancel because going out $\theta = 0$, and coming in $\theta = 2\pi$. For this figure, (22) reduces to

$$\int_0^{2\pi} \tfrac{1}{2}a^2 \, d\theta = \pi a^2,$$

and (23) becomes

$$\int_a^0 -2\pi r \, dr = \pi a^2.$$

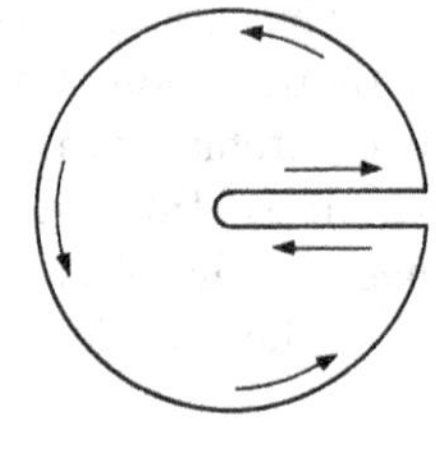

FIGURE 10–15

Note that to take only the circle as ∂A would yield zero on the left side of (23).

3. Let A be the interior of the unit circle, and consider

$$\int_{\partial A} \frac{x \, dy - y \, dx}{x^2 + y^2}.$$

Observe that

$$\frac{\partial}{\partial x}\frac{x}{x^2 + y^2} = \frac{y^2 - x^2}{(x^2 + y^2)^2}, \qquad \frac{\partial}{\partial y}\frac{-y}{x^2 + y^2} = \frac{y^2 - x^2}{(x^2 + y^2)^2};$$

thus purely formally, Green's theorem yields the answer zero. However, the variables involved are discontinuous at the origin; hence Green's theorem does not apply. On ∂A, $x^2 + y^2 = 1$, and

$$x \, dy - y \, dx = \cos \theta \, (\sin \theta \, dr + \cos \theta \, d\theta)$$
$$- \sin \theta \, (\cos \theta \, dr - \sin \theta \, d\theta) = d\theta;$$

thus

$$\int_{\partial A} \frac{x \, dy - y \, dx}{x^2 + y^2} = \int_0^{2\pi} d\theta = 2\pi.$$

The single discontinuity in the interior affects the value of the line integral.

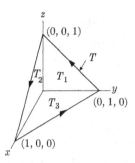

FIGURE 10–16

4. Let T be the plane triangle in three-space having vertices at $(1, 0, 0)$, $(0, 1, 0)$, and $(0, 0, 1)$. Let T be oriented so that on ∂T these vertices come in this order (Fig. 10–16). Use Stokes' theorem to evaluate

$$\int_{\partial T} (z - y)\, dx + (x - z)\, dy + (y - x)\, dz.$$

The first term in the surface integral will be

$$\left[\frac{\partial}{\partial y} (y - x) - \frac{\partial}{\partial z} (x - z)\right] dy \wedge dz = 2\, dy \wedge dz.$$

Other terms follow a similar pattern, and the surface integral is

$$2 \iint_T (dy \wedge dz + dz \wedge dx + dx \wedge dy)$$
$$= 2 \iint_{T_1} dy \wedge dz + 2 \iint_{T_2} dz \wedge dx + 2 \iint_{T_3} dx \wedge dy$$
$$= 2\alpha(T_1) + 2\alpha(T_2) + 2\alpha(T_3) = 3.$$

5. Let V be the interior of the unit sphere. Use the divergence theorem to find

$$\iint_{\partial V} (x^2\, dy \wedge dz + 2xy\, dz \wedge dx + 2xz\, dx \wedge dy).$$

Note that

$$\frac{\partial}{\partial x} x^2 + \frac{\partial}{\partial y} 2xy + \frac{\partial}{\partial z} 2xz = 6x;$$

hence the triple integral is

$$\iiint_V 6x\, dx \wedge dy \wedge dz = 6\bar{x}\tau(V) = 0$$

because $\bar{x} = 0$ for the sphere, by symmetry.

EXERCISES

1. Evaluate each of the following integrals by Green's theorem.

(a) $\int_C [(x^2 + y^2)\, dx + 2xy\, dy]$, where C is any simple closed curve in the plane

(b) $\int_C (ay\, dx + bx\, dy)$, where C is any counterclockwise, simple closed curve in the plane

(c) $\int_C [(x^2 + y^2)\, dx - 2xy\, dy]$, where C is the circle on which $x^2 + y^2 = 1$

(d) The integral in part (c), where C is the circle on which $x^2 + y^2 = 2x$

(e) The integral in part (c), where C is the circle on which $x^2 + y^2 = 2y$ with counterclockwise orientation

(f) $\int_C [2y^2 \cos^2 x\, dx - y(2x - \sin 2x)\, dy]$, where C is any counterclockwise, simple closed curve in the plane

(g) $\int_C [(\sin xy + xy \cos xy)\, dx + x^2 \cos xy\, dy]$, where C is a simple closed curve in the plane

(h) $\int_C (y^3\, dx - x^3\, dy)$, where C is the circle on which $x^2 + y^2 = 1$ with counterclockwise orientation

(i) $\int_C [xy^6\, dx + (3x^2y^5 + 6x)\, dy]$, where C is any counterclockwise, simple closed curve in the plane

2. Let C be the intersection of the cylinder on which $x^2 + y^2 = 1$ and the plane on which $z = x$, oriented so that y increases for positive x. Use Stokes' theorem to evaluate each of the following integrals.

(a) $\int_C (-3y\, dx + 3x\, dy + dz)$

(b) $\int_C [2xy^2z\, dx + 2x^2yz\, dy + (x^2y^2 - 2z)\, dz]$

(c) $\int_C (xy\, dx + x^2\, dy + yz\, dz)$

3. Let V be the interior of the unit sphere with right-handed orientation. Use the divergence theorem to evaluate each of the following exercises.

(a) $\iint_{\partial V} (x\, dy \wedge dz + y\, dz \wedge dx + z\, dx \wedge dy)$

(b) $\iint_{\partial V} (e^x \cos y\, dy \wedge dz - e^x \sin y\, dz \wedge dx)$

(c) $\iint_{\partial V} (xy\, dy \wedge dz + yz\, dz \wedge dx + xz\, dx \wedge dy)$

4. Show that each of the following surface integrals gives the volume of a right-handed V that is suitably well behaved with respect to the coordinates involved.

(a) $\iint_{\partial V} \frac{1}{2}r^2\, d\theta \wedge dz$ (b) $\iint_{\partial V} r\theta\, dz \wedge dr$

(c) $\iint_{\partial V} rz\, dr \wedge d\theta$ (d) $\iint_{\partial V} \frac{1}{3}\rho^3 \sin \phi\, d\phi \wedge d\theta$

(e) $\iint_{\partial V} \rho^2 \cos \phi\, d\rho \wedge d\theta$ (f) $\iint_{\partial V} \rho^2\theta \sin \phi\, d\rho \wedge d\phi$

5. Use each of formulas (a), (b), and (c) in Exercise 4 to find the volume of the figure described by

$$0 \le r \le a, \qquad 0 \le \theta \le \pi/2, \qquad 0 \le z \le b.$$

See Fig. 10–17.

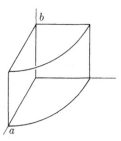

FIGURE 10–17

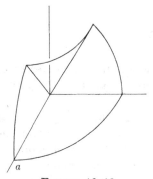

FIGURE 10–18

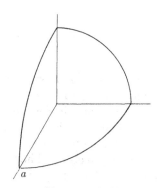

FIGURE 10–19

6. Use each of formulas (d), (e), and (f) in Exercise 4 to find the volume of the figure described by

$$0 \leq \rho \leq a, \qquad \frac{\pi}{4} \leq \phi \leq \frac{\pi}{2}, \qquad 0 \leq \theta \leq \frac{\pi}{2}.$$

See Fig. 10–18.

7. Apply formulas (d) and (f) of Exercise 4 to find the volume of an octant of a sphere (Fig. 10–19):

$$0 \leq \rho \leq a, \qquad 0 \leq \phi \leq \frac{\pi}{2}, \qquad 0 \leq \theta \leq \frac{\pi}{2}.$$

Discuss the difficulty that arises in applying formula (e) to this problem.

10–5 Vector integral calculus. The Stokes type theorems of Section 10–4 deal in transformations of differential forms. These differentials come from coordinate variables, so Section 10–4 could be classified under "theory of coordinates." Exactly which coordinates are used does not matter; for example,

$$\iint_{\partial V} (u \, d\phi \wedge d\theta + v \, d\theta \wedge d\rho + w \, d\rho \wedge d\phi)$$

$$= \iiint_V \left(\frac{\partial u}{\partial \rho} + \frac{\partial v}{\partial \phi} + \frac{\partial w}{\partial \theta} \right) d\rho \wedge d\phi \wedge d\theta. \qquad (24)$$

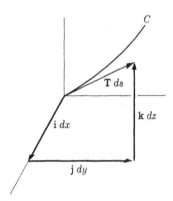

FIGURE 10–20

This is a perfectly good specialization of the generalized Stokes theorem, but its physical significance is not so important as that of a slightly different transformation on spherical-coordinate integrals.

In particular, the classical Stokes theorem and the divergence theorem applied to rectangular coordinates may be written as integral transformations involving vectors. The strictly vector formulas are, of course, coordinate-free, and it is in this form that the physical significance of the theorems stands out more clearly. Thus, for example, (24) above is a perfectly good Stokes type theorem, but it would not ordinarily be called the divergence theorem. The theory of Section 10–4 may be applied to spherical-coordinate differential forms other than

$$u \, d\phi \wedge d\theta + v \, d\theta \wedge d\rho + w \, d\rho \wedge d\phi.$$

The form that would yield the divergence theorem is the one with the same physical significance as

$$r \, dy \wedge dz + s \, dz \wedge dx + t \, dx \wedge dy,$$

and the best way to see this form is to put the whole thing into vectors from the rectangular form, and then convert back to spherical coordinates if desired.

First, note that if $\mathbf{T}$ is the unit tangent variable on a curve C, then (Fig. 10–20)

$$\mathbf{T} \, ds = \mathbf{i} \, dx + \mathbf{j} \, dy + \mathbf{k} \, dz. \qquad (25)$$

Next, suppose that $\mathbf{N}$ is the unit normal variable on a surface S, then

$$\mathbf{N} \, d\sigma = \mathbf{i} \, dy \wedge dz + \mathbf{j} \, dz \wedge dx + \mathbf{k} \, dx \wedge dy. \qquad (26)$$

To understand this result, recall that $d\sigma = ab \, du \wedge dv$ whenever $a \, du$ and

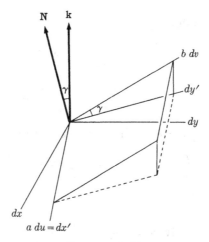

FIGURE 10–21

$b\,dv$ are orthonormal components on the planes tangent to S. Any such orthonormal components will do. Choose them as in Fig. 10–21 so that

$$a\,du = dx'$$

measures components parallel to the xy-plane. Then (note Fig. 10–21 again)

$$\mathbf{N} \cdot \mathbf{k} = \cos \gamma,$$

but

$$dy' = b\,dv \cos \gamma;$$

hence

$$\mathbf{N} \cdot \mathbf{k}\,d\sigma = \cos \gamma\, ab\,du \wedge dv = (a\,du) \wedge (b\,dv \cos \gamma) = dx' \wedge dy'.$$

However, $dx' \wedge dy' = dx \wedge dy$ because a rotation does not affect exterior products. Thus,

$$\mathbf{k} \cdot (\mathbf{N}\,d\sigma) = dx \wedge dy,$$

which is to say that $dx \wedge dy$ is the $\mathbf{k}$-component of $\mathbf{N}\,d\sigma$, as stated in (26). The other components obviously follow the same pattern.

Now, let

$$\mathbf{w} = w_1\mathbf{i} + w_2\mathbf{j} + w_3\mathbf{k}.$$

It follows from (25) that

$$w_1\,dx + w_2\,dy + w_3\,dz = \mathbf{w} \cdot \mathbf{T}\,ds.$$

Furthermore,

$$\mathbf{curl}\,\mathbf{w} = \left(\frac{\partial w_3}{\partial y} - \frac{\partial w_2}{\partial z}\right)\mathbf{i} + \left(\frac{\partial w_1}{\partial z} - \frac{\partial w_3}{\partial x}\right)\mathbf{j} + \left(\frac{\partial w_2}{\partial x} - \frac{\partial w_1}{\partial y}\right)\mathbf{k};$$

thus in the light of (26),

$$\left(\frac{\partial w_3}{\partial y} - \frac{\partial w_2}{\partial z}\right) dy \wedge dz + \left(\frac{\partial w_1}{\partial z} - \frac{\partial w_3}{\partial x}\right) dz \wedge dx$$

$$+ \left(\frac{\partial w_2}{\partial x} - \frac{\partial w_1}{\partial y}\right) dx \wedge dy = \mathbf{curl\ w} \cdot \mathbf{N}\ d\sigma.$$

Therefore, the classical Stokes theorem (see Section 10–4) may be written

$$\int_{\partial S} \mathbf{w} \cdot \mathbf{T}\ ds = \iint_S \mathbf{curl\ w} \cdot \mathbf{N}\ d\sigma. \tag{27}$$

Recall that

$$\mathrm{div}\ \mathbf{w} = \frac{\partial w_1}{\partial x} + \frac{\partial w_2}{\partial y} + \frac{\partial w_3}{\partial z}\ ;$$

and note that by (26) above,

$$w_1\ dy \wedge dz + w_2\ dz \wedge dx + w_3\ dx \wedge dy = \mathbf{w} \cdot \mathbf{N}\ d\sigma.$$

Thus, the divergence theorem (see Section 10–4) takes the form

$$\iint_{\partial V} \mathbf{w} \cdot \mathbf{N}\ d\sigma = \iiint_V \mathrm{div}\ \mathbf{w}\ d\tau. \tag{28}$$

EXAMPLES

1. *Physical interpretation.* Let **v** be the velocity variable in a fluid dynamics problem. That is, at a given instant of time, the fluid at the point p of space is moving with velocity $\mathbf{v}(p)$. The dependence of **v** on time will not be considered here.

If the fluid moves with constant velocity **v** for one unit of time, then in Fig. 10–22 the flow through the parallelogram A will just fill the parallelepiped P.

The volume of the parallelepiped is clearly

$$\mathbf{v} \cdot \mathbf{N}\alpha(A),$$

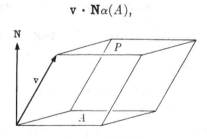

FIGURE 10–22

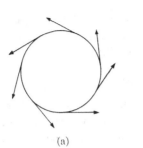 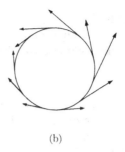

(a)   (b)

FIGURE 10–23

where **N** is the unit normal to A. Thus, it would appear that

$$\iint_{\partial V} \mathbf{v} \cdot \mathbf{N} \, d\sigma$$

gives the net flow per unit time through the surface V. This can be confirmed (the details are omitted). Thus, by (28),

$$\iiint_V \operatorname{div} \mathbf{v} \, d\tau$$

gives net flow per unit time out of V. Hence, just as mass density is a factor whose integral with respect to $d\tau$ gives mass, so div **v** is a *flow-density* variable (net flow per unit time per unit volume) inasmuch as its integral with respect to $d\tau$ yields the flow.

If the fluid is incompressible, the quantity flowing into any region V is equal to that flowing out. So

$$\operatorname{div} \mathbf{v} = 0$$

characterizes *incompressible flow*. In general, div **v** > 0 indicates expansion, and div **v** < 0 indicates compression.

Now, look at Fig. 10–23, and regard the arrows as indicating **v** at various points. In Fig. 10–23(a) there is obviously counterclockwise circulation around C. The issue is not so clear-cut in Fig. 10–23(b), but there seems to be an overall net circulation counterclockwise in the sense that the tangential components of velocity in the counterclockwise direction outweigh those in the clockwise direction. This observation suggests

$$\int_C \mathbf{v} \cdot \mathbf{T} \, ds$$

as a definition of *circulation* around C. This being the case, it follows from (27) that

$$\iint_S \operatorname{curl} \mathbf{v} \cdot \mathbf{N} \, d\sigma$$

gives the circulation around ∂S. Thus, **curl v** $\cdot$ **N** is a *circulation-density* variable. It gives circulation per unit area in a plane normal to **N**. *Irrotational flow* is thus characterized by the condition

$$\text{curl } \mathbf{v} = 0.$$

This example indicates a reason for the terminology, divergence and curl. However, a more careful look shows that the terms are misapplied. One speaks of the divergence or curl of the *velocity*, meaning measures of the local tendency of the *flow* to diverge or to curl.

2. *Other coordinate systems.* The descriptions of **T** ds and **N** $d\sigma$ given by (25) and (26) above are readily generalized. Let

$$(a\, dt,\ b\, du,\ c\, dv)$$

be any orthonormal component system on the tangents to three-space with basis vectors

$$(\mathbf{b}_t,\ \mathbf{b}_u,\ \mathbf{b}_v).$$

It follows by exactly the arguments used to derive (25) and (26) that

$$\mathbf{T}\, ds = a\, dt\, \mathbf{b}_t + b\, du\, \mathbf{b}_u + c\, dv\, \mathbf{b}_v,$$

and

$$\mathbf{N}\, d\sigma = bc\, du \wedge dv\, \mathbf{b}_t + ca\, dv \wedge dt\, \mathbf{b}_u + ab\, dt \wedge du\, \mathbf{b}_v.$$

In words, the **b**-component of **T** ds is the arc-length differential in the **b**-direction; the **b**-component of **N** $d\sigma$ is the area differential in a plane perpendicular to **b**.

Thus, for example, in spherical coordinates,

$$\mathbf{N}\, d\sigma = \rho^2 \sin\phi\, d\phi \wedge d\theta\, \mathbf{b}_\rho + \rho \sin\phi\, d\theta \wedge d\rho\, \mathbf{b}_\phi + \rho\, d\rho \wedge d\phi\, \mathbf{b}_\theta;$$

hence if

$$\mathbf{w} = w_1\mathbf{b}_\rho + w_2\mathbf{b}_\phi + w_3\mathbf{b}_\theta,$$

then

$$\mathbf{w} \cdot \mathbf{N}\, d\sigma = w_1\rho^2 \sin\phi\, d\phi \wedge d\theta + w_2\rho \sin\phi\, d\theta \wedge d\rho + w_3\rho\, d\rho \wedge d\phi,$$

and the divergence theorem concerns a transformation of this form. The transformation proceeds according to (25) of Section 10–4, however; thus the volume integral is

$$\iiint_V \left[\frac{\partial}{\partial\rho}(\rho^2 \sin\phi w_1) + \frac{\partial}{\partial\phi}(\rho \sin\phi w_2) + \frac{\partial}{\partial\theta}(\rho w_3) \right] d\rho \wedge d\phi \wedge d\theta$$

$$= \iiint_V \frac{1}{\rho^2 \sin\phi} \left[\frac{\partial}{\partial\rho}(\rho^2 \sin\phi w_1) + \frac{\partial}{\partial\phi}(\rho \sin\phi w_2) + \frac{\partial}{\partial\theta}(\rho w_3) \right] d\tau.$$

Thus,

$$\text{div } \mathbf{w} = \frac{1}{\rho^2 \sin \phi} \left[\frac{\partial}{\partial \rho} (\rho^2 \sin \phi w_1) + \frac{\partial}{\partial \phi} (\rho \sin \phi w_2) + \frac{\partial}{\partial \theta} (\rho w_3) \right].$$

The reader should expand this equation and compare it with the expression for div $\mathbf{w}$ derived by another method in Example 1, Section 7–5.

The generalized Stokes theorem together with the vector forms of the classical theorems are a powerful tool for transforming vector-derivative operators. Note the steps taken above. The problem was to find div $\mathbf{w}$ in spherical coordinates.

(i) Write the divergence theorem in vector form:

$$\iint_{\partial V} \mathbf{w} \cdot \mathbf{N} \, d\sigma = \iiint_V \text{div } \mathbf{w} \, d\tau.$$

(ii) Find $\mathbf{w} \cdot \mathbf{N} \, d\sigma$ in spherical coordinates and thus set up the integral over ∂V.

(iii) Transform to an integral over V by the generalized Stokes theorem.

(iv) From this last integrand, factor out $d\tau$; the remaining factor is div $\mathbf{w}$.

3. *Solid angles.* Let $\mathbf{r}$ be the position vector and $\rho = |\mathbf{r}|$ be the spherical coordinate in three-space. Direct computation (see Example 1, Section 7–3) shows that

$$\text{div } \frac{\mathbf{r}}{\rho^3} = 0.$$

Thus, one would expect from the divergence theorem that

$$\iint_{\partial V} \frac{\mathbf{r} \cdot \mathbf{N} \, d\sigma}{\rho^3} = 0. \tag{29}$$

However, one must be careful about continuity in these theorems, and $\mathbf{r}/\rho^3$ is discontinuous at the origin. Thus, (29) holds provided that V does not contain the origin.

Let S_0 be the unit sphere. On S_0, $\rho = 1$ and $\mathbf{r} \cdot \mathbf{N} = 1$; hence

$$\iint_{S_0} \frac{\mathbf{r} \cdot \mathbf{N} \, d\sigma}{\rho^3} = \iint_{S_0} d\sigma = \sigma(S_0) = 4\pi.$$

With the origin in V, (29) does not necessarily hold.

Let S be an arbitrary piecewise smooth surface. Let L be a piecewise smooth conical surface with vertex at the origin, cutting off portions S_0^1 and S^1 of S_0 and S, respectively (Fig. 10–24). With orientations as shown in Fig. 10–24, S_0^1, L, and S^1 make up the boundary of a region in which

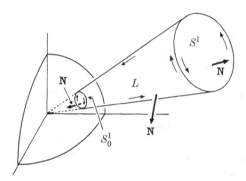

FIGURE 10–24

$\mathbf{r}/\rho^3$ is continuous; hence (29) applies, and

$$\iint_{S^1} \frac{\mathbf{r} \cdot \mathbf{N}\, d\sigma}{\rho^3} + \iint_{L} \frac{\mathbf{r} \cdot \mathbf{N}\, d\sigma}{\rho^3} + \iint_{S_0^1} \frac{\mathbf{r} \cdot \mathbf{N}\, d\sigma}{\rho^3} = 0. \qquad (30)$$

Now, $\mathbf{r}$ is tangent to L; thus $\mathbf{r} \cdot \mathbf{N} = 0$ on L, and the middle integral is zero. Now, the $\mathbf{N}$ in (30) is the exterior normal to the region bounded by S_0^1, S^1, and L; hence $\mathbf{r} \cdot \mathbf{N} = -1$ on S_0^1. Therefore, (30) becomes

$$\iint_{S^1} \frac{\mathbf{r} \cdot \mathbf{N}\, d\sigma}{\rho^3} - \iint_{S_0^1} d\sigma = 0,$$

or

$$\iint_{S^1} \frac{\mathbf{r} \cdot \mathbf{N}\, d\sigma}{\rho^3} = \sigma(S_0^1). \qquad (31)$$

Now, S_0^1 is the conical projection of S^1 onto the unit sphere, and the surface area of this projection is called the solid angle subtended by S^1 at the origin. So, (31) gives an integral formula for solid angles. Note that the understood convention in (31) is that $\mathbf{N}$ is the unit normal to S^1 away from the origin.

EXERCISES

1. Let $\mathbf{w}$ be given in spherical coordinate components as in Example 2.
   (a) Show that on a curve C,

   $$\mathbf{w} \cdot \mathbf{T}\, ds = w_1\, d\rho + w_2\rho\, d\phi + w_3\rho \sin \phi\, d\theta.$$

   (b) Follow the plan of Example 2 to find **curl w** in spherical coordinates. That is,

   $$\iint_{\partial S} \mathbf{w} \cdot \mathbf{T}\, ds = \iint_{S} \text{curl } \mathbf{w} \cdot \mathbf{N}\, d\sigma;$$

apply (18) to the coordinate form of $\mathbf{w} \cdot \mathbf{T} \, ds$ in part (a); note the expression for $\mathbf{N} \, d\sigma$ in Example 2, and recover **curl w** from the surface integral. Compare Example 1, Section 7–5.

2. Follow the plan of Example 2 and Exercise 1 to find div **w** and **curl w** in cylindrical coordinates. Compare Exercise 1(b) and 1(c), Section 7–5.

3. Let (t, u, v) be an orthogonal—though possibly curvilinear—coordinate system in three-space. Let a, b, and c be the multipliers such that $(a \, dt, b \, du, c \, dv)$ gives orthonormal differential-component variables. Let $\mathbf{b}_t$, $\mathbf{b}_u$, $\mathbf{b}_v$ be unit vectors in the t, u, v directions, respectively; and let $\mathbf{w} = w_1 \mathbf{b}_t + w_2 \mathbf{b}_u + w_3 \mathbf{b}_v$.

(a) Show that on a curve C, $\mathbf{w} \cdot \mathbf{T} \, ds = w_1 a \, dt + w_2 b \, du + w_3 c \, dv$.

(b) Show that on a surface S,

$$\mathbf{N} \, d\sigma = \mathbf{b}_t bc \, du \wedge dv + \mathbf{b}_u ca \, dv \wedge dt + \mathbf{b}_v ab \, dt \wedge du.$$

(c) Use part (a) and proceed, as in Exercise 1, to show that

$$\mathbf{curl \ w} = \frac{\mathbf{b}_t}{bc}\left[\frac{\partial}{\partial u}(cw_3) - \frac{\partial}{\partial v}(bw_2)\right]$$
$$+ \frac{\mathbf{b}_u}{ca}\left[\frac{\partial}{\partial v}(aw_1) - \frac{\partial}{\partial t}(cw_3)\right]$$
$$+ \frac{\mathbf{b}_v}{ab}\left[\frac{\partial}{\partial t}(bw_2) - \frac{\partial}{\partial u}(aw_1)\right].$$

(d) Use part (b) and proceed, as in Example 2, to show that

$$\text{div } \mathbf{w} = \frac{1}{abc}\left[\frac{\partial}{\partial t}(bcw_1) + \frac{\partial}{\partial u}(caw_2) + \frac{\partial}{\partial v}(abw_3)\right].$$

4. The procedure being studied here will also find Laplacians. Let w be a scalar variable.

(a) Apply the divergence theorem to **grad** w to show that

$$\iint_{\partial V} \mathbf{grad} \ w \cdot \mathbf{N} \, d\sigma = \iiint_V \nabla^2 w \, d\tau.$$

(b) Show that in spherical coordinates,

$$\mathbf{grad} \ w = \frac{\partial w}{\partial \rho} \mathbf{b}_\rho + \frac{1}{\rho} \frac{\partial w}{\partial \phi} \mathbf{b}_\phi + \frac{1}{\rho \sin \phi} \frac{\partial w}{\partial \theta} \mathbf{b}_\theta;$$

then use part (a) to find $\nabla^2 w$, and compare Example 2, Section 7–5.

(c) Show that in cylindrical coordinates,

$$\mathbf{grad} \ w = \frac{\partial w}{\partial r} \mathbf{b}_r + \frac{1}{r} \frac{\partial w}{\partial \theta} \mathbf{b}_\theta + \frac{\partial w}{\partial z} \mathbf{b}_z;$$

find $\nabla^2 w$ and compare with Exercise 1(d), Section 7–5.

(d) Show that for the coordinates of Exercise 3,

$$\mathbf{grad}\ w = \frac{1}{a}\frac{\partial w}{\partial t}\,\mathbf{b}_t + \frac{1}{b}\frac{\partial w}{\partial u}\,\mathbf{b}_u + \frac{1}{c}\frac{\partial w}{\partial v}\,\mathbf{b}_v.$$

Apply part (a) to get the general formula for a Laplacian in any orthogonal coordinate system:

$$\nabla^2 w = \frac{1}{abc}\left[\frac{\partial}{\partial t}\left(\frac{bc}{a}\frac{\partial w}{\partial t}\right) + \frac{\partial}{\partial u}\left(\frac{ca}{b}\frac{\partial w}{\partial u}\right) + \frac{\partial}{\partial v}\left(\frac{ab}{c}\frac{\partial w}{\partial v}\right)\right].$$

(e) Apply the general formula of part (d) to the spherical and cylindrical systems, and recover the results of parts (b) and (c).

5. Let

$$x = t^2 - u^2, \qquad y = 2tu, \qquad z = v.$$

(a) Show that

$$dx^2 + dy^2 + dz^2 = a^2\,dt^2 + b^2\,du^2 + c^2\,dv^2.$$

Conclude that (t, u, v) is an orthogonal system, and that a, b, and c, thus found in terms of t, u, and v, are the required multipliers for this system.

(b) Find $\mathbf{w} \cdot \mathbf{T}\,ds$, $\mathbf{N}\,d\sigma$, and $\mathbf{grad}\ \mathbf{w}$ in the (t, u, v)-system, and use appropriate Stokes type theorems to find div $\mathbf{w}$, curl $\mathbf{w}$, and $\nabla^2 w$.

(c) Find div $\mathbf{w}$, curl $\mathbf{w}$, and $\nabla^2 w$ by the general formulas of Exercise 3(c), 3(d), and 4(d). Compare part (b) and also Exercise 5, Section 7–5.

6. Let V be a convex, three-dimensional, oriented manifold in three-space having a piecewise smooth boundary. Show that the value of

$$\iint_{\partial V} \frac{\mathbf{r} \cdot \mathbf{N}\,d\sigma}{\rho^3}$$

is (a) 4π if V contains the origin, (b) 2π if ∂V contains the origin and has a tangent plane there, and (c) 0 if the origin is outside V and ∂V.

7. Let V be an oriented, three-dimensional manifold in three-space with a piecewise smooth boundary.

(a) Show that

$$\iint_{\partial V} \mathbf{r} \cdot \mathbf{N}\,d\sigma = 3\tau(V).$$

(b) Show that

$$\iint_{\partial V} (x^2\mathbf{i} + 2xy\mathbf{j} + 2xz\mathbf{k}) \cdot \mathbf{N}\,d\sigma = 6\tau(V)\bar{x},$$

where $(\bar{x}, \bar{y}, \bar{z})$ is the centroid of V.

(c) Find surface integrals for $\bar{y}$ and $\bar{z}$.

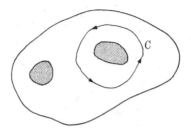

FIGURE 10-25

10-6 Integrals independent of the path.   Let A be a two-dimensional manifold in the plane. The differential form $u\,dx + v\,dy$ is said to *have line integrals independent of the path in A* provided that for every pair of points p and q in A,

$$\int_{C_1} (u\,dx + v\,dy) = \int_{C_2} (u\,dx + v\,dy), \tag{32}$$

whenever C_1 and C_2 are oriented, piecewise smooth, one-dimensional manifolds in A, each having p as a beginning point and q as an end point.

Of the paths C_1 and C_2 in (32), reverse the orientation of one of them and join them together. This forms an oriented closed curve C, and (32) may be written

$$\int_C (u\,dx + v\,dy) = 0. \tag{33}$$

If C in (33) bounds the two-dimensional manifold B, then by Green's theorem in the plane, (33) may be written

$$\iint_B \left(\frac{\partial v}{\partial x} - \frac{\partial u}{\partial y} \right) d\alpha = 0; \tag{34}$$

thus independence of the path may be studied in terms of these partial derivatives.

Now, one speaks of line integrals being independent of the path *in a given region A* (see definition above), and the current project is to characterize this phenomenon in terms of properties of u and v and their partial derivatives in A. This characterization will be done by reducing (32) to (34) as above, but for (34) to give any information from the type data just mentioned, B must be contained in A. If A has a hole in it (Fig. 10-25), and C goes around the hole, then B covers the hole, and hypotheses concerning u and v in A only, give no information about (34). Thus, the discussion of independence of path will be limited to *simply connected* regions. Intuitively, this means regions without holes. Precisely, A is simply connected if every simple closed curve in A bounds a two-dimensional manifold that is contained in A.

FIGURE 10–26

LINE INTEGRAL THEOREM (*two dimensions*). Let A be a simply connected, two-dimensional manifold in the plane, and let u and v be variables on A, having continuous partial derivatives with respect to x and y. Then, the following three statements are equivalent.

(a) $u\,dx + v\,dy$ has line integrals independent of the path in A.

(b) $\partial u/\partial y = \partial v/\partial x$ on A.

(c) There exists a function ϕ on pairs having continuous second partial derivatives and such that

$$u = \phi_1(x, y), \qquad v = \phi_2(x, y)$$

on A.

Proof. Equivalence of the three statements will be proved by showing that (c) → (b) → (a) → (c).

Statement (c) implies (b). Given (c), the equation in (b) merely states the equality of the second-order cross partial derivatives of ϕ.

Statement (b) implies (a). Here the simple connectedness of A is important. Given an oriented, piecewise smooth, simple closed curve C in A, let it bound the oriented, two-dimensional manifold B which will be contained in A. Now, apply Green's theorem:

$$\int_C (u\,dx + v\,dy) = \iint_B \left(\frac{\partial v}{\partial x} - \frac{\partial u}{\partial y}\right) d\alpha = 0$$

because, given (b), the integrand in the double integral is zero on A, hence on B. Note that for the most general piecewise smooth curve C, the extended form of Green's theorem (not proved in Section 10–4) is required.

This proves that (32) above holds provided that C_1 and C_2 have a common beginning point and a common end point, lie in A, and do not cross. Paths that cross form a number (perhaps infinite) of simple closed curves (Fig. 10–26), and the above argument yields (32) for such a pair of paths by the formation of a sum (perhaps an infinite series) of zeros.

Statement (a) implies (c). Let (a, b) be an arbitrary but fixed point of A. Define ϕ by setting

$$\phi(x_0, y_0) = \int_{(a,b)}^{(x_0, y_0)} (u\,dx + v\,dy).$$

This is unambiguous because of hypothesis (a). Then,

$$\phi(x_0 + h, y_0) - \phi(x_0, y_0) = \int_{(x_0,y_0)}^{(x_0+h,y_0)} (u\,dx + v\,dy),$$

and, because of (a), the path may be taken as a horizontal line. On this path, $dy = 0$; thus

$$\phi_1(x_0, y_0) = \lim_{h \to 0} \frac{\phi(x_0 + h, y_0) - \phi(x_0, y_0)}{h}$$

$$= \lim_{h \to 0} \frac{1}{h} \int_{x_0}^{x_0+h} u\,dx = u[(x_0, y_0)]$$

by the fundamental theorem of calculus. The result $\phi_2(x, y) = v$ is proved in a similar manner, using a vertical path from (x_0, y_0) to $(x_0, y_0 + h)$.

There is an analogous theorem in three dimensions, the tool used in the proof being Stokes' theorem instead of Green's theorem. To this end, it will be required that the region V considered have the property that every simple closed curve in V bound a surface that lies entirely in V. A solid figure will be called *simply connected* if it has this property. Note that in three-space simple connectedness does not preclude the possibility of holes, provided that they are of the right type. The solid figure between concentric spheres is simply connected; note that if a circle goes around the hole, it still bounds a hemispherical surface that lies in the given figure. On the other hand, the solid figure between coaxial cylinders is not simply connected.

It is more informative to look at the three-dimensional case in vector form. Let **w** be a vector variable on a simply connected region V. One investigates independence of the path for integrals of

$$\mathbf{w} \cdot \mathbf{T}\,ds.$$

Through Stokes' theorem this leads to surface integrals of

$$\operatorname{curl} \mathbf{w};$$

hence one wants this to be zero. Finally, the idea that components of **w** are partial derivatives of some one variable u is described by saying that

$$\mathbf{w} = \operatorname{grad} u.$$

LINE INTEGRAL THEOREM (*three dimensions*). Let V be a simply connected, three-dimensional manifold in three-space; and let **w** be a vector variable on V whose components have continuous partial derivatives.

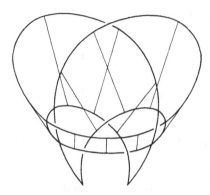

FIGURE 10–27

Then the following statements are equivalent.

(a) $\mathbf{w} \cdot \mathbf{T}\, ds$ has line integrals independent of the path in V.

(b) $\mathbf{curl}\ \mathbf{w} = 0$ on V.

(c) There is a scalar variable u on V with continuous second partial derivatives and such that

$$\mathbf{w} = \mathbf{grad}\ u$$

on V.

Proof is left to the reader with the following comment. In the proof that (b) implies (a), Stokes' theorem will be employed to show that the integral of $u\, dx + v\, dy + w\, dz$ around a piecewise smooth, simple closed curve in V is zero. The usual extension to cover a curve with an infinite number of "kinks" will be required, but there is an even deeper problem in this case. A simple closed curve in three-space may be tied in a knot, and it is not at all clear that such a curve may be made to bound an *orientable* two-dimensional manifold as will be required for the application of Stokes' theorem. To prove independence of the path (including knotted ones) one may consider the process of "shrinking a knot continuously to a point." Intuitively, pull on the ends, and tighten the knot into an arbitrarily small one. A knotted curve and another looped through it, making a similar, smaller knot (see Fig. 10–27), may be made to bound an orientable surface that does not cross itself; thus by Stokes' theorem the line integral is the same around large and small (loose and tight) knots. Finally, the absolute value of a line integral is less than or equal to the maximum absolute value of the integrand times the length of the curve (why?). Thus, the integral around a sufficiently small knot contributes an arbitrarily small amount to the total line integral, and the desired result may be obtained.

EXAMPLES

1. *Exact differentials.* Consider the differential equation

$$u\,dx + v\,dy = 0, \tag{35}$$

where u and v have continuous partial derivatives with respect to x and y. If

$$\frac{\partial v}{\partial x} = \frac{\partial u}{\partial y}, \tag{36}$$

then by the line-integral theorem, there is a function ϕ such that

$$\phi_1(x, y) = u, \qquad \phi_2(x, y) = v.$$

Thus, if $w = \phi(x, y)$, then

$$dw = u\,dx + v\,dy,$$

and the differential equation reads

$$dw = 0.$$

This equation has the obvious solution $w = $ constant, or

$$\phi(x, y) = \text{constant.} \tag{37}$$

If (36) is satisfied, then (35) is called an *exact differential equation.* In this case, once the function ϕ is found, (37) gives the solution of (35).

In many simple examples, ϕ may be found by inspection; but if not, the proof of the line-integral theorem shows how to find it. Given (36), line integrals of $u\,dx + v\,dy$ are independent of the path, and

$$\phi(x_0, y_0) = \int_{(a,b)}^{(x_0, y_0)} (u\,dx + v\,dy)$$

defines ϕ. Here the path is arbitrary; hence it is easiest to take a rectangular one from (a, b) to (x_0, b), then from (x_0, b) to (x_0, y_0). This procedure yields

$$\phi(x_0, y_0) = \int_a^{x_0} f(x, b)\,dx + \int_b^{y_0} g(x_0, y)\,dy,$$

where $u = f(x, y)$ and $v = g(x, y)$.

For example,

$$xy^2\,dx + x^2 y\,dy = 0$$

is exact because

$$\frac{\partial}{\partial y}(xy^2) = 2xy = \frac{\partial}{\partial x}(x^2 y).$$

Perhaps $xy^2\, dx + x^2 y\, dy$ is recognizable as

$$\tfrac{1}{2} d(x^2 y^2);$$

if so, the solution is obvious. Otherwise, let $(a, b) = (0, 0)$ and write

$$\phi(x_0, y_0) = \int_0^{x_0} x \cdot 0 \, dx + \int_0^{y_0} x_0^2 y \, dy = 0 + \tfrac{1}{2} x_0^2 y_0^2.$$

2. *Forces and potentials.* Let **w** give the force at each point in three-space due to a certain force field. Then

$$\int_C \mathbf{w} \cdot \mathbf{T} \, ds$$

is the work done by the force field along with the oriented curve C. If this work is zero on every closed path, the force field is called *conservative* (energy is conserved on a trip that ends where it started). By the line-integral theorem, if a force is conservative, then its components are the partial derivatives of some one scalar variable. That is, the vector variable giving force is the gradient of some scalar variable. This scalar variable is called the *potential*. The potential at p is the work done (on any path) from some fixed point p_0 to p. Conservative forces and only conservative forces admit potentials. Such forces are characterized by the fact that they have zero curl.

3. *Velocity potentials.* Let **v** be the velocity variable in fluid dynamics, and apply the line-integral theorem to $\mathbf{v} \cdot \mathbf{T} \, ds$. If the flow is irrotational, **curl v** $= 0$, and **v** appears as the gradient of a scalar variable. This scalar variable is called the *velocity potential*. Only irrotational flows admit velocity potentials.

EXERCISES

1. Show that each of the following is an exact differential equation, and solve it.
  (a) $2xy \, dx + x^2 \, dy = 0$ (b) $ye^{xy} \, dx + xe^{xy} \, dy = 0$
  (c) $2xy \, dx + (x^2 - y^2) \, dy = 0$ (d) $\sin y \, dx + x \cos y \, dy = 0$

2. Show that condition (b) of the line-integral theorem in two dimensions is satisfied for each of the following differential forms. Then, check continuity conditions and describe in each case restrictions on A so that the given form will have line integrals independent of the path in A.

  (a) $\dfrac{x \, dx + y \, dy}{(x^2 + y^2)^{3/2}}$ (b) $\dfrac{y \, dx - x \, dy}{x^2}$

  (c) $\dfrac{3x^2}{y} \, dx - \dfrac{x^3}{y^2} \, dy$ (d) $\dfrac{x \, dy - y \, dx}{x^2 + y^2}$

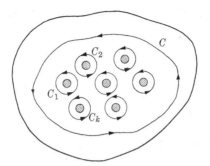

FIGURE 10–28

3. By direct computation, evaluate

$$\int_C \frac{x \, dy - y \, dx}{x^2 + y^2},$$

where C is the square with vertices $(+1, +1)$ and has counterclockwise orientation. Why does this not contradict the line-integral theorem?

4. Show that

$$\int_C (y^2 \, dx + x^2 \, dy) = 0,$$

where C is the square with vertices $(\pm 1, \pm 1)$. Show that condition (b) of the theorem is not satisfied. Why does this not contradict the theorem?

5. Let the domain A have k holes in it as in Fig. 10–28. For each i, let C_i be a simple closed curve in A enclosing the ith hole and no others. Let u and v have continuous partial derivatives with respect to x and y in A, and let $\partial v/\partial x = \partial u/\partial y$ in A. Suppose that for each i,

$$\int_{C_i} (u \, dx + v \, dy) = a_i.$$

(a) Let C be a simple closed curve in A enclosing all the holes. Evaluate

$$\int_C (u \, dx + v \, dy).$$

(b) Find all possible values of the integral

$$\int_p^q (u \, dx + v \, dy),$$

for paths in A, given that this integral has the value K for some one path.

6. Show that each of the following is independent of the path, and find its value.

(a) $\displaystyle\int_{(1,1,2)}^{(3,5,0)} (yz\,dx + xz\,dy + xy\,dz)$

(b) $\displaystyle\int_{(1,0,0)}^{(1,0,2)} (\sin yz\,dx + xz\cos yz\,dy + xy\cos yz\,dz)$

7. Consider the differential form

$$\frac{x\,dx + y\,dy + z\,dz}{(x^2 + y^2 + z^2)^{3/2}}.$$

(a) Show that this form has line integrals independent of the path in all of three-space with the origin removed.
(b) Show that line integrals of this form yield the work done by the gravitational force due to a point mass at the origin.
(c) Show that this gravitational force field is conservative, and find a potential variable for it.

8. Consider the differential form

$$\frac{-y\,dx}{x^2 + y^2} + \frac{x\,dy}{x^2 + y^2} + z\,dz.$$

(a) Show by a symmetry argument that its integral around the circle on which $x^2 + y^2 = 1$ and $z = 0$ is different from zero.
(b) Check condition (b) of the line integral theorem.
(c) Check continuity conditions, and describe a domain in which this form has line integrals independent of the path.
(d) In Exercise 7(a) above, the domain for line integrals independent of the path was the domain of continuity of the variables. Why does this not hold in part (c) of this exercise?

Section 1–1

1. (a)

0	4
1	−1
2	0
4	−3

(b)

0	4
1	−5
2	−2
4	3

(c)

0	0
1	−6
2	−1
4	0

(d)

1	$-\frac{3}{2}$
2	−1
4	0

(e)

0	0
1	$-\frac{2}{3}$
2	−1

(f) 4

(g) 0

(h)

2	1
0	1
−3	1
1	1

(i)

4	1
1	1
−1	1
2	1

(j)

2	1
0	1
−3	1
4	1
1	1

(k)

4	1
1	1
0	1
−1	1
2	1

2. (a) $-2, 10, -\frac{5}{4}, \frac{22}{9}$
   (b) $\frac{1}{3}, -\frac{4}{3}, -1, \frac{1}{9999}$
   (c) $11, -24, \frac{23}{8}, 3$
   (d) $-\frac{1}{2}, -\frac{1}{5}, -\frac{9}{10}, -\frac{25}{29}$
   (e) $1 - a^9, (1 - a^3)^3, (a^3 - 1)/a^3, 1/(1 - a^3)$
   (f) $-a - (1/a), (1/a) + a, a/(a^2 + 1), -a - (1/a)$
   (g) $4 - a^4, (4 - a^2)^2, (4a^2 - 1)/a^2, a^2$
   (h) $4a - [1/(4a)], -2a + [1/(2a)], (2/a) - (a/2), 2a/(4a^2 - 1)$

3. (a) $I^2 + I^0$
   (b) $(I^0 + I)^2$
   (c) $(I^2 + I^0)/(I^3 - I^0)$
   (d) $(I^2 + I^0)^{1/2}$
   (e) $[(I + I^0)^{1/2} + I^0]^2$
   (f) $I^0 + (I^0/I)$
   (g) $I + (I^0/I)$
   (h) $I^0/(I + I^0)$
   (i) $I^0/(I^0 + I^{1/2})$
   (j) $[I^0 + (I^0/I)]^{1/2}$

4. (a)

−1	−1
0	4
1	−1
2	−3
4	2

(b)

−3	1
0	−3
2	2
4	0

(c)

−3	−3
0	2
2	−1
4	4

(d)

−1	2
0	−3
1	2
4	1

(e)

−3	−1
0	0
1	2
4	4

(f)

−1	1
0	0
1	1
2	2

(g) 4

(h) −3

(i) 4

(j) −3

5. (a) $1 + (1 + \sqrt{a})^2$, $1 + \sqrt{1 + a^2}$ (b) $1 + 2a^{-3/2}$, $1/\sqrt{1 + 2a^3}$
   (c) $[(1 + 2a^2)/a^2]^2$, $1/(2 + a^2)^2$ (d) $2 + a$, a
   (e) $\sqrt{1 - \sqrt{a}}$, $1 - \sqrt{1 - a}$ (f) $1 + 3a^2$, $3(a + 1)^2$
   (g) $(1 - \sqrt{a})/(1 + \sqrt{a})$, $\sqrt{(1 - a)/(1 + a)}$
   (h) $\sqrt{1 + \sqrt{1 - a}}$, $\sqrt{1 - \sqrt{1 + a}}$

6. (a) $(I^2 - I^0)^{1/2}$ (b) $(I + I^{1/2})^3$
   (c) $[I^0 + (I^2 + I^0)^{1/2}]^2$ (d) $[-2I^0 - I^{1/2}]^4$
   (e) $3(I^2 - 2I)^5$ (f) $(I^3 + 3I^2)^{1/2}$
   (g) $[I^0 - (I^0 - I)^{1/2}]^{1/2}$ (h) $[I^0 - 2(3I^0 + I)^{1/2}]^3$

7. (a) $I^5 \circ (I + I^2)$ (b) $I^3 \circ (I - I^{1/2})$
   (c) $I^{1/2} \circ (I - I^3)$
   (d) $I^3 \circ [I^{1/2} \circ (I - I^0) - I^{1/2} \circ (I + I^0)]$
   (e) $I^2 \circ (I^0 + I^2) \circ (I^0 + I)$ (f) $I^{1/2} \circ (I + I^{1/2}) \circ (I - I^0)$
   (g) $I^3 \circ (I^0 + I^{-1}) \circ (I^0 - I^2)$
   (h) $I^{1/2} \circ (I + I^{1/2}) \circ (I + I^{1/2}) \circ (I - I^0)$

12. (a) $n\pi$ (b) 0   (c) $n\pi/2$
    (d) ± 2 (e) 0   (f) $(2n + 1)\pi/2$
    (g) None   (h) 0   (i) $0, \pm 1$
    (j) $(4n - 1)\pi/2$ (k) None   (l) 0
    (m) $0, e$ (n) 0   (o) $n\pi/2$
    (p) $\pm 1, 0$ (q) None   (r) 1
    (s) None   (t) $\pm 1, 0$ (u) None
    (v) 0

Section 1-2

1. (a) $I^2 \circ (I^0 + I^{1/2}) \circ x$ (b) $I^{1/2} \circ (I^0 - I^3) \circ x$
   (c) $(3I^0 + I^4) \circ (2I^0 - I) \circ x$ (d) $I^4 \circ (3I^0 + I^{1/2}) \circ (2I^0 - I) \circ x$
   (e) $(2I^0 + I^3) \circ (I^0 - I^{1/2}) \circ (I + 4I^0) \circ x$
   (f) $I^{1/2} \circ (I^0 - I^3) \circ (4I^0 + I^2) \circ x$ (g) $I^2 \circ (2I^0 - I^{1/2}) \circ (I^0 - I^3) \circ x$
   (h) $I^2 \circ \{I^0 - I[I^{1/2} \circ (I^0 - I^2)]\} \circ x$
   (i) $I^{1/2} \circ (I - I^{1/2}) \circ (I^0 - I) \circ x$ (j) $I[I^{1/2} \circ (I^0 - 2I^{1/2})] \circ x$

2. (a) $\sqrt{1 + x}$ (b) $(\sqrt{x} - x^2)^3$
   (c) $[2(1 - x) - (1 - x)^2]^2$ (d) $\sqrt{1 + \sqrt{x}}$
   (e) $1 - (x - x^2)^2$ (f) $x - \sqrt{x} - (x - \sqrt{x})^3$
   (g) $(1 - x)^{3/2}$ (h) $x\sqrt{1 - x}$
   (i) $(x^3 - x^2)^3(3x^2 - 2x)$ (j) $\sqrt{1 + \sqrt{1 + x}}$

3. (a) v (b) n (c) m (d) m (e) n
   (f) m (g) m (h) n (i) m (j) v
   (k) m (l) m (m) f (n) m (o) n
   (p) m (q) v (r) v (s) m (t) n
   (u) m (v) m (w) n (x) v (y) m
   (z) f

4. (a) $x + y$ (b) $(xy)(p)$ (c) $x(p)y(p)$ (d) x^3
   (e) $(I^3 + I^2)(a)$ (f) $(x + y)^2$ (g) $x^2 + y^2$ (h) $(I^0 + I)^{1/2}$
   (i) $I^0 + I^{1/2}$

Section 2–1

1. (a) $2\,dx/y,\ 2\,dy/x$ (b) $\sqrt{1 + y^4}\,dx,\ \sqrt{1 + x^4}\,dy$
   (c) $\sqrt{1 + y^4}\,dx,\ \sqrt{1 + x^4}\,dy$ (d) $\sqrt{1 + 9x^4}\,dx,\ \sqrt{1 + 9x^4}\,dy/3x^2$
   (e) $\sqrt{5 - 4x + 4x^2}\,dx,\ \sqrt{5 - 4x + 4x^2}\,dy/(2 - 2x)$
   (f) $\sqrt{10 - 6x^2 + 9x^4}\,dx,\ \sqrt{10 - 6x^2 + 9x^4}\,dy/(3 - 3x^2)$
   (g) $\sqrt{x^2 + y^2}\,dx/y,\ \sqrt{x^2 + y^2}\,dy/x$ (h) $\sqrt{1 + \cos^2 x}\,dx,\ \sqrt{1 + \sec^2 x}\,dy$
   (i) $\sqrt{1 + (1/4x)}\,dx,\ \sqrt{1 + 4y^2}\,dy$ (j) $\sqrt{1 + (1/4x)}\,dx,\ \sqrt{1 + 4y^2}\,dy$
   (k) $\sqrt{1 + \sec^2 y}\,dx,\ \sqrt{1 + \cos^2 y}\,dy$ (l) $\sqrt{1 + \sec^2 y}\,dx,\ \sqrt{1 + \cos^2 y}\,dy$

2. (a) 1   (b) 4   (c) $\sqrt{17}$ (d) 3
   (e) -1 (f) 0   (g) -1 (h) -2
   (i) -2 (j) $2\sqrt{17}$ (k) -2 (l) -4
   (m) $2\sqrt{5}$ (n) $-2\sqrt{5}$ (o) -4 (p) 4

3. (a) $(3, 5)$ (b) $(-\tfrac{5}{2}, 4)$
   (c) $(2 + \sqrt{17}, 4 + 4\sqrt{17})$ (d) $(-\tfrac{3}{2}, 2)$
   (e) $(-5, 0)$ (f) $(1 - \sqrt{5}, 1 - 2\sqrt{5})$
   (g) $(\tfrac{5}{2}, 4)$ (h) $(0, -4)$
   (i) $(4, 0)$ (j) $(-1 - \sqrt{5}, 1 + 2\sqrt{5})$

Section 2–2

1. (a) $\tfrac{1}{2}$ (b) 2   (c) $1/\sqrt{5}$ (d) -2
   (e) $\sqrt{17}$ (f) 1   (g) 1   (h) $1/\sqrt{17}$
   (i) $-\tfrac{1}{2}$ (j) $4/\sqrt{17}$ (k) $-\sqrt{17}/4$ (l) $\sqrt{17}$
   (m) 1   (n) 1   (o) $-2\sqrt{5}$ (p) $-\sqrt{5}/2$

3. (a) $-x/y$ (b) $-2y/(3x)$

   (c) $\dfrac{-y \cos xy - \sin (x + y)}{x \cos xy + \sin (x + y)}$ (d) $\dfrac{\sqrt{y}(1 - 2y\sqrt{x})}{\sqrt{x}(2x\sqrt{y} - 1)}$

   (e) $1/(x + y - 1)$ (f) $\dfrac{\sqrt{y/x} - 2xy}{x^2 - \sqrt{x/y}}$

   (g) $\dfrac{x^3 - 2x^4 y - 2y}{x^5 - x}$ (h) $\dfrac{y \cos xy - \sin y}{x \cos y - x \cos xy}$

   (i) $\dfrac{2y^2 + y}{y\sqrt{x} + x - 2xy}$ (j) $\dfrac{1 + xy}{e^{-xy} - x^2}$

Section 2–4

1. (a) var, M (b) var, M (c) no.
   (d) var, T_p (e) no.   (f) var, T_p

(g) var, T_p (h) var, M (i) var, tan bundle

(j) var, tan bundle   (k) var, tan bundle   (l) var, M

(m) no.   (n) var, M (o) no.

(p) var, T_p (q) var, tan bundle   (r) no.

2. (a) $-1/y^3$ (b) $-6/(x + 2y)^3$ (c) $(2xy - 3)/x^3$

(d) $\csc x(y \csc x - \cot x) + \cot x(y \cot x - \csc x)$

(e) $y(2y - y^2 - 2)/[x^2(y - 1)^3]$

(f) $[6x(1 - 2y)^2 - 3x^2(1 - 2y - 3x^2)]/(1 - 2y)^3$

(g) $4 \sec^2 (x + y) \tan (x + y)$

(h) $[(2y^4 - y^3 + 2xy^2)(xy - x^2) - xy^4]/(xy - x^2)^3$

(i) $-y^2/(x + y)^3$

(j) $\dfrac{4y - 12x^2}{2(\sqrt{xy} - x)} + \dfrac{(8x + 3)(4xy + y - 4x^3)}{4(\sqrt{xy} - x)^2} + \dfrac{(4xy + y - 4x^3)^2}{4(\sqrt{xy} - x)^3}$

3. $-D_x^2 y/(D_x y)^3$

Section 2–5

1. (a) $\frac{64}{15}$ (b) $\frac{64}{15}$ (c) $\frac{3}{32} - \frac{1}{10} \ln 2$ (d) $2\sqrt{3} - \pi/3$

(e) $\frac{8}{9}$ (f) $\frac{4}{15}$

2. (a) 8 (b) $(51\sqrt{17} - 3)/4$ (c) $(2\sqrt{2} - 1)/3$

(d) $\pi/2$ (e) $\pi/2$ (f) 0

(g) 10 (h) $\frac{22}{3}$

Section 3–1

1. (a) $\mathbf{i} + 4\mathbf{j}$ (b) $3\mathbf{i} - 2\mathbf{j}$ (c) 1 (d) $\sqrt{5}$

(e) $\sqrt{10}$ (f) $1/\sqrt{10}$ (g) $1/\sqrt{5}$

2. (a) $(\mathbf{A} \cdot \mathbf{B})\mathbf{C}$ (b) $\mathbf{A} \cdot (\mathbf{B} + \mathbf{C})$ (c) $(\mathbf{A} \cdot \mathbf{B}) + a$ (d) $\mathbf{A}/(\mathbf{B} \cdot \mathbf{C})$

(e) $\mathbf{A}(\mathbf{B} \cdot \mathbf{C})$ (f) $(\mathbf{A}/a) \cdot \mathbf{B}$ (g) $(\mathbf{A} + \mathbf{B}) \cdot \mathbf{C}$ (h) $(\mathbf{A}/a) + \mathbf{B}$

(i) $a + (\mathbf{A} \cdot \mathbf{B})$

6. (b) $\mathbf{u} = \frac{3}{5}\mathbf{i} - \frac{4}{5}\mathbf{j}$, $\mathbf{v} = \frac{4}{5}\mathbf{i} + \frac{3}{5}\mathbf{j}$ (c) $-\frac{6}{5}\mathbf{u} + \frac{17}{5}\mathbf{v}$

Section 3–2

5. (a) $\kappa = s/2$ (b) $\kappa = 2s$

6. (a) $\frac{1}{2}$; $(0, \sqrt{2})$ (b) $17\sqrt{17}/4$; $(-15, \frac{83}{12})$

(c) $13\sqrt{13}/6$; $(-\frac{11}{2}, \frac{16}{3})$ (d) $-(e^2 + 1)^{3/2}/e$; $[2e + (1/e), -e^2]$

(e) -1; $(0, -1)$ (f) $2\sqrt{2}$; $(-2, 3)$

(g) $5\sqrt{5}/4$; $[(\pi/4) - \frac{5}{2}, \frac{9}{4}]$ (h) 1; $(0, 2)$

(i) -1; $(\pi/2, 0)$ (j) $-\frac{1}{8}$, $(\pi/4, \frac{15}{8})$

10. (a) $X = \frac{1}{2}x(1 - 18x^2)$, $Y = x^3 + (1 + 9x^2)/(6x)$

(b) $X = x - \tan x$, $Y = 1 + \ln \sec x$

(c) $X = x - 1 + e^{2x}$, $Y = 2e^x + e^{-x}$

(d) $X = t + \sin t$, $Y = \cos t - 1$

(e) $X = \sec t + \sec^2 t(\tan^2 t + \sec^2 t)$,
$Y = \tan t - \sec t \tan t(\tan^2 t + \sec^2 t)$

(f) $X = \cos^3 t + (\frac{1}{2}) \sin t \tan 2t$, $Y = \sin^3 t + (\frac{1}{2}) \cos t \tan 2t$

(g) $X = \cos^4 t + (\frac{1}{2}) \sin^2 t(\cos^4 t + \sin^4 t)$,
$Y = \sin^4 t + (\frac{1}{2}) \cos^2 t(\cos^4 t + \sin^4 t)$

(h) $X = x - \tanh x(1 + \sinh^2 x)$, $Y = \cosh x + \text{sech}(1 + \sinh^2 x)$

(i) $X = \sin^3 t(1 - 16 \cos^2 t)$, $Y = \cos 2t - (\frac{1}{4}) \cos^2 t(1 + 16 \sin^2 t)$

(j) $X = -x - (\frac{9}{2})x^2$, $Y = (\frac{4}{3})x^{1/2} + 4x^{3/2}$

12. (a) $X^2 + Y^2 = 0$ (b) $Y = (\frac{1}{2}) + 3X^{2/3}/\sqrt[3]{16}$
(c) $(3X/7)^{2/3} + (4Y/7)^{2/3} = 1$ (d) $Y = \cosh X$
(e) $(X^2 + Y^2 - 1)^2 = 54 + 108X$

Section 3-3

1. (a) πab (b) πab

2. (a) π (b) $\pi/8$ (c) $\pi/2$ (d) $\frac{3}{2}$
(e) $\frac{8}{15}$ (f) $\frac{8}{3}$ (g) $8\pi/3$

3. (a) $2\pi + (3\sqrt{3}/2)$ (b) $\pi - (3\sqrt{3}/2)$ (c) $\pi + 3\sqrt{3}$

4. (a) $1 - (\pi/8)$ (b) $\pi + 3\sqrt{3}$ (c) $(3\pi + 1)/2$
(d) $(\pi/3) + (\sqrt{3}/2)$ (e) $(\pi/8) - (\frac{1}{4})$

Section 3-4

1. (a) $-\sin t, \cos t, -\cos t, -\sin t, 1, 1, 0, 1$
(b) $-3 \sin t, 4 \cos t, -3 \cos t, -4 \sin t, \sqrt{9 \sin^2 t + 16 \cos^2 t}$,
$\sqrt{9 \cos^2 t + 16 \sin^2 t}, -7 \sin t \cos t/\sqrt{9 \sin^2 t + 16 \cos^2 t}$,
$12/\sqrt{9 \sin^2 t + 16 \cos^2 t}$
(c) $-e^{-t}(\sin t + \cos t), e^{-t}(\cos t - \sin t), 2e^{-t} \sin t, -2e^{-t} \cos t, \sqrt{2} e^{-t}$,
$2e^{-t}, -\sqrt{2} e^{-t}, \sqrt{2} e^{-t}$
(d) $e^t, 2e^{-t}, e^t, 2e^{-t}, \sqrt{e^{2t} + 4e^{-2t}}, \sqrt{e^{2t} + 4e^{-2t}}$,
$(e^{2t} - 4e^{-2t})/\sqrt{e^{2t} + 4e^{-2t}}, 4/\sqrt{e^{2t} + 4e^{-2t}}$
(e) $-\text{sech } t \tanh t, \text{sech}^2 t, \text{sech } t(\tanh^2 t - \text{sech}^2 t), -2 \text{sech}^2 t \tanh t$,
$\text{sech } t, \text{sech } t, -\text{sech } t \tanh t, \text{sech}^2 t$
(f) $40, 30 - 32t, 0, -32, \sqrt{1024t^2 - 1920t + 2500}, 32$,
$(1024t - 960)/\sqrt{1024t^2 - 1920t + 2500}$,
$-1280/\sqrt{1024t^2 - 1920t + 2500}$
(g) $-\sin t, 2 \cos 2t, -\cos t, -4 \sin 2t, \sqrt{\sin^2 t + 4 \cos^2 2t}$,
$\sqrt{\cos^2 t + 16 \sin^2 2t}, (\sin 2t - 8 \sin 4t)/\sqrt{4 \sin^2 t + 16 \cos^2 2t}$,
$(4 \sin t \sin 2t + 2 \cos t \cos 2t)/\sqrt{\sin^2 t + 4 \cos^2 2t}$
(h) $1, -t/\sqrt{1 - t^2}, 0, -1/(1 - t^2)^{3/2}, 1/\sqrt{1 - t^2}, 1/(1 - t^2)^{3/2}$,
$t/(1 - t^2)^{3/2}, -1/(1 - t^2)$

2. (a) $v_0 \cos \theta(v_0 \sin \theta + \sqrt{v_0^2 \sin^2 \theta - 2gh})/g$
(b) Will just clear edge of cliff if cliff is $(v_0^2 \sin 2\theta)/(2g)$ horizontally from gun.
(c) Will not clear cliff.

3. (a) $v/\sqrt{4x^2+1}$, $2xv/\sqrt{4x^2+1}$, $-4v^2x/(1+4x^2)^2$, $2v^2/(1+4x^2)^2$

   (b) $v/\sqrt{1+e^{-2x}}$, $-v/\sqrt{1+e^{2x}}$, $v^2e^{-2x}/(1+e^{-2x})^2$, $v^2e^{3x}/(1+e^{2x})^2$

   (c) $v\,\mathrm{sech}\,x$, $v\tanh x$, $-v^2\,\mathrm{sech}^2\,x\tanh x$, $v^2\,\mathrm{sech}^3\,x$

   (d) $v\cos x$, $v\sin x$, $-v^2\sin x\cos x$, $v^2\cos^2 x$

   (e) $3v/\sqrt{9+4x^{-2/3}}$, $2vx^{-1/3}/\sqrt{9+4x^{-2/3}}$, $-12v^2x^{-5/3}/(9+4x^{-2/3})^2$,
   $18x^{-4/3}/(9+4x^{-2/3})^2$

   (f) $2v/\sqrt{9x+4}$, $3v\sqrt{x}/\sqrt{9x+4}$, $-18v^2/(9x+4)^2$, $12v^2x^{-1/2}/(9x+4)^2$

Section 4–1

1. (b)
$$AB = \begin{bmatrix} 1 & 3 & -2 \\ 1 & 2 & 0 \end{bmatrix}, \quad BC = \begin{bmatrix} 1 & 1 \\ 4 & -1 \end{bmatrix}, \quad BD = \begin{bmatrix} 1 & 1 & 4 \\ 4 & -3 & 0 \end{bmatrix},$$

$$CA = \begin{bmatrix} 0 & -1 \\ 1 & 0 \\ 2 & -2 \end{bmatrix}, \quad CB = \begin{bmatrix} 1 & 3 & -2 \\ 0 & -1 & 2 \\ 2 & 4 & 0 \end{bmatrix}, \quad DC = \begin{bmatrix} 1 & -2 \\ 4 & 1 \\ 4 & -3 \end{bmatrix}$$

   (c)
$$A' = \begin{bmatrix} 1 & 1 \\ -1 & 0 \end{bmatrix} \quad B' = \begin{bmatrix} 1 & 0 \\ 2 & -1 \\ 0 & 2 \end{bmatrix} \quad C' = \begin{bmatrix} 1 & 0 & 2 \\ -1 & 1 & 0 \end{bmatrix}$$

$$D' = \begin{bmatrix} 1 & 0 & 2 \\ -1 & 1 & -1 \\ 0 & 2 & -1 \end{bmatrix}$$

   (d)
$$A^{-1} = \begin{bmatrix} 0 & 1 \\ -1 & 1 \end{bmatrix} \quad D^{-1} = \begin{bmatrix} -3 & -1 & 2 \\ -4 & -1 & 2 \\ 2 & 1 & -1 \end{bmatrix}$$

2. (a) $\begin{bmatrix} u \\ v \end{bmatrix} = \begin{bmatrix} 2 & -3 \\ 4 & -1 \end{bmatrix}\begin{bmatrix} x \\ y \end{bmatrix}$ (b) $\begin{bmatrix} u \\ v \end{bmatrix} = \begin{bmatrix} 1 & 1 \\ 0 & -2 \end{bmatrix}\begin{bmatrix} x \\ y \end{bmatrix}$

   (c) $\begin{bmatrix} u \\ v \\ w \end{bmatrix} = \begin{bmatrix} 2 & -3 & 1 \\ 1 & -2 & 3 \\ 4 & -1 & 5 \end{bmatrix}\begin{bmatrix} x \\ y \\ z \end{bmatrix}$

   (d) $\begin{bmatrix} u \\ v \\ w \end{bmatrix} = \begin{bmatrix} 1 & -1 & 0 \\ 0 & 1 & -1 \\ -1 & 0 & 1 \end{bmatrix}\begin{bmatrix} x \\ y \\ z \end{bmatrix}$

3. $u = 3x - 2y$, $v = 4x$

Section 4–2

4. (b) $\begin{bmatrix} 3 \\ -1 \end{bmatrix}$ (c) $\begin{bmatrix} 0 \\ -\frac{1}{2} \end{bmatrix}$

5. (a) $\frac{1}{2}\mathbf{i} - \frac{1}{4}\mathbf{j}$ (b) $-5, 11$

   (c) $\frac{5}{16}r(\mathbf{u})r(\mathbf{v}) + \frac{3}{16}r(\mathbf{u})s(\mathbf{v}) + \frac{3}{16}s(\mathbf{u})r(\mathbf{v}) + \frac{5}{16}s(\mathbf{u})s(\mathbf{v})$

8. $\begin{bmatrix} 1 & 2 \\ 1 & -1 \end{bmatrix}$

9. $\begin{bmatrix} -\frac{7}{4} & -\frac{1}{4} \\ \frac{1}{4} & \frac{7}{4} \end{bmatrix}$

Section 4–3

1. (a) 13   (b) -3 (c) 1   (d) -4 (e) 6   (f) 1   (g) 118   (h) 0

Section 4–4

1. 53

3. (a) $\begin{bmatrix} \frac{1}{7} & -\frac{3}{14} \\ \frac{2}{7} & \frac{1}{14} \end{bmatrix}$ (b) $\begin{bmatrix} -\frac{2}{5} & \frac{3}{5} \\ -\frac{1}{5} & \frac{4}{5} \end{bmatrix}$

5. (a) $x = -\frac{5}{7}, y = -\frac{4}{7}$ (b) $x = \frac{31}{53}, y = \frac{5}{53}, z = -\frac{30}{53}$

Section 4–5

1. (a) -5 (b) $-19\mathbf{i} - 9\mathbf{j} + 11\mathbf{k}$ (c) $19\mathbf{i} + 9\mathbf{j} - 11\mathbf{k}$

   (d) -5 (e) -17 (f) $-23\mathbf{i} - 10\mathbf{j} + 16\mathbf{k}$

2. $\dfrac{\mathbf{i} + \mathbf{j} + \mathbf{k}}{\sqrt{3}}, \dfrac{3\mathbf{i} - 2\mathbf{j} - \mathbf{k}}{\sqrt{14}}, \dfrac{\mathbf{i} + 4\mathbf{j} - 5\mathbf{k}}{\sqrt{42}}; \quad 0, \dfrac{9}{\sqrt{14}}, \dfrac{3}{\sqrt{42}}$

Section 4–6

1. (a) $\mathbf{i} + 2\mathbf{j}, -(\frac{8}{5})\mathbf{i} + (\frac{4}{5})\mathbf{j}$ (b) $2\mathbf{i} + \mathbf{j}, \mathbf{i} - 2\mathbf{j}$ (c) $\mathbf{i}, \mathbf{j}$

   (d) $2\mathbf{i} + \mathbf{j} + \mathbf{k}, (\frac{2}{3})\mathbf{i} - (\frac{5}{3})\mathbf{j} + (\frac{1}{3})\mathbf{k}, (\frac{4}{5})\mathbf{i} - (\frac{8}{5})\mathbf{k}$

   (e) $\mathbf{i} + \mathbf{j} + \mathbf{k}, (\frac{2}{3})\mathbf{i} + (\frac{2}{3})\mathbf{j} - (\frac{4}{3})\mathbf{k}, \mathbf{i} - \mathbf{j}$

Section 4–7

1. (a) $\begin{bmatrix} \sqrt{3}/2 & -\frac{1}{2} \\ \frac{1}{2} & \sqrt{3}/2 \end{bmatrix}$

   (b) $(\sqrt{3}/2, \frac{1}{2}), (-\frac{1}{2}, \sqrt{3}/2), (-\sqrt{3} + \frac{1}{2}, -1 - \sqrt{3}/2),$
   $(3\sqrt{3}/2 + 1, -\sqrt{3})$

Section 4–8

1. (a) Parabola   (b) Ellipse   (c) Hyperbola   (d) Ellipse
   (e) Parabola   (f) Hyperbola   (g) Hyperbola

2. (a) Locus of $4x^2 - \dfrac{\sqrt{3}}{2}x - \frac{1}{2}y = 0$ rotated by $-\pi/6$

   (b) Locus of $\frac{3}{2}x^2 + \frac{1}{2}y^2 - 3 = 0$ rotated by $\pi/4$
   (c) Locus of $-\frac{1}{2}x^2 + \frac{5}{2}y^2 - 5 = 0$ rotated by $\pi/4$
   (d) Locus of $4x^2 + 2y^2 - 19 = 0$ rotated by $\pi/4$

   (e) Locus of $2x^2 + \dfrac{1}{\sqrt{2}}x + \dfrac{1}{\sqrt{2}}y - 1 = 0$ rotated by $\pi/4$

   (f) Locus of $3x^2 - y^2 - 4 = 0$ rotated by $\pi/6$
   (g) Locus of $x^2 - y^2 - 1 = 0$ rotated by $\pi/4$

Section 4–9

1. (a) $4(x - 3) - 5(y + 5) - 7(z - 2) = 0$
   (b) $33(x + 1) - 10(y - 3) + 9(z + 5) = 0$
   (c) $4(x - 3) - 30y + 23(z + 1) = 0$
   (d) $3x + y + 2(z - 1) = 0$
   (e) $7(x + 1) - 5(y + 3) - 11(z + 5) = 0$

2. (a) $\dfrac{x - 3}{4} = \dfrac{y + 5}{-5} = \dfrac{z - 2}{-7}$ (b) $\dfrac{x + 1}{33} = \dfrac{y - 3}{-10} = \dfrac{z + 5}{9}$

   (c) $\dfrac{x - 3}{4} = \dfrac{y}{-30} = \dfrac{z + 1}{23}$ (d) $\dfrac{x}{3} = y = \dfrac{z - 1}{2}$

   (e) $\dfrac{x + 1}{7} = \dfrac{y + 3}{-5} = \dfrac{z + 5}{-11}$

3. (a) $2(x - 3) + 3(y + 5) - (z - 2) = 0$
   (b) $x + 1 - 3(y - 3) - 7(z + 5) = 0$
   (c) $3(x - 3) + 5y + 6(z + 1) = 0$
   (d) $x - y - (z - 1) = 0$
   (e) $3(x + 1) + 2(y + 3) + z + 5 = 0$

4. (a) $\dfrac{x - 3}{2} = \dfrac{y + 5}{3} = \dfrac{z - 2}{-1}$ (b) $x + 1 = \dfrac{y - 3}{-3} = \dfrac{z + 5}{-7}$

   (c) $\dfrac{x - 3}{3} = \dfrac{y}{5} = \dfrac{z + 1}{6}$ (d) $x = \dfrac{y}{-1} = \dfrac{z - 1}{-1}$

   (e) $\dfrac{x + 1}{3} = \dfrac{y + 3}{2} = z + 5$

5. (a) $13(x - 3) - 2(y - 4) - 5(z + 2) = 0$
   (b) $2(x - 3) - y + z + 2 = 0$
   (c) $11x - 27y + 2z = 0$
   (d) $x - 3 + 5y - 9(z - 1) = 0$
   (e) $4(x + 1) - 7(y - 2) - 2(z + 5) = 0$

6. (a) $\dfrac{x-3}{13} = \dfrac{y-4}{-2} = \dfrac{z+2}{-5}$ (b) $\dfrac{x-3}{2} = \dfrac{y}{-1} = z+2$

(c) $\dfrac{x}{11} = \dfrac{y}{-27} = \dfrac{z}{2}$ (d) $x-3 = \dfrac{y}{5} = \dfrac{z-1}{-9}$

(e) $\dfrac{x+1}{4} = \dfrac{y-2}{-7} = \dfrac{z+5}{-2}$

7. (a) $2(x-3) + 3(y-4) + 4(z+2) = 0$
   (b) $2(x-3) + 3y - (z+2) = 0$
   (c) $4x + 2y + 5z = 0$
   (d) $2(x-3) + 5y + 3(z-1) = 0$
   (e) $-(x+1) - 2(y-2) + 5(z+5) = 0$

8. (a) $\dfrac{x-3}{2} = \dfrac{y-4}{3} = \dfrac{z+2}{4}$ (b) $\dfrac{x-3}{2} = \dfrac{y}{3} = \dfrac{z+2}{-1}$

(c) $\dfrac{x}{4} = \dfrac{y}{2} = \dfrac{z}{5}$ (d) $\dfrac{x-3}{2} = \dfrac{y}{5} = \dfrac{z-1}{3}$

(e) $\dfrac{x+1}{-1} = \dfrac{y-2}{-2} = \dfrac{z+5}{5}$

Section 5-1

2. (a) $x = y = 0$ (b) $x = -y$ (c) $x = y$ (d) $x = 0$
   (e) $y = 0$ (f) $y = 0$ and $|x| \geq 1$ (g) $x = y = 0$
   (h) $y = 0$ (i) $x = y = 0$

Section 5-2

1. (a) $2xy - 2y^2$, $x^2 - 4xy$ (b) $2e^{-y}$, $-2xe^{-y}$
   (c) $2e^{-\theta}$, $-2re^{-\theta}$ (d) $-e^{-r}\cos\theta$, $-e^{-r}\sin\theta$
   (e) $1/v$, $-u/v^2$ (f) $-u/v^2$, $1/v$
   (g) $\cos\theta$, $-r\sin\theta$ (h) $\sin\theta$, $r\cos\theta$
   (i) $x/\sqrt{x^2+y^2}$, $y/\sqrt{x^2+y^2}$
   (j) $-y/(x^2+y^2)$, $x/(x^2+y^2)$ (k) $2x/(x^2+y^2)$, $2y/(x^2+y^2)$
   (l) $-x/(x^2+y^2)^{3/2}$, $-y/(x^2+y^2)^{3/2}$
   (m) $v^{1.4}$, $1.4pv^{.4}$
   (n) $Be^{-t}\cos At \cos Bx$, $-e^{-t}\sin Bx(A\sin At + \cos At)$
   (o) $x/\sqrt{x^2+y^2+z^2}$, $y/\sqrt{x^2+y^2+z^2}$, $z/\sqrt{x^2+y^2+z^2}$
   (p) $-x/(x^2+y^2+z^2)^{3/2}$, $-y/(x^2+y^2+z^2)^{3/2}$, $-z/(x^2+y^2+z^2)^{3/2}$
   (q) $-y/(x^2+y^2)$, $x/(x^2+y^2)$, 0
   (r) $x/\sqrt{x^2+y^2}$, $y/\sqrt{x^2+y^2}$, 0
   (s) $xz/(x^2+y^2+z^2)\sqrt{x^2+y^2}$, $yz/(x^2+y^2+z^2)\sqrt{x^2+y^2}$,
   $-\sqrt{x^2+y^2}/(x^2+y^2+z^2)$
   (t) $\sin\phi\cos\theta$, $\rho\cos\phi\cos\theta$, $-\rho\sin\phi\sin\theta$
   (u) $\sin\phi\sin\theta$, $\rho\cos\phi\sin\theta$, $\rho\sin\phi\cos\theta$
   (v) $\cos\phi$, $-\rho\sin\phi$, 0

(w) $r/\sqrt{z^2 + r^2}$, 0, $z/\sqrt{z^2 + r^2}$

(x) $-z/(z^2 + r^2)$, 0, $r/(z^2 + r^2)$

(y) $\sin \phi$, $\rho \cos \phi$, 0   (z) $z/(r^2 + z^2)$, 0, $-r(r^2 + z^2)$

5. (a) $1 + x + \frac{1}{2}x^2 - \frac{1}{2}y^2 \cdots$ (b) $x + xy \cdots$

    (c) $x + y \cdots$ (d) $1 - \frac{1}{2}x^2 + xy - \frac{1}{2}y^2 \cdots$

Section 5-3

1. (a) $\cos \theta$ (b) $\sec \theta$ (c) $-r \sin \theta$ (d) $-r \csc \theta$

    (e) $-\tan \theta$ (f) $\cot \theta$ (g) $\sin \theta$ (h) $\csc \theta$

    (i) $r \cos \theta$ (j) $r \sec \theta$ (k) $-\cot \theta$ (l) $\tan \theta$

    (m) $\cos \theta$ (n) $\sec \theta$ (o) $\sin \theta$ (p) $\csc \theta$

    (q) $r \tan \theta$ (r) $-r \cot \theta$ (s) $-(\sin \theta)/r$ (t) $-1/(r \sin \theta)$

    (u) $(\cos \theta)/r$ (v) $1/(r \cos \theta)$ (w) $(\cot \theta)/r$ (x) $-(\tan \theta)/r$

Section 5-4

1. (a) $4e^{2t}$ (b) $\text{sech}^2 2t$

    (c) $(6t^5 + 8t^3 + 2t) \cos 4t - 4(t^6 + 2t^4 + t^2) \sin 4t$

    (d) $t \sin 2t + t^2 \cos 2t$ (e) $e^{t^2 - t^3}(2t^2 - 3t^3 - 3)/t^4$

    (f) $(\sin t)^{\tan t}(1 + \sec^2 t \ln \sin t)$ (g) $(t \cos t - \sin t \log_t \sin t)/(t \sin t \ln t)$

2. $\left(\dfrac{\partial z}{\partial x}\right)_y = y$, $\left(\dfrac{\partial z}{\partial y}\right)_x = x$, $\left(\dfrac{\partial z}{\partial r}\right)_\theta = r \sin 2\theta$, $\left(\dfrac{\partial z}{\partial \theta}\right)_r = r^2 \cos 2\theta$

$\left(\dfrac{\partial z}{\partial x}\right)_r = (r^2 - 2x^2)/y$, $\left(\dfrac{\partial z}{\partial r}\right)_x = rx/y$,

$\left(\dfrac{\partial z}{\partial y}\right)_r = (r^2 - 2y^2)/x$, $\left(\dfrac{\partial z}{\partial r}\right)_y = ry/x$,

$\left(\dfrac{\partial z}{\partial x}\right)_\theta = 2x \tan \theta$, $\left(\dfrac{\partial z}{\partial \theta}\right)_x = x^2 \sec^2 \theta$,

$\left(\dfrac{\partial z}{\partial y}\right)_\theta = 2y \cot \theta$, $\left(\dfrac{\partial z}{\partial \theta}\right)_y = -y^2 \csc^2 \theta$

3. (a) $e^{-xy}(y \, dx + x \, dy)$ (b) $(x \, dx - y \, dy)/(x^2 - y^2)^{1/2}$

    (c) $2(x \, dx + y \, dy)/(x^2 + y^2)$ (d) $\cos (x + y)(dx + dy)$

    (e) $(x \, dy - y \, dx)/(x^2 + y^2)$ (f) $(y \, dx + x \, dy) \sinh (xy)$

    (g) $e^{-y}(dx - x \, dy)$

    (h) $\cos\left(\dfrac{y}{x}\right)\left(\dfrac{dy}{y} - \dfrac{dx}{x}\right) + \sin\left(\dfrac{y}{x}\right)\left(\dfrac{dy}{y} - \dfrac{x \, dy}{y^2}\right)$

    (i) $(2x + y) \, dx + (x + 2y) \, dy$ (j) $\ln y \, dx + (x/y) \, dy$

    (k) $y^x \ln y \, dx + xy^{x-1} \, dy$ (l) $e^{-x/y}(x \, dy - y \, dx)/y^2$

    (m) $2x \sin y^2 \, dx + 2x^2y \cos y^2 \, dy$ (n) $2(y \, dx - x \, dy)/(x^2 - y^2)$

    (o) $e^x(\sin y \, dx + \cos y \, dy)$

    (p) $e^{x+y}[\sin y(\cos x - \sin x) \, dx + \cos x(\cos y + \sin y) \, dy]$

4. (a) $2r\,dr + 2z\,dz$

   (b) $2\rho\,d\rho$

   (c) $(6t + 2u - 2v)\,dt + (6u + 2t + 2v)\,du + (6v - 2t + 2u)\,dv$

   (d) 0

   (e) $4u(1 + u^2 + v^2)\,du + 4v(1 + u^2 + v^2)\,dv$

   (f) $2t(u^2 + v^2)\,dt + 2u(v^2 + t^2)\,du + 2v(t^2 + u^2)\,dv$

   (g) $\sin 2(u - v)(dv - du)$

   (h) $2t\,dt$

   (i) $(4t^3 - 6t^2 + 6t^5 - 2e^{-2t})\,dt$

   (j) $4t(3t^2 + u^2 - v^2)\,dt + 4u(3u^2 + t^2 + v^2)\,du + 4v(3v^2 - t^2 + u^2)\,dv$

   (k) $\left(2u - \dfrac{2v}{u^2 + v^2}\arctan\dfrac{v}{u}\right)du + \left(2v + \dfrac{2u}{u^2 + v^2}\arctan\dfrac{v}{u}\right)dv + 2z\,dz$

6. (a) M (b) K (c) K

   (d) tan bundle M (e) L (f) tan bundle M

   (g) tan bundle K (h) tan bundle L

7. (a) tan bundle K (b) tan bundle L (c) tan bundle L

Section 5–5

1. Rect, comp: dr, $r\,d\theta$, dz

2. $2\,du\sqrt{u^2 + v^2}$, $2\,dv\sqrt{u^2 + v^2}$

4. (a) $\left(\dfrac{\partial\rho}{\partial x}\right)_{yz} = \sin\phi\cos\theta$, $\left(\dfrac{\partial\rho}{\partial y}\right)_{zx} = \sin\phi\sin\theta$, $\left(\dfrac{\partial\rho}{\partial z}\right)_{xy} = \cos\phi$

   $\left(\dfrac{\partial\phi}{\partial x}\right)_{yz} = \dfrac{\cos\phi\cos\theta}{\rho}$, $\left(\dfrac{\partial\phi}{\partial y}\right)_{zx} = \dfrac{\cos\phi\sin\theta}{\rho}$, $\left(\dfrac{\partial\phi}{\partial z}\right)_{xy} = \dfrac{-\sin\phi}{\rho}$

   $\left(\dfrac{\partial\theta}{\partial x}\right)_{yz} = \dfrac{-\sin\theta}{\rho\sin\phi}$, $\left(\dfrac{\partial\theta}{\partial y}\right)_{zx} = \dfrac{\cos\theta}{\rho\sin\phi}$, $\left(\dfrac{\partial\theta}{\partial z}\right)_{xy} = 0$

   (b) $\left(\dfrac{\partial\rho}{\partial x}\right)_{\phi\theta} = \csc\phi\sec\theta$, $\left(\dfrac{\partial\rho}{\partial y}\right)_{\phi\theta} = \csc\phi\csc\theta$, $\left(\dfrac{\partial\rho}{\partial z}\right)_{\phi\theta} = \sec\phi$

   $\left(\dfrac{\partial\phi}{\partial x}\right)_{\theta\rho} = \dfrac{\sec\phi\sec\theta}{\rho}$, $\left(\dfrac{\partial\phi}{\partial y}\right)_{\theta\rho} = \dfrac{\sec\phi\csc\theta}{\rho}$, $\left(\dfrac{\partial\phi}{\partial z}\right)_{\theta\rho} = \dfrac{-\csc\phi}{\rho}$

   $\left(\dfrac{\partial\theta}{\partial x}\right)_{\rho\phi} = \dfrac{-\csc\theta}{\rho\sin\phi}$, $\left(\dfrac{\partial\theta}{\partial y}\right)_{\rho\phi} = \dfrac{\sec\theta}{\rho\sin\phi}$, $\left(\dfrac{\partial\theta}{\partial z}\right)_{\rho\phi} = \infty$

   (c) Take reciprocals from part (b).   (d) Take reciprocals from part (a).

5. (a) $\left(\dfrac{\partial\rho}{\partial r}\right)_{\theta z} = \sin\phi$, $\left(\dfrac{\partial\rho}{\partial\theta}\right)_{zr} = 0$, $\left(\dfrac{\partial\rho}{\partial z}\right)_{r\theta} = \cos\phi$

   $\left(\dfrac{\partial\phi}{\partial r}\right)_{\theta z} = \dfrac{\cos\phi}{\rho}$, $\left(\dfrac{\partial\phi}{\partial\theta}\right)_{zr} = 0$, $\left(\dfrac{\partial\phi}{\partial z}\right)_{r\theta} = \dfrac{-\sin\phi}{\rho}$

   $\left(\dfrac{\partial\theta}{\partial r}\right)_{\theta z} = 0$, $\left(\dfrac{\partial\theta}{\partial\theta}\right)_{zr} = 1$, $\left(\dfrac{\partial\theta}{\partial z}\right)_{r\theta} = 0$

(b) $\left(\dfrac{\partial \rho}{\partial r}\right)_{\phi \theta} = \csc \phi, \left(\dfrac{\partial \rho}{\partial \theta}\right)_{\phi \theta} = \infty, \left(\dfrac{\partial \rho}{\partial z}\right)_{\phi \theta} = \sec \phi$

$\left(\dfrac{\partial \phi}{\partial r}\right)_{\theta \rho} = \dfrac{\sec \phi}{\rho}, \left(\dfrac{\partial \phi}{\partial \theta}\right)_{\theta \rho} = \infty, \left(\dfrac{\partial \phi}{\partial z}\right)_{\theta \rho} = \dfrac{-\csc \phi}{\rho}$

$\left(\dfrac{\partial \theta}{\partial r}\right)_{\rho \phi} = \infty, \left(\dfrac{\partial \theta}{\partial \theta}\right)_{\rho \phi} = 1, \left(\dfrac{\partial \theta}{\partial z}\right)_{\rho \phi} = \infty$

(c) Take reciprocals from part (b). (d) Take reciprocals from part (a).

Section 6–1.

2. (a) arctan $(\tfrac{3}{4})$ (b) 0 and arctan $(3\sqrt{3}/5)$
   (c) $\pi/4$ (d) $\pi/3$ and $\pi/2$
   (e) $\pi/6$ (f) 0 and arctan $3\sqrt{3}$
   (g) $\pi/3$ (h) $\pi/2$

7. (a) $2\pi(2 - \sqrt{2})$ (b) $2\pi\sqrt{2}$

8. (a) 4π (b) 4π

9. (a) $a\pi$ (b) $8a$ (c) $a(4\pi - 3\sqrt{3})/8$ (d) $\pi a\sqrt{2}$ (e) 5π

12. (a) $2/a$ (b) $1/(e^{a\theta}\sqrt{1 + a^2})$ (c) $(\theta^2 + 2)/[a(\theta^2 + 1)^{3/2}]$
    (d) $(1 - \cos \theta + 2 \sin^2 \theta)/(8a \sin^3 \theta/2)$

15. $h\sqrt{2 - 2\epsilon \cos \theta}/m$

16. $\omega \sin \phi$

Section 6–2

1. (a) $0(x = 0)$ (b) $-b^2 x/(a^2 y)(\pm a, 0)$ (c) $b^2 x/(a^2 y)(\pm a, 0)$
   (d) $-y/x$ (none) (e) $-\sqrt{y/x}\ (0, a)$ (f) $-\sqrt[3]{y/x}\ (0, \pm a)$
   (g) $(ay - x^2)/(y^2 - ax)(0, 0), (a\sqrt[3]{4}, a\sqrt[3]{2})$
   (h) $-(x^2 + 2xy)/(x^2 + y^2)$ (none)
   (i) $(y - x^2)/(y^2 - x)\ (\sqrt[3]{3} \pm 2\sqrt{2}, \sqrt[3]{1 \pm \sqrt{2}})$
   (j) $-(x^3 + 3x^2 y)/(x^3 + y^3)$ (none)
   (k) $-\sqrt{y/x}\ (4, 0)$
   (l) $-(x\sqrt{xy} + y)/(x + y\sqrt{xy})\ (\pm\sqrt{6}, 0), (-1, -1)$
   (m) $(y - 3x^2)/(2y - x)(0, 0), (\tfrac{1}{4}, \tfrac{1}{8})$
   (n) $(2x - 3\sqrt{xy})/(2y + x\sqrt{x/y})\ (0, 0)$
   (o) $(2x - y^2 - 3x^2)/(2y + 2xy)\ (0, 0), (1, 0)$
   (p) $2x/(3y^2 - 2y)\ (0, 0)$
   (q) $x(1 - x^2 - y^2)/[y(1 + x^2 + y^2)]\ (0, 0), (+\sqrt{2}, 0)$
   (r) $(x^2 - y)/(y^2 + x)\ (0, 0), (\sqrt[3]{4}, -\sqrt[3]{2})$
   (s) $(5x^4 + 8xy^2 - 16x^3)/[4y(y^2 - 2x^2)]\ (0, 0), (4, 0)$
   (t) $(x^2 + ye^{1/x})/(x^2 + x^2 e^{1/x})$ (none)
   (u) $(y^2 + 3x^2)/(4y - 2xy)\ (0, 0)$ (v) $(5x^4 + 4y - 8x)/(2y - 4x)\ (0, 0)$
   (w) $\cos x/\sin y\ (n\pi, 2n\pi)$ (x) $-y/x$ (none)

2. (a) 8 (b) $-\tfrac{23}{26}$ (c) -2 (d) $-\pi/2$ (e) -1
   (f) -1 (g) 1 (h) 0 (i) -1 (j) -1

3.

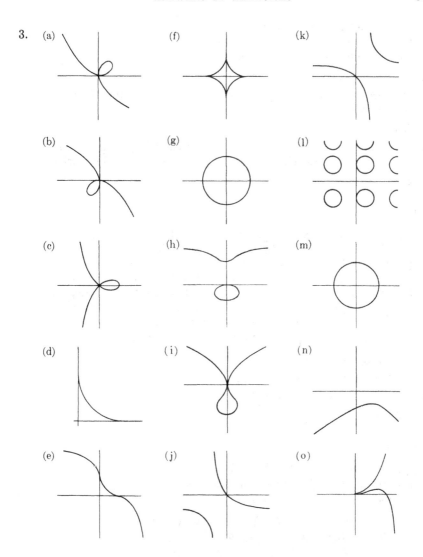

Section 6-3

1. For max: square only. For min: side of triangle $= \sqrt{3}$ times side of square.
2. For max: square only. For min: hypotenuse of triangle $= (1 + \sqrt{2})$ times side of square.
3. For max: equilateral only. For min: hypotenuse of right isosceles $= (1 + \sqrt{2})/\sqrt{3}$ times side of equilateral.
4. (a) $bx = ay$ for min; $x = 0$ for max
   (b) $x = y$ for min; $x = 0$ or $y = 0$ for max
   (c) $x = 2y$ for min; $x = 0$ for max

(d) $x = 0$ for min; $y = 0$ for max

(e) $x = 0$ for min; $y = 0$ for max

(f) $y = 0$ for min; $x = 0$ for max

5. (a) $(C - \frac{1}{2}, \sqrt{C - \frac{1}{2}})$ (b) $(0, 0)$

6. Distance from source of strength $a = \sqrt[3]{a/b}$ times that from source of strength b.

7. Height $= \frac{1}{2}$ side of base

8. (a) square   (b) Height $= \sqrt{2}$ times radius

     (c) Base $= 2$ times height   (d) Radius $= \sqrt{2}$ times height

9. $2 \arcsin \sqrt{\frac{2}{3}}$

Section 6-4

1. (a) Max $(1/\sqrt{2}, 1/\sqrt{2}, 0)$, min $(-1/\sqrt{2}, -1\sqrt{2}, 0)$

     (b) Max $(\sqrt{\frac{1}{3}}, \pm\sqrt{\frac{2}{3}}, \pm\sqrt{\frac{2}{3}}$, min $(-\sqrt{\frac{1}{3}}, \pm\sqrt{\frac{2}{3}}, \pm\sqrt{\frac{2}{3}})$

     (c) Max $(\pm1/\sqrt{2}, \pm1/\sqrt{2})$, min $(\pm1/\sqrt{2}, \mp1/\sqrt{2})$

     (d) Max $(1/\sqrt{3}, 1/\sqrt{3}, 1/\sqrt{3})$, min $(-1/\sqrt{3}, -1/\sqrt{3}, -1/\sqrt{3})$

     (e) Max $(\pm1/\sqrt{3}, \pm1/\sqrt{3}, 1/\sqrt{3})$ or $(\pm1/\sqrt{3}, \mp1/\sqrt{3}, -1/\sqrt{3})$,

         min $(\pm1/\sqrt{3}, \pm1/\sqrt{3}, -1/\sqrt{3})$ or $(\pm1/\sqrt{3}, \mp1/\sqrt{3}, 1/\sqrt{3})$

     (f) Max $(\frac{4}{3}, \frac{4}{3}, \frac{4}{3})$, min $(1, 1, 2)$ or $(1, 2, 1)$ or $(2, 1, 1)$

2. (a) Cube   (b) Square base, height $= \frac{1}{2}$ side of base

3. $(0, \pm1, 0)$, $(\pm1, 0, 0)$

5. $\sqrt{5}\,h = (1 + \sqrt{5})r$, $\sec\alpha = 3/2$

6. $\sqrt{3}\,h = (1 + \sqrt{3})r$, $\alpha = \pi/3$

7. \$89, \$94

Section 6-5

2. (a) $e^{2r}/\sqrt{e^{2r} + e^{2s}}$, $e^{2s}/\sqrt{e^{2r} + e^{2s}}$

     (b) $2r(2 + s^2)/(2r^2 + 2s^2 + r^2s^2)$, $2s(2 + r^2)/(2r^2 + 2s^2 + r^2s^2)$

     (c) $(3r^2 - s^2)/s$, $-r(r^2 + s^2)/s^2$

3. (a) $\dfrac{dw}{dt} = \left(\dfrac{\partial w}{\partial x}\right)_{yz} \dfrac{dx}{dt} + \left(\dfrac{\partial w}{\partial y}\right)_{zz} \dfrac{dy}{dt} + \left(\dfrac{\partial w}{\partial z}\right)_{xy} \dfrac{dz}{dt}$

     (b) $\dfrac{dw}{dx} = \left(\dfrac{\partial w}{\partial x}\right)_{yz} + \left(\dfrac{\partial w}{\partial y}\right)_{zz} \dfrac{dy}{dx} + \left(\dfrac{\partial w}{\partial z}\right)_{xy} \dfrac{dz}{dx}$

     (c) $\left(\dfrac{\partial w}{\partial t}\right)_{x} = \left(\dfrac{\partial w}{\partial y}\right)_{zz} \left(\dfrac{\partial y}{\partial t}\right)_{x} + \left(\dfrac{\partial w}{\partial z}\right)_{xy} \left(\dfrac{\partial z}{\partial t}\right)_{x}$

         $\left(\dfrac{\partial w}{\partial x}\right)_{t} = \left(\dfrac{\partial w}{\partial x}\right)_{yz} + \left(\dfrac{\partial w}{\partial y}\right)_{zz} \left(\dfrac{\partial y}{\partial x}\right)_{t} + \left(\dfrac{\partial w}{\partial z}\right)_{xy} \left(\dfrac{\partial z}{\partial x}\right)_{t}$

     (d) $\left(\dfrac{\partial w}{\partial u}\right)_{v} = \left(\dfrac{\partial w}{\partial x}\right)_{yz} \left(\dfrac{\partial x}{\partial u}\right)_{v} + \left(\dfrac{\partial w}{\partial y}\right)_{zz} \left(\dfrac{\partial y}{\partial u}\right)_{v} + \left(\dfrac{\partial w}{\partial z}\right)_{xy} \left(\dfrac{\partial z}{\partial u}\right)_{v}$

         $\left(\dfrac{\partial w}{\partial v}\right)_{u} = \left(\dfrac{\partial w}{\partial x}\right)_{yz} \left(\dfrac{\partial x}{\partial v}\right)_{u} + \left(\dfrac{\partial w}{\partial y}\right)_{zz} \left(\dfrac{\partial y}{\partial v}\right)_{u} + \left(\dfrac{\partial w}{\partial z}\right)_{xy} \left(\dfrac{\partial z}{\partial v}\right)_{u}$

(e) $\left(\dfrac{\partial w}{\partial x}\right)_y = \left(\dfrac{\partial w}{\partial x}\right)_{yz} + \left(\dfrac{\partial w}{\partial z}\right)_{xy} \left(\dfrac{\partial z}{\partial x}\right)_y$

$\left(\dfrac{\partial w}{\partial y}\right)_x = \left(\dfrac{\partial w}{\partial y}\right)_{zx} + \left(\dfrac{\partial w}{\partial z}\right)_{xy} \left(\dfrac{\partial z}{\partial y}\right)_x$

(f) $\left(\dfrac{\partial w}{\partial x}\right)_{yt} = \left(\dfrac{\partial w}{\partial x}\right)_{yz} + \left(\dfrac{\partial w}{\partial z}\right)_{xy} \left(\dfrac{\partial z}{\partial x}\right)_{yt}$

$\left(\dfrac{\partial w}{\partial y}\right)_{tx} = \left(\dfrac{\partial w}{\partial y}\right)_{zx} + \left(\dfrac{\partial w}{\partial z}\right)_{xy} \left(\dfrac{\partial z}{\partial y}\right)_{tx}$

$\left(\dfrac{\partial w}{\partial t}\right)_{xy} = \left(\dfrac{\partial w}{\partial z}\right)_{xy} \left(\dfrac{\partial z}{\partial t}\right)_{xy}$

(g) $\left(\dfrac{\partial w}{\partial x}\right)_{uv} = \left(\dfrac{\partial w}{\partial x}\right)_{yz} + \left(\dfrac{\partial w}{\partial y}\right)_{zx} \left(\dfrac{\partial y}{\partial x}\right)_{uv} + \left(\dfrac{\partial w}{\partial z}\right)_{xy} \left(\dfrac{\partial z}{\partial x}\right)_{uv}$

$\left(\dfrac{\partial w}{\partial u}\right)_{vx} = \left(\dfrac{\partial w}{\partial y}\right)_{zx} \left(\dfrac{\partial y}{\partial u}\right)_{vx} + \left(\dfrac{\partial w}{\partial z}\right)_{xy} \left(\dfrac{\partial z}{\partial u}\right)_{vx}$

$\left(\dfrac{\partial w}{\partial v}\right)_{xu} = \left(\dfrac{\partial w}{\partial y}\right)_{zx} \left(\dfrac{\partial y}{\partial v}\right)_{xu} + \left(\dfrac{\partial w}{\partial z}\right)_{xy} \left(\dfrac{\partial z}{\partial v}\right)_{xu}$

(h) $\left(\dfrac{\partial w}{\partial t}\right)_{uv} = \left(\dfrac{\partial w}{\partial x}\right)_{yz} \left(\dfrac{\partial x}{\partial t}\right)_{uv} + \left(\dfrac{\partial w}{\partial y}\right)_{zx} \left(\dfrac{\partial y}{\partial t}\right)_{uv} + \left(\dfrac{\partial w}{\partial z}\right)_{xy} \left(\dfrac{\partial z}{\partial t}\right)_{uv}$

$\left(\dfrac{\partial w}{\partial u}\right)_{vt} = \left(\dfrac{\partial w}{\partial x}\right)_{yz} \left(\dfrac{\partial x}{\partial u}\right)_{vt} + \left(\dfrac{\partial w}{\partial y}\right)_{zx} \left(\dfrac{\partial y}{\partial u}\right)_{vt} + \left(\dfrac{\partial w}{\partial z}\right)_{xy} \left(\dfrac{\partial z}{\partial u}\right)_{vt}$

$\left(\dfrac{\partial w}{\partial v}\right)_{tu} = \left(\dfrac{\partial w}{\partial x}\right)_{yz} \left(\dfrac{\partial x}{\partial v}\right)_{tu} + \left(\dfrac{\partial w}{\partial y}\right)_{zx} \left(\dfrac{\partial y}{\partial v}\right)_{tu} + \left(\dfrac{\partial w}{\partial z}\right)_{xy} \left(\dfrac{\partial z}{\partial v}\right)_{tu}$

4. (a) $\left(\dfrac{\partial x_1}{\partial x_2}\right)_{x_3 x_4} = \left(\dfrac{\partial x_1}{\partial x_2}\right)_{x_3 x_5} + \left(\dfrac{\partial x_1}{\partial x_5}\right)_{x_2 x_3} \left(\dfrac{\partial x_5}{\partial x_2}\right)_{x_3 x_4}$

(b) $\left(\dfrac{\partial x_1}{\partial x_2}\right)_{x_3 x_4} = \left(\dfrac{\partial x_1}{\partial x_2}\right)_{x_5 x_6} + \left(\dfrac{\partial x_1}{\partial x_5}\right)_{x_2 x_6} \left(\dfrac{\partial x_5}{\partial x_2}\right)_{x_3 x_4} + \left(\dfrac{\partial x_1}{\partial x_6}\right)_{x_2 x_5} \left(\dfrac{\partial x_6}{\partial x_2}\right)_{x_3 x_4}$

Section 6–6

1. (a) $\dfrac{1}{2}\left[\left(\dfrac{\partial w}{\partial u}\right)^2 + \left(\dfrac{\partial w}{\partial v}\right)^2\right]$ (b) $\left[\left(\dfrac{\partial w}{\partial u}\right)^2 + \left(\dfrac{\partial w}{\partial v}\right)^2\right] \Big/ [4(u^2 + v^2)]$

(c) $e^{-2u}\left[\left(\dfrac{\partial w}{\partial u}\right)^2 + \left(\dfrac{\partial w}{\partial v}\right)^2\right]$

(d) $\dfrac{1}{484}\left[34\left(\dfrac{\partial w}{\partial u}\right)^2 + 28\,\dfrac{\partial w}{\partial u}\,\dfrac{\partial w}{\partial v} + 20\left(\dfrac{\partial w}{\partial v}\right)^2\right]$

(e) $\dfrac{1}{\cosh 2v}\left[\left(\dfrac{\partial w}{\partial u}\right)^2 - \dfrac{2\sinh 2v}{u\cosh 2v}\,\dfrac{\partial w}{\partial u}\,\dfrac{\partial w}{\partial v} + \dfrac{1}{u^2}\left(\dfrac{\partial w}{\partial v}\right)^2\right]$

2. (a) $\left(\dfrac{\partial w}{\partial r}\right)^2 + \dfrac{1}{r^2}\left(\dfrac{\partial w}{\partial \theta}\right)^2 + \left(\dfrac{\partial w}{\partial z}\right)^2$

(b) $\left(\dfrac{\partial w}{\partial \rho}\right)^2 + \dfrac{1}{\rho^2}\left(\dfrac{\partial w}{\partial \phi}\right)^2 + \dfrac{1}{\rho^2 \sin^2 \phi}\left(\dfrac{\partial w}{\partial \theta}\right)^2$

(c) $\dfrac{1}{2}\left[\left(\dfrac{\partial w}{\partial t}\right)^2 + \left(\dfrac{\partial w}{\partial u}\right)^2 + \left(\dfrac{\partial w}{\partial v}\right)^2 + \dfrac{\partial w}{\partial t}\dfrac{\partial w}{\partial v} - \dfrac{\partial w}{\partial u}\dfrac{\partial w}{\partial v} - \dfrac{\partial w}{\partial t}\dfrac{\partial w}{\partial u}\right]$

3. (a) $t - p\left(\dfrac{\partial t}{\partial p}\right)_v + \begin{vmatrix} \left(\dfrac{\partial t}{\partial v}\right)_p & \left(\dfrac{\partial t}{\partial p}\right)_v \\[2ex] \left(\dfrac{\partial u}{\partial v}\right)_p & \left(\dfrac{\partial u}{\partial p}\right)_v \end{vmatrix} = 0$

(b) $\left(\dfrac{\partial u}{\partial p}\right)_t + t\left(\dfrac{\partial v}{\partial t}\right)_p + p\left(\dfrac{\partial v}{\partial p}\right)_t = 0$

(c) $\left(\dfrac{\partial t}{\partial p}\right)_u - t\left(\dfrac{\partial v}{\partial u}\right)_p + p\begin{vmatrix} \left(\dfrac{\partial v}{\partial u}\right)_p & \left(\dfrac{\partial v}{\partial p}\right)_u \\[2ex] \left(\dfrac{\partial t}{\partial u}\right)_p & \left(\dfrac{\partial t}{\partial p}\right)_u \end{vmatrix} = 0$

(d) $t\begin{vmatrix} \left(\dfrac{\partial p}{\partial t}\right)_u & \left(\dfrac{\partial p}{\partial u}\right)_t \\[2ex] \left(\dfrac{\partial v}{\partial t}\right)_u & \left(\dfrac{\partial v}{\partial u}\right)_t \end{vmatrix} - p\left(\dfrac{\partial v}{\partial u}\right)_t - 1 = 0$

Section 6-7

2. (a) $\dfrac{\partial x}{\partial u} = -\dfrac{5}{3}, \dfrac{\partial x}{\partial v} = \dfrac{-4}{3}, \dfrac{\partial y}{\partial u} = \dfrac{7}{3}, \dfrac{\partial y}{\partial v} = \dfrac{5}{3}$

(b) $\dfrac{\partial x}{\partial u} = \dfrac{-u}{x}, \dfrac{\partial x}{\partial v} = \dfrac{v}{2x}, \dfrac{\partial y}{\partial u} = 0, \dfrac{\partial y}{\partial v} = \dfrac{3v}{2y}$

(c) $\dfrac{\partial x}{\partial u} = \dfrac{ve^v}{e^{u+v} - (1-v)(xe^y + 1)}, \dfrac{\partial x}{\partial v} = \dfrac{ue^v + (ye^v - x)(xe^y + 1)}{e^{u+v} - (1-v)(xe^y + 1)},$

$\dfrac{\partial y}{\partial u} = \dfrac{v^2 - v}{e^{u+v} - (1-v)(xe^y + 1)}, \dfrac{\partial y}{\partial v} = \dfrac{e^v(x - ye^v) + u(v-1)}{e^{u+v} - (1-v)(xe^y + 1)}$

(d) $\dfrac{\partial x}{\partial u} = \dfrac{-2y}{4xy + 1}, \dfrac{\partial x}{\partial v} = \dfrac{-1}{4xy + 1}, \dfrac{\partial y}{\partial u} = \dfrac{-1}{4xy + 1}, \dfrac{\partial y}{\partial v} = \dfrac{2x}{4xy + 1}$

3. (a) $\dfrac{\partial w}{\partial u} = \dfrac{7x - 5y}{3}, \dfrac{\partial w}{\partial v} = \dfrac{5x - 4y}{3}$

(b) $\dfrac{\partial w}{\partial u} = \dfrac{-uy}{x}, \dfrac{\partial w}{\partial v} = \dfrac{v(y^2 + 3x^2)}{2xy}$

(c) $\dfrac{\partial w}{\partial u} = \dfrac{yve^v + xv(v-1)}{e^{y+v} - (1-v)(xe^y+1)}$

$\dfrac{\partial w}{\partial v} = \dfrac{yue^v + y(ye^v - x)(xe^y+1) + xe^y(x - ye^v) + xu(v-1)}{e^{y+v} - (1-v)(xe^y+1)}$

(d) $\dfrac{\partial w}{\partial u} = \dfrac{-2y^2 - x}{4xy+1},\ \dfrac{\partial w}{\partial v} = \dfrac{2x^2 - y}{4xy+1}$

4. (a) $\dfrac{\partial x}{\partial y} = \dfrac{-f_2(x,y,z)}{f_1(x,y,z)},\ \dfrac{\partial x}{\partial z} = \dfrac{-f_3(x,y,z)}{f_1(x,y,z)}$

(b)

$$\frac{dx}{dz} = \frac{-\begin{vmatrix} f_3(x,y,z) & f_2(x,y,z) \\ g_3(x,y,z) & g_2(x,y,z) \end{vmatrix}}{\begin{vmatrix} f_1(x,y,z) & f_2(x,y,z) \\ g_1(x,y,z) & g_2(x,y,z) \end{vmatrix}}$$

$$\frac{dy}{dz} = \frac{-\begin{vmatrix} f_1(x,y,z) & f_3(x,y,z) \\ g_1(x,y,z) & g_3(x,y,z) \end{vmatrix}}{\begin{vmatrix} f_1(x,y,z) & f_2(x,y,z) \\ g_1(x,y,z) & g_2(x,y,z) \end{vmatrix}}$$

(c)

$$\frac{dx}{dz} = \frac{\begin{vmatrix} 1 & f_2(x,y) \\ -g_2(y,z) & g_1(y,z) \end{vmatrix}}{\begin{vmatrix} f_1(x,y) & f_2(x,y) \\ -1 & g_1(y,z) \end{vmatrix}},\ \frac{dy}{dz} = \frac{\begin{vmatrix} f_1(x,y) & 1 \\ -1 & -g_2(y,z) \end{vmatrix}}{\begin{vmatrix} f_1(x,y) & f_2(x,y) \\ -1 & g_1(y,z) \end{vmatrix}}$$

(d)

$$\frac{\partial w}{\partial y} = \frac{\begin{vmatrix} f_2(w,x,y,z) & f_3(w,x,y,z) \\ g_2(x,y) & g_2(x,y) \end{vmatrix}}{f_1(w,x,y,z)g_1(x,y)},\ \frac{\partial x}{\partial y} = \frac{-g_2(x,y)}{g_1(x,y)}$$

$$\frac{\partial w}{\partial z} = \frac{\begin{vmatrix} f_2(w,x,y,z) & f_4(w,x,y,z) \\ g_1(x,y) & -1 \end{vmatrix}}{f_1(w,x,y,z)g_1(x,y)},\ \frac{\partial x}{\partial z} = \frac{1}{g_1(x,y)}$$

(e)

$$\frac{\partial x}{\partial y} = \frac{1}{g_1(x,y)\phi_1(u,v)} \begin{vmatrix} f_1(y,z) & 0 & 1 \\ -g_2(x,y) & -1 & 0 \\ 0 & \phi_1(u,v) & \phi_2(u,v) \end{vmatrix},$$

$$\frac{\partial u}{\partial y} = \frac{-f_1(y,z)\phi_2(u,v)}{\phi_1(u,v)},\ \frac{\partial v}{\partial y} = f_1(y,z)$$

(f)

$$\frac{\partial u}{\partial y} = \frac{\begin{vmatrix} f_2(u, v, x, y, z) & f_3(u, v, x, y, z) & f_4(u, v, x, y, z) \\ g_2(u, v, x, y, z) & g_3(u, v, x, y, z) & g_4(u, v, x, y, z) \\ \phi_2(u, v, x, y, z) & \phi_3(u, v, x, y, z) & \phi_4(u, v, x, y, z) \end{vmatrix}}{\begin{vmatrix} f_1(u, v, x, y, z) & f_2(u, v, x, y, z) & f_3(u, v, x, y, z) \\ g_1(u, v, x, y, z) & g_2(u, v, x, y, z) & g_3(u, v, x, y, z) \\ \phi_1(u, v, x, y, z) & \phi_2(u, v, x, y, z) & \phi_3(u, v, x, y, z) \end{vmatrix}}$$

Other answers by cyclic permutation of symbols

(g)

$$\frac{\partial u}{\partial x} = \frac{\begin{vmatrix} g_2(u, v, w) & g_3(u, v, w) \\ f_2(u, v, w) & f_3(u, v, w) \end{vmatrix}}{\begin{vmatrix} \phi_1(u, v, w) & \phi_2(u, v, w) & \phi_3(u, v, w) \\ g_1(u, v, w) & g_2(u, v, w) & g_3(u, v, w) \\ f_1(u, v, w) & f_2(u, v, w) & f_3(u, v, w) \end{vmatrix}}$$

Other answers by cyclic permutation of symbols

(h) Let $\Delta = \begin{vmatrix} g_1(u, v) & g_2(u, v) \\ \phi_1(u, v) & \phi_2(u, v) \end{vmatrix}$;

$$\frac{\partial x}{\partial y} = \frac{1}{\Delta} \begin{vmatrix} f_1(u, v) & f_2(u, v) \\ \phi_1(u, v) & \phi_2(u, v) \end{vmatrix}, \frac{\partial u}{\partial y} = \frac{\phi_2(u, v)}{\Delta},$$

$$\frac{\partial v}{\partial y} = \frac{-\phi_1(u, v)}{\Delta}, \frac{\partial x}{\partial z} = \frac{-1}{\Delta} \begin{vmatrix} f_1(u, v) & f_2(u, v) \\ g_1(u, v) & g_2(u, v) \end{vmatrix},$$

$$\frac{\partial u}{\partial z} = \frac{-g_2(u, v)}{\Delta}, \frac{\partial v}{\partial z} = \frac{g_1(u, v)}{\Delta}$$

(i)

$$\frac{\partial u}{\partial x} = \frac{\begin{vmatrix} f_2(u, v, x, y, z) & f_3(u, v, x, y, z) \\ g_2(u, v, x, y, z) & g_3(u, v, x, y, z) \end{vmatrix}}{\begin{vmatrix} f_1(u, v, x, y, z) & f_2(u, v, x, y, z) \\ g_1(u, v, x, y, z) & g_2(u, v, x, y, z) \end{vmatrix}}$$

Other answers by cyclic permutation of symbols.

(j) $\dfrac{du}{dz} = \dfrac{1}{\phi'(u)}, \dfrac{dx}{dz} = \dfrac{f'(u)}{\phi'(u)}, \dfrac{dy}{dz} = \dfrac{g'(u)}{\phi'(u)}$

Section 7–1

1. (a) Scalar   (b) Matrix (3×1) (c) Scalar
   (d) Vector   (e) Linear transformation (vector to scalar)
   (f) Matrix (1×3) (g) Matrix (3×1) (h) Matrix (3×1)
   (i) Matrix (3×1) (j) Vector
   (k) Linear transformation (vector to vector)   (l) Vector
   (m) Matrix (3×3) (n) Matrix (3×1) (o) Scalar

Section 7–2

1. (a) $yz\mathbf{i} + zx\mathbf{j} + xy\mathbf{k}$, $\sqrt{y^2z^2 + z^2x^2 + x^2y^2}$, $3xyz/\sqrt{x^2 + y^2 + z^2}$

   (b) $(x\mathbf{i} + y\mathbf{j} + z\mathbf{k})/\sqrt{x^2 + y^2 + z^2}$, 1, 1

   (c) $(-y\mathbf{i} + x\mathbf{j})/(x^2 + y^2)$, $1/\sqrt{x^2 + y^2}$, 0

   (d) $(y/z)\mathbf{i} + (x/z)\mathbf{j} - (xy/z^2)\mathbf{k}$, $\sqrt{y^2z^2 + x^2z^2 + x^2y^2}/z^2$,
   $xy/(z\sqrt{x^2 + y^2 + z^2})$

   (e) $(2x\mathbf{i} + 2y\mathbf{j} + 2z\mathbf{k})/(x^2 + y^2 + z^2)$, $2/\sqrt{x^2 + y^2 + z^2}$,
   $2/\sqrt{x^2 + y^2 + z^2}$

   (f) $-(x\mathbf{i} + y\mathbf{j} + z\mathbf{k})/(x^2 + y^2 + z^2)^{3/2}$, $1/(x^2 + y^2 + z^2)$,
   $-1/(x^2 + y^2 + z^2)$

   (g) $[xz\mathbf{i} + yz\mathbf{j} - (x^2 + y^2)\mathbf{k}]/[(x^2 + y^2 + z^2)\sqrt{x^2 + y^2}]$,
   $1/\sqrt{x^2 + y^2 + z^2}$, 0

   (h) $(y + z)\mathbf{i} + (x + z)\mathbf{j} + (x + y)\mathbf{k}$, $\sqrt{(y + z)^2 + (x + z)^2 + (x + y)^2}$,
   $2(xy + yz + zx)/\sqrt{x^2 + y^2 + z^2}$

2. (a) $6t^5/\sqrt{1 + 4t^2 + 9t^4}$ (b) 0
   (c) $1/\sqrt{2}$ (d) $4t^5/\sqrt{4t^6 + t^4 + 1}$
   (e) $2/(t\sqrt{14})$ (f) $\sqrt{3}\,(t - 2)/(3t^2 - 12t + 14)^{3/2}$
   (g) 0   (h) $2x\sqrt{3}$

3. (a) $6/\sqrt{14}$ (b) $-\sqrt{5}/3$ (c) 0   (d) $3\sqrt{2}/8$
   (e) $-3/(7\sqrt{2})$ (f) $\frac{1}{9}$ (g) $1/\sqrt{3}$ (h) $3\sqrt{2}/2$

4. (a) $x + y + z - 3 = 0$ (b) $x + 2y + z - 6 = 0$
   (c) $x + y + z - 3 = 0$ (d) $x + y - z - 1 = 0$
   (e) $2x + 4y - z - 3 = 0$ (f) $z - 1 = 0$
   (g) $ex + ey - z - e = 0$ (h) $x - y + 2z - \pi/2 = 0$

5. (a) $x - 1 = y - 1 = z - 1$ (b) $x - 1 = \dfrac{y - 2}{2} = z - 1$

   (c) $x - 1 = y - 1 = z - 1$ (d) $x - 1 = y - 1 = 1 - z$

   (e) $\dfrac{x - 1}{2} = \dfrac{y - 1}{4} = 3 - z$ (f) $x = y = \pi/4$

   (g) $\dfrac{x - 1}{e} = \dfrac{y - 1}{e} = e - z$ (h) $x - 1 = 1 - y = \dfrac{z - \pi/4}{2}$

Section 7–3

1. (a) $0, 0$

   (b) $\dfrac{-3(x+y+z)}{(x^2+y^2+z^2)^{5/2}}, \dfrac{-3[(y-z)\mathbf{i}+(z-x)\mathbf{j}+(x-y)\mathbf{k}]}{(x^2+y^2+z^2)^{5/2}}$

   (c) $3e^x \cos y + e^z, 3e^x \sin y \mathbf{k}$

2. Same as Exercise 1

Section 7–4

3.

$$D_{X(\mathbf{r})}X(\mathbf{w}) = \begin{bmatrix} \dfrac{2x^2-y^2-z^2}{(x^2+y^2+z^2)^{5/2}} & \dfrac{3xy}{(x^2+y^2+z^2)^{5/2}} & \dfrac{3xz}{(x^2+y^2+z^2)^{5/2}} \\[3mm] \dfrac{3xy}{(x^2+y^2+z^2)^{5/2}} & \dfrac{2y^2-x^2-z^2}{(x^2+y^2+z^2)^{5/2}} & \dfrac{3yz}{(x^2+y^2+z^2)^{5/2}} \\[3mm] \dfrac{3xz}{(x^2+y^2+z^2)^{5/2}} & \dfrac{3yz}{(x^2+y^2+z^2)^{5/2}} & \dfrac{2z^2-x^2-y^2}{(x^2+y^2+z^2)^{5/2}} \end{bmatrix}$$

5. (a) $\mathbf{v}_0 = \mathbf{i} + 2\mathbf{j} + 3\mathbf{k}$ (b) $\mathbf{v}_0 = 2\mathbf{i} + \mathbf{j} - 3\mathbf{k}$

6. (a) $R_\alpha^+ = \begin{bmatrix} \cos\alpha & 0 \\ 0 & \cos\alpha \end{bmatrix}, R_\alpha^- = \begin{bmatrix} 0 & -\sin\alpha \\ \sin\alpha & 0 \end{bmatrix}$

   $T^+\mathbf{i} = \cos\alpha\mathbf{i}, T^-\mathbf{i} = \sin\alpha\mathbf{j}.$

   (b) $T^+ : \begin{bmatrix} 2 & -1 \\ -1 & 1 \end{bmatrix}, T^- : \begin{bmatrix} 0 & 4 \\ -4 & 0 \end{bmatrix}$

   $T(\mathbf{i}+2\mathbf{j}) = 8\mathbf{i} - 3\mathbf{j}, T^+(\mathbf{i}+2\mathbf{j}) = \mathbf{j}, T^-(\mathbf{i}+2\mathbf{j}) = 8\mathbf{i} - 4\mathbf{j}.$

7. $\lambda^2 - 3\lambda - 4 = 0$

Section 7–5

2. (b) $\begin{bmatrix} \dfrac{2}{\rho^3} & 0 & 0 \\[3mm] 0 & -\dfrac{1}{\rho^3} & 0 \\[3mm] 0 & 0 & -\dfrac{1}{\rho^3} \end{bmatrix}$

$\begin{bmatrix} \dfrac{2r^2-z^2}{(r^2+z^2)^{5/2}} & 0 & \dfrac{-3rz}{(r^2+z^2)^{5/2}} \\[3mm] 0 & \dfrac{-r^2-z^2}{(r^2+z^2)^{5/2}} & 0 \\[3mm] \dfrac{-3rz}{(r^2+z^2)^{5/2}} & 0 & \dfrac{2z^2-r^2}{(r^2+z^2)^{5/2}} \end{bmatrix}$

3. (b)
$$\begin{bmatrix} \dfrac{2\csc\phi}{\rho^3} & \dfrac{2\csc\phi\cot\phi}{\rho^3} & 0 \\[3mm] \dfrac{2\csc\phi\cot\phi}{\rho^3} & \dfrac{\csc\phi(\csc^2\phi+\cot^2\phi-1)}{\rho^3} & 0 \\[3mm] 0 & 0 & \dfrac{-\csc\phi(1+\cot^2\phi)}{\rho^3} \end{bmatrix}$$

$$\begin{bmatrix} \dfrac{2}{r^3} & 0 & 0 \\[3mm] 0 & \dfrac{-1}{r^3} & 0 \\[3mm] 0 & 0 & 0 \end{bmatrix}$$

4. (b) $\mathbf{w} = -y\mathbf{i} + x\mathbf{j} = \rho\sin\phi\,\mathbf{b}_\theta$
   (c) $\operatorname{div}\mathbf{w} = 0$, $\operatorname{curl}\mathbf{w} = 2\mathbf{k} = 2\cos\phi\,\mathbf{b}_\rho - 2\sin\phi\,\mathbf{b}_\phi$

5. (a) $\dfrac{1}{\sqrt{s^2+t^2}}\left(\dfrac{\partial u}{\partial s}\,\mathbf{b}_s + \dfrac{\partial u}{\partial t}\,\mathbf{b}_t\right) + \dfrac{\partial u}{\partial z}\,\mathbf{b}_z$

   (b) $\dfrac{1}{\sqrt{s^2+t^2}}\left(\dfrac{\partial w_1}{\partial s} + \dfrac{\partial w_2}{\partial t} + tw_2 + sw_1\right) + \dfrac{\partial w_3}{\partial z}$

   (c) $\left(\dfrac{1}{\sqrt{s^2+t^2}}\dfrac{\partial w_3}{\partial t} - \dfrac{\partial w_2}{\partial z}\right)\mathbf{b}_s + \left(\dfrac{\partial w_1}{\partial z} - \dfrac{1}{\sqrt{s^2+t^2}}\dfrac{\partial w_3}{\partial s}\right)\mathbf{b}_t$

   $\quad + \dfrac{1}{\sqrt{s^2+t^2}}\left(\dfrac{\partial w_2}{\partial s} - \dfrac{\partial w_1}{\partial t} - tw_1 + sw_2\right)\mathbf{b}_z$

   (d) $\dfrac{1}{\sqrt{s^2+t^2}}\left[\dfrac{\partial}{\partial s}\left(\dfrac{1}{\sqrt{s^2+t^2}}\dfrac{\partial u}{\partial s}\right) + \dfrac{\partial}{\partial t}\left(\dfrac{1}{\sqrt{s^2+t^2}}\dfrac{\partial u}{\partial t}\right) + s\dfrac{\partial u}{\partial s} + t\dfrac{\partial u}{\partial t}\right]$

   $\quad + \dfrac{\partial^2 u}{\partial z^2}$

Section 8–1

1. (a) $(\pi^2/2) + 2$ (b) $2 - e$ (c) $\pi/4$ (d) $\frac{5}{6}$
   (e) $-\frac{1}{20}$ (f) $(e/2) - 1$ (g) $\frac{423}{28}$ (h) 2
   (i) $\frac{14}{15}$ (j) $\pi^2/8$ (k) $-\frac{6}{35}$ (l) $\frac{3}{20}$
   (m) $\frac{1}{6}$ (n) 1 (o) $\frac{1}{4}$ (p) 1

2. (a) $\displaystyle\int_1^{e^2}\int_{\ln y}^2 dx\,dy$ (b) $\displaystyle\int_0^1\int_0^{x^2} dy\,dx$

   (c) $\displaystyle\int_{-1}^1\int_0^{\sqrt{1-x^2}} dy\,dx$ (d) $\displaystyle\int_{-4}^5\int_{(y-2)/3}^{\sqrt{y+4}-2} dx\,dy$

(e) $\displaystyle\int_0^{1/2}\int_{(1-\sqrt{1-4y^2})/2}^{(1+\sqrt{1-4y^2})/2}dx\,dy$ (f) $\displaystyle\int_0^1\int_{x^2}^x dy\,dx$

(g) $\displaystyle\int_0^1\int_{-1}^2 dy\,dx$ (h) $\displaystyle\int_1^2\int_{\ln y}^{\ln 2}dx\,dy$

(i) $\displaystyle\int_0^1\int_{\arcsin y}^{\pi/2}dx\,dy$ (j) $\displaystyle\int_0^1\int_{e^x}^e dy\,dx$

(k) $\displaystyle\int_0^1\int_x^{\sqrt{x}}dx\,dy$ (l) $\displaystyle\int_0^4\int_{y/2}^{\sqrt{y}}dx\,dy$

Section 8–2

1. In order $\alpha(A)$, $\mu_1(x,0)$, $\mu_1(y,0)$ $\mu_2(x,0)$, $\mu_2(y,0)$.
   (a) $\frac{1}{6}, \frac{1}{12}, \frac{1}{15}, \frac{1}{20}, \frac{1}{28}$
   (b) $\frac{1}{3}, \frac{3}{20}, \frac{3}{20}, \frac{3}{35}, \frac{3}{35}$
   (c) $\frac{4}{3}, \frac{4}{3}, \frac{4}{5}, \frac{8}{5}, \frac{88}{105}$
   (d) $\frac{8}{3}, \frac{12}{5}, \frac{24}{5}, \frac{96}{35}, \frac{384}{35}$
   (e) $\frac{1}{3}, \frac{1}{6}, \frac{1}{6}, \frac{11}{105}, \frac{1}{10}$
   (f) $\pi, \pi, 0, 5\pi/4, \pi/4$
   (g) $(\pi-2)/4, \frac{1}{6}, \frac{1}{6}, (\pi/16)-(\frac{1}{12}), (\pi/16)-(\frac{1}{12})$
   (h) $\frac{1}{3}, \frac{2}{15}, \frac{2}{15}, \frac{1}{14}, \frac{1}{14}$
   (i) $4, 0, 8, \frac{4}{5}, \frac{96}{5}$
   (j) $(4-\pi)/2\pi, 0, 0, [32(\pi-2)-\pi^3]/16\pi^3, (16-3\pi)/36\pi$
   (k) $1, (e-2)/2, (e^2+1)/4, (6-2e)/3, (2e^3+1)/9$
   (l) $e-2, (e^3-3)/4, \frac{1}{2}, (e^3-4)/9, (3e-7)/3$
   (m) $\frac{5}{2}\ln 2, 0, \frac{17}{16}\ln 2 - \frac{15}{32}, \frac{5}{6}(\ln 2)^3 - \frac{3}{2}(\ln 2)^2 + 5\ln 2 - 3, \frac{125}{96}\ln 2 - \frac{57}{64}$
   (n) $(\pi-2)/2, \pi/8, (\pi^2-8)/8, (3\pi-4)/18, (\pi^3-24\pi+48)/24$
   (o) $\frac{1}{2}, \frac{1}{3}, \frac{1}{2}, \frac{1}{4}, \frac{7}{12}$
   (p) $\frac{1}{12}, \frac{1}{20}, \frac{11}{35}, \frac{1}{30}, \frac{1}{70}$
   (q) $(\frac{15}{8})-2\ln 2, \frac{9}{16}, \frac{9}{16}, \frac{45}{64}, \frac{45}{64}$
   (r) $(3-e)/2, (2e-5)/6, (5+2e-e^2)/12, (25-9e)/12, (3e^2+3e-e^3+7)/36$
   (s) $\frac{1}{4}, \frac{1}{8}, \frac{1}{6}, \frac{7}{96}, \frac{1}{8}$
   (t) $\ln 2, 1, \frac{1}{4}, \frac{3}{2}, \frac{1}{8}$

Section 8–3

1. (a) Sphere, center origin, radius 1
   (b) Hyperboloid 1 sheet, center origin, axis z
   (c) Hyperboloid 2 sheet, center origin, axis x
   (d) Hyperboloid 2 sheet, center origin, axis z
   (e) Hyperboloid 1 sheet, center origin, axis x
   (f) Paraboloid, vertex $(0, 0, -1)$, opens $+z$
   (g) Paraboloid, vertex $(0, 0, -1)$, opens $-z$
   (h) Cone, vertex origin, axis z
   (i) Cone, vertex origin, axis x
   (j) Cylinder, center $(\frac{1}{2}, 0, 0)$, axis z, radius $\frac{1}{2}$
   (k) Cylinder, center $(0, -\frac{1}{2}, 0)$, axis z, radius $\frac{1}{2}$
   (l) 2 planes, intersect on y-axis
   (m) Parabolic cylinder, elements parallel y-axis, parabola opens $+x$

(n) Cone, vertex origin, axis z

(o) Paraboloid, vertex origin, opens $+x$

(p) Paraboloid, vertex origin, opens $+y$

(q) Paraboloid, vertex origin, opens $-y$

(r) Sphere, center $(1, -2, 0)$, radius 3

(s) Hyperboloid 2 sheet, center $(1, 2, 0)$, axis x

(t) Cone, vertex $(1, 2, 0)$, axis x

(u) Hyperboloid 1 sheet, center $(1, 2, 0)$, axis x

(v) Paraboloid, vertex $(1, -2, -9)$, opens $+z$

(w) Paraboloid, vertex $(1, -2, 9)$, opens $-z$

(x) Cylinder, center $(1, -2, 0)$, axis z, radius 3

(y) Hyperbolic cylinder, center $(1, 2, 0)$, elements parallel z axis, axis of hyperbola x

(z) Hyperbolic cylinder, center $(1, 2, 0)$, elements parallel z-axis, axis of hyperbola y

Section 8-4

1. (a) $\frac{1}{90}$ (b) $\frac{11}{60}$ (c) $\pi/2$
   (d) $(8e - e^2 - 15)/8$ (e) $\frac{4}{35}$ (f) $\frac{49}{20}$
   (g) $\frac{77}{120}$ (h) $\frac{1}{15}$

2. (a) $\pi/2$ (b) $3\pi/2$ (c) 4 (d) 8π
   (e) $\frac{16}{3}$ (f) $\frac{16}{3}$ (g) 4π (h) $\frac{1}{3}$
   (i) 8π (j) $99\pi/2$ (k) $\frac{11}{12}$ (l) $\frac{27}{2}$
   (m) $abc/6$ (n) $\frac{4}{3}$

3. (a) $(0, 0, \frac{1}{3})$ (b) $(0, \frac{4}{3}, \frac{10}{9})$
   (c) $(\frac{8}{3}, \frac{16}{15}, \frac{8}{15})$ (d) $(0, 0, 4)$
   (e) $(0, 0, 0)$ (f) $(0, 0, 1)$
   (g) $(\frac{1}{2}, 0, \frac{5}{4})$ (h) $\left(\dfrac{22 - 3\pi}{16}, \dfrac{22 - 3\pi}{16}, \dfrac{44 - 3\pi}{32} \right)$
   (i) $(1, 0, \frac{11}{3})$ (j) $(0, 0, \frac{999}{55})$
   (k) $(\frac{97}{55}, \frac{114}{55}, \frac{76}{55})$ (l) $(\frac{3}{2}, \frac{1}{2}, \frac{12}{5})$
   (m) $(a/4, b/4, c/4)$ (n) $(\frac{3}{5}, -\frac{1}{5}, \frac{2}{5})$

Section 9-1

2. (a) Right (b) Left

4. Those with the origin in the interior

5. Polar, $r < 0$: clockwise
   Spherical (all ρ), $0 < \phi < \pi$: right, $\pi < \phi < 2\pi$: left

8. $D_x y = 2x$ changes sign at $x = 0$

9. $D_\theta x = -r \sin \theta$ changes sign at $\theta = 0, \pi, 2\pi$.
   $D_\theta y = r \cos \theta$ changes sign at $\theta = \pi/2, 3\pi/2$.

10. $\operatorname{Det} D_{\begin{bmatrix} \phi \\ \theta \end{bmatrix}} \begin{bmatrix} x \\ y \end{bmatrix} = \sin \phi \cos \phi$ changes sign at $\phi = \pi/2$

Section 9–2

1. (a) $\begin{vmatrix} 1 & 1 \\ 0 & 2 \end{vmatrix} = 2$ (b) $\sqrt{5}\begin{vmatrix} 1/\sqrt{5} & \sqrt{5} \\ -2/5 & 0 \end{vmatrix} = 2$

(c) $4\sqrt{3}\begin{vmatrix} 1/3 & 2/3 \\ -\sqrt{3}/6 & \sqrt{3}/6 \end{vmatrix} = 2$

2. (a) $4(u^2 + v^2)\ du \wedge dv$ (b) $u\ du \wedge dv$ (c) $3\ dv \wedge du$

5. (a) $\sqrt{2}\begin{vmatrix} 1/\sqrt{2} & 1/\sqrt{2} & 0 \\ -1/2 & 1/2 & 0 \\ 0 & 0 & 1 \end{vmatrix} = 1$

(b) $3\dfrac{\sqrt{2}}{\sqrt{3}}\begin{vmatrix} 1/\sqrt{3} & 1/\sqrt{3} & 1/\sqrt{3} \\ 1/(3\sqrt{2}) & 1/(3\sqrt{2}) & -\sqrt{2}/3 \\ -1/2 & 1/2 & 0 \end{vmatrix} = 1$

Section 9–3

1.	Region	Differentials	Integrand	Integral
(a)	cc	cc	+	+
(b)	cc	c	−	+
(c)	c	c	+	+
(d)	cc	c	−	+
(e)	c	c	−	−
(f)	cc	cc	+	+
(g)	c	c	+	+
(h)	c	cc	−	+
(i)	L	L	+	+
(j)	L	R	−	+
(k)	R	R	−	−
(l)	R	R	+	+
(m)	L	L	+	+
(n)	R	L	−	+
(o)	R	R	+	+
(p)	R	L	+	−
(q)	R	R	+	+
(r)	L	L	+	+
(s)	R	R	+	+
(t)	L	L	−	−
(u)	R	R	+	+
(v)	L	L	+	+

2. (a) $\displaystyle\int_0^1 \int_{\sqrt{y}}^y dx\, dy$

(b) $\displaystyle\int_0^1 \int_{x^2}^x dy\, dx$

(c) $\displaystyle\int_0^1 \int_y^1 dx\, dy$

(d) $\displaystyle\int_1^0 \int_0^x dy\, dx$

(e) $\displaystyle\int_0^{\pi/2} \int_0^{\sin\theta} r\, dr\, d\theta$

(f) $\displaystyle\int_0^1 \int_{\pi/2}^{\arcsin r} r\, d\theta\, dr$

(g) $\displaystyle\int_{\pi/2}^0 \int_0^1 r\, dr\, d\theta$

(h) $\displaystyle\int_0^1 \int_0^{\pi/2} r\, d\theta\, dr$

(i) $\displaystyle\int_0^1 \int_0^{1-z} \int_0^{1-y-z} dx\, dy\, dz$

(j) $\displaystyle\int_1^0 \int_0^{1-x} \int_0^{1-x-y} dz\, dy\, dx$

(k) $\displaystyle\int_{-1}^1 \int_{-\sqrt{1-x^2}}^{\sqrt{1-x^2}} \int_{x^2+y^2}^1 dz\, dy\, dx$

(l) $\displaystyle\int_1^{-1} \int_{x^2}^1 \int_{-\sqrt{z-x^2}}^{\sqrt{z-x^2}} dy\, dz\, dx$

(m) $\displaystyle\int_0^1 \int_0^{2\pi} \int_0^z r\, dr\, d\theta\, dz$

(n) $\displaystyle\int_0^{2\pi} \int_0^1 \int_r^1 r\, dz\, dr\, d\theta$

(o) $\displaystyle\int_0^{2\pi} \int_0^1 \int_{\sqrt{z}}^1 r\, dr\, dz\, d\theta$

(p) $\displaystyle\int_{2\pi}^0 \int_0^1 \int_0^{r^2} r\, dz\, dr\, d\theta$

(q) $\displaystyle\int_0^{2\pi} \int_{\pi/4}^{\pi/2} \int_0^1 \rho^2 \sin\phi\, d\rho\, d\phi\, d\theta$

(r) $\displaystyle\int_{2\pi}^0 \int_0^1 \int_{\pi/4}^{\pi/2} \rho^2 \sin\phi\, d\phi\, d\rho\, d\theta$

(s) $\displaystyle\int_0^{2\pi} \int_0^1 \int_0^{\arccos\rho} \rho^2 \sin\phi\, d\phi\, d\rho\, d\theta$

(t) $\displaystyle\int_{2\pi}^0 \int_0^{\pi/2} \int_0^{\cos\phi} \rho^2 \sin\phi\, d\rho\, d\phi\, d\theta$

Section 9–4

1. (a) $\pi/2$ (b) $-\pi^4/3$ (c) -1 (d) $\frac{112}{45}$

2. (a) $\displaystyle\int_0^{1/2} \int_v^{1-v} 2\sqrt{1 + 2u^2 + 2v^2}\, du\, dv$

(b) $\displaystyle\int_0^2 \int_0^{1-(u^2/4)} \frac{1}{4\sqrt[4]{u^2 + v^2}}\, dv\, du$

(c) $\displaystyle\int_0^1 \int_{1-2u}^1 \sqrt{u^2 + (u+v)^2}\, dv\, du$

(d) $\displaystyle\int_0^1 \int_0^{2\pi} \int_0^1 r^4 \cos^2 \theta \; dr \; d\theta \; dz$

(e) $\displaystyle\int_0^1 \int_0^{\pi/2} \int_0^{1+r(\cos\theta+\sin\theta)} zr^3 \sin\theta \cos\theta \; dz \; d\theta \; dr$

(f) $\displaystyle\int_0^{2\pi} \int_0^1 \int_{-\sqrt{4-r^2}}^{\sqrt{4-r^2}} rz \; dz \; dr \; d\theta$

(g) $\displaystyle\int_0^{2\pi} \int_0^{\pi} \int_0^1 \rho^3 \sin\phi \; d\rho \; d\phi \; d\theta$

(h) $\displaystyle\int_0^{2\pi} \int_0^{\pi/4} \int_0^{\sec\phi} \rho^3 \sin\phi \cos\phi \; d\rho \; d\phi \; d\theta$

(i) $\displaystyle\int_0^{2\pi} \int_0^{\pi/4} \int_0^2 \rho^4 \sin\phi \cos^2\phi \; d\rho \; d\phi \; d\theta$

(j) $\displaystyle\int_0^{2\pi} \int_{\pi/4}^{3\pi/4} \int_0^1 \rho^4 \sin^3\phi \; d\rho \; d\phi \; d\theta$

(k) $\displaystyle\int_0^{2\pi} \int_0^{\pi/4} \int_0^1 \rho^3 \sin\phi \cos\phi \; d\rho \; d\phi \; d\theta$

(l) $\displaystyle\int_0^{2\pi} \int_0^1 \int_r^1 2rz \; dz \; dr \; d\theta$

(m) $\displaystyle\int_0^1 \int_{-\sqrt{y-y^2}}^{\sqrt{y-y^2}} \int_0^y y \; dz \; dx \; dy$

(n) $\displaystyle\int_{-1/\sqrt{2}}^{1/\sqrt{2}} \int_{-\sqrt{(1/2)-y^2}}^{\sqrt{(1/2)-y^2}} \int_{\sqrt{x^2-y^2}}^{\sqrt{1-x^2-y^2}} z \; dz \; dx \; dy$

3. (a) $2\pi/3$, $(0, 0, \frac{3}{8})$ (b) $\pi/3$, $(0, 0, \frac{3}{4})$
   (c) π, $(0, 0, \frac{1}{2})$ (d) $4\pi/3$, $(0, 0, 0)$
   (e) $2\pi\sqrt{2}/3$, $(0, 0, 0)$ (f) $28\pi/3$, $(0, 0, 0)$
   (g) $\frac{8}{9}$, $(\frac{9}{15}, 0, 0)$ (h) $\pi/3$, $(0, 0, \frac{1}{2})$
   (i) $\pi/2$, $(\frac{1}{2}, 0, \frac{5}{12})$ (j) $\pi/2$, $(0, 0, \frac{1}{3})$
   (k) $\pi/6$, $(0, 0, \frac{1}{2})$ (l) $\pi/2$, $(0, 0, \frac{1}{3})$
   (m) $3\pi/2$. $(0, \frac{4}{3}, \frac{10}{9})$ (n) $99\pi/2$, $(0, 0, \frac{282}{55})$
   (o) $\frac{64}{9}$, $(\frac{6}{5}, 0, 0)$ (p) $\pi/6$, $(0, 0, \frac{1}{2})$
   (q) $\pi(2 - \sqrt{2})/3$, $(0, 0, 3/(16 - 8\sqrt{2}))$
   (r) $\pi(8\sqrt{2} - 7)/6$, $(0, 0, 7/(16\sqrt{2} - 14))$
   (s) $4\pi(8 - \sqrt{3})/3$, $(0, 0, 0)$ (t) 8π, $(0, 0, 4)$

Section 9–5

1. (a) $2, \pi/4$
   (b) $\frac{1}{6}, \frac{1}{15}$
   (c) $6, 18$
   (d) $6, 26$
   (e) $ab/2, ab(3a^2 + b^2)/12$
   (f) $\frac{4}{3}, \frac{8}{15}$
   (g) $\pi a^2/2, \pi a^3/3$
   (h) $\pi^3/6, \pi^4/12$
   (i) $\pi a^2, \pi a^4/2$
   (j) $3\pi/2, 5\pi/3$
   (k) $\pi/8, \frac{8}{75}$
   (l) $1, \pi/8$

2. (a) $\frac{19}{30}, \frac{67}{105}$
   (b) $(\frac{32}{3}) + (\pi/2), \frac{56}{15}$
   (c) $a^3, 3a^4/2$
   (d) $\frac{2}{3}, \frac{28}{45}$
   (e) a^3, a^5
   (f) $2\pi/3, \frac{1}{12}$
   (g) $\pi a^2 h, m a^2 h^2/2$
   (h) $4\pi\sqrt{3}, 9\pi/2$
   (i) $8\pi/3, 16\pi/5$
   (j) $(2 - \sqrt{2})\pi/3, \pi/8$
   (k) $2\pi\sqrt{2}/3, (\pi^2/8) + \pi$
   (l) $4\pi a^3/3, 32\pi a^4/5$

3. (a) $2\pi(b + \sqrt{a^2 + h^2} - a - \sqrt{b^2 + h^2}$
   (b) $2\pi(h + \sqrt{a^2 + m^2} - \sqrt{a^2 + (m + h)^2})$
   (c) $2\pi h[1 - (h/\sqrt{a^2 + h^2})]$
   (d) $4\pi a^3/(3b^2)$
   (e) $2\pi[1 - (b/\sqrt{a^2 + b^2})]$

4. (c) $2\pi(\sqrt{a^2 + h^2} - h)$
   (e) $2\pi(\sqrt{a^2 + b^2} - b)$

Section 9–6

1. (a) $u^3/3 (0 \le u \le 1); u^2 - (u^3/3) - (\frac{1}{3})(1 \le u \le 2)$
   (b) $2u - u^2 (0 \le u \le 1)$
   (c) $u - u \ln u (0 < u \le 1)$
   (d) $u^2/2 (0 \le u \le 1); 2u - (u^2/2) - 1 (1 \le u \le 2)$
   (e) $u^2 (0 \le u \le 1)$
   (f) $2u - u^2 (0 \le u \le 1)$
   (g) $(ae^{-bu} - be^{-au} + b - a)/(b - a)(u \ge 0)$
   (h) $1 - aue^{-au} - e^{-au}(u \ge 0)$

2. (a) $(1/\sqrt{2\pi})e^{-a^2/2}$
   (b) $(1/\sqrt{2\pi})e^{-a^2/2}$
   (c) $1 - e^{-a^2/2} \quad (a \ge 0)$
   (d) $1 - e^{-a/2} \quad (a \ge 0)$

Section 9–7

1. From Exercise 1, Section 9–5
   (a) $(\pi/2, 16/9\pi)$
   (b) $(\frac{5}{8}, \frac{15}{28})$
   (c) $(2, 1)$
   (d) $(\frac{105}{52}, \frac{15}{13})$
   (e) $\left(\dfrac{4a}{5}, \dfrac{3b(2a^2 + b^2)}{5(3a^2 + b^2)} \right)$
   (f) $(0, \frac{4}{7})$
   (g) $(0, 3a/2\pi)$
   (h) $(12(6\pi - \pi^3)/\pi^4, 3(\pi^4 - 12\pi^2 + 48)/\pi^4)$
   (i) Center
   (j) $(-\frac{21}{20}, 0)$

From Exercise 2, Section 9–5
   (d) $(0, 15\pi/64, 675\pi/1792)$
   (e) $(7a/12, 7a/12, 7a/12)$ with respect to corner at which $D_r m = 0$.
   (f) $(0, 0, \frac{16}{35})$
   (g) $(0, 0, 2h/3)$
   (h) $(0, 0, 0)$
   (i) $(0, 0, \frac{16}{7})$
   (j) $(0, 0, (16 - 4\sqrt{2})/15)$
   (k) $(0, 0, 0)$
   (l) $(0, 0, 10a/7)$

3. (a) $\frac{7}{6}$ (b) $\frac{4}{3}$ (c) $\frac{1}{4}$ (d) 1
   (e) $\frac{2}{3}$ (f) $\frac{1}{3}$ (g) $(1/a) + (1/b)$ (h) $2/a$
6. (a) $\sigma\sqrt{\pi/2}$ (b) σ^2 (c) $2\sigma^2$ (d) $\sigma^2(4 - \pi)/2$

Section 10–1

1. (a) 8   (b) 0   (c) -2π (d) $-\frac{1}{4}$ (e) $-\frac{5}{3}$ (f) $-\frac{1}{2}$
2. One on each path

Section 10–2

1. (a) π (b) $\frac{1}{2}$ (c) $\pi/2$ (d) 0
2. Same as Exercise 1

Section 10–3

1. (a) $\pi a^2(2 - \sqrt{2})$, $\left(0, 0, \dfrac{a}{4 - 2\sqrt{2}}\right)$

   (b) $2a^2$, $(a/3, 0, \pi a/8)$ (c) $2\pi a^2/3$, $(0, 0, 0)$ (d) $\pi a^2\sqrt{2}$, $\left(a, 0, \dfrac{32a}{9\pi}\right)$
2. Same as Exercise 1
3. (a) $F = \pi$, $P = 2\pi a$
   (b) $F = 2\pi(1 - a/\sqrt{a^2 + h^2})$, $P = 2\pi a \operatorname{arcsinh}(h/a)$

5. (a) $d\sigma = \sqrt{r^2 + r^2\left(\dfrac{\partial z}{\partial r}\right)^2 + \left(\dfrac{\partial z}{\partial \theta}\right)^2}\ dr \wedge d\theta$

Section 10–4

1. (a) 0   (b) $(b - a)\alpha(A)$ (c) 0   (d) 0
   (e) -4π (f) $-4\bar{y}\alpha(A)$ (g) 0   (h) $-3\pi/2$ (i) $6\alpha(A)$
2. (a) 6π (b) 0   (c) 0
3. (a) 4π (b) 0   (c) 0
5. $\pi a^2 b/4$
6. $\pi a^3/(6\sqrt{2})$
7. $\pi a^3/6$

Section 10–6

1. (a) $x^2 y = C$ (b) $e^{xy} = C$ (c) $x^2 y = \dfrac{y^3}{3} = C$ (d) $x \sin y = C$
2. (a) A neither contains nor surrounds the origin.
   (b) A lies completely on one side of the y-axis.
   (c) A lies completely on one side of the x-axis
   (d) A neither contains nor surrounds the origin.
3. Integral $= 2\pi$
5. (a) $\sum_{i=1}^{k} a_i$
   (b) $K + \sum_{i=1}^{k} n_i a_i$ where each n_i is a positive or negative integer or zero.
6. (a) -2 (b) 0

INDEX